Combinatorial Theory

A BLAISDELL BOOK IN PURE AND APPLIED MATHEMATICS

MARSHALL HALL, JR.
California Institute of Technology

Combinatorial Theory

BLAISDELL PUBLISHING COMPANY
A Division of Ginn and Company
WALTHAM, MASSACHUSETTS · TORONTO · LONDON

To Sally, Marshall, and Jonathan

LIBRARY OF CONGRESS CATALOG CARD NUMBER: 67–11108
PRINTED IN THE UNITED STATES OF AMERICA.

Preface

Combinatorial theory is the name now given to the subject formerly called "combinatorial analysis" or "combinatorics," though these terms are still used by many people. Like many branches of Mathematics, its boundaries are not clearly defined, but the central problem may be considered that of arranging objects according to specified rules and finding out in how many ways this may be done. If the specified rules are very simple, then the chief emphasis is on the enumeration of the number of ways in which the arrangement may be made. If the rules are subtle or complicated, the chief problem is whether or not such arrangements exist, and to find methods for constructing the arrangements. An intermediate area is the relationship between related choices, and a typical theorem will assert that the maximum for one kind of choice is equal to the minimum for another kind.

The text is divided into three major segments. The first four chapters deal with problems of enumeration. Chapters 5 through 9 deal with the intermediate area of theorems on choice. Chapters 10 through 16 are concerned with the existence and construction of designs.

The theory of enumeration is covered extensively in the classical work of Major P. A. MacMahon, *Combinatorial Analysis*, London, Vol. I, 1915, Vol. II, 1916, and in the recent book by John Riordan, *An Introduction to Combinatorial Analysis*, John Wiley & Sons, Inc., New York, 1958. The treatment of this subject in the first four chapters of this book is relatively brief, and does not attempt to match the scope of these books. H. J. Ryser in the Carus monograph, *Combinatorial Mathematics*, 1963, gives a brief but elegant account of the theorems on choice and the construction and existence of block designs.

Many people have been helpful to me in preparing this book. These include Dr. Leonard Baumert, Professor Robert Dilworth, Dr. Karl Goldberg, Professor Donald Knuth, Dr. Morris Newman, and Professor A. W. Tucker. Special thanks are due to Professors Garrett Birkhoff, Robert Greenwood, and Herbert Ryser, who read the entire manuscript and gave me many helpful suggestions. In preparation of the manuscript and correction of clerical errors, the assistance of Mrs. Kay Hardt, Dr. Allen Pfeffer, and Mr. Robert McEliece was invaluable.

I am indebted to the Literary Executor of the late Sir Ronald A. Fisher, F.R.S., Cambridge, to Dr. Frank Yates, F.R.S., Rothamsted, and to Messrs.

Oliver & Boyd Ltd., Edinburgh, for permission to quote a portion of text from their book *Statistical Tables for Biological, Agricultural, and Medical Research.* I also wish to thank Dr. C. R. Rao of the Indian Statistical Institute and the editors of *Sankhyā* for permission to quote from the paper "A Study of BIB Designs with Replications 11 to 15."

MARSHALL HALL, JR.

Pasadena, California

Contents

1

Permutations and Combinations

1.1 Definitions

A permutation is an ordered selection of objects from a set S.
A combination is an unordered selection of objects from a set S.

We may or may not permit repetition in our permutations and combinations. Thus, selecting two letters from the three letters a, b, c, we have nine permutations with repetitions permitted:

$$aa,\ ab,\ ac,\ ba,\ bb,\ bc,\ ca,\ cb,\ cc\,.$$

We have six permutations without repetitions:

$$ab,\ ac,\ ba,\ bc,\ ca,\ cb\,.$$

We have six combinations with repetitions permitted:

$$aa,\ bb,\ cc,\ ab,\ ac,\ bc\,,$$

and three combinations without repetitions:

$$ab,\ ac,\ bc\,.$$

The number of permutations of n things taken r at a time, without repetition, written ${}_nP_r$, is easily evaluated. For in a permutation $a_1a_2 \cdots a_r$ we may choose a_1 as any of the n objects, a_2 as any one of the remaining $(n - 1)$ objects, and having chosen $a_1a_2 \cdots a_i$, we may take a_{i+1} as any one of the $(n - i)$ remaining objects. Hence,

$$ {}_nP_r = n(n-1)\cdots(n-r+1) = \frac{n!}{(n-r)!} = (n)_r. \tag{1.1.1} $$

A combination of n things taken r at a time without repetition, say, $a_1a_2 \cdots a_r$, will lead to $r!$ different permutations, namely, all $r!$ permutations of $a_1, \ldots, a_r$. Hence the number of combinations of n things taken r at a time, written ${}_nC_r$, is given by

$$ {}_nC_r = \frac{{}_nP_r}{r!} = \frac{n!}{(n-r)!\,r!} = \binom{n}{r}, \tag{1.1.2} $$

which is the familiar binomial coefficient. Indeed, in the product $(x + y)^n = (x + y) \cdots (x + y)$, the coefficient of the term, x^ry^{n-r} is the number of ways of choosing r of the factors $x + y$, from which we take an x, and then y from the remaining $n - r$ factors $x + y$. We note that

$$ {}_nC_r = {}_nC_{n-r}. \tag{1.1.3} $$

The number of permutations of n things taken r at a time, repeats permitted, is n^r, since in $a_1a_2 \cdots a_r$ there are n choices for each of $a_1, a_2, \ldots, a_r$ in turn.

To find the number of combinations of n things taken r at a time with repeats permitted, we cannot simply divide n^r by an appropriate factor, since different combinations may yield a different number of permutations. Thus, taking combinations of a, b, c, d, e three at a time, the combination *abc* gives six permutations, the combination *aab* gives three permutations, and the combination *aaa* gives only one permutation. Here we use the device of counting a different set, which is in one-to-one correspondence with our given set. To a given combination (say, *bbd*), let us adjoin the entire set *abcde* and write the whole in order *abbbcdde* and then insert marks separating the different letters thus: $a \mid bbb \mid c \mid dd \mid e$. In general, to a combination of r letters, repeats permitted, from a set of n letters adjoin all n letters and write in order the set of $(n + r)$ letters and then insert $(n - 1)$ marks between the different letters. Thus, with $(n + r)$ positions to be filled and $(n + r - 1)$ spaces between these positions, we are to insert $(n - 1)$ marks. The number of ways of doing this is

$$ \binom{n+r-1}{n-1} = \binom{n+r-1}{r}. \tag{1.1.4} $$

There is a one-to-one correspondence between the ways of inserting the $(n - 1)$ marks in the $(n + r - 1)$ spaces and the combinations with repeats of n things taken r at a time. Hence, this number is $\binom{n+r-1}{r}$. The expression for the number of combinations, n things taken r at a time, without repeats and with repeats, are similar in form. Thus, for five things taken three at a time, the numbers are respectively

$$\frac{5 \cdot 4 \cdot 3}{1 \cdot 2 \cdot 3} \quad \text{and} \quad \frac{5 \cdot 6 \cdot 7}{1 \cdot 2 \cdot 3},$$

where the factors in the numerators decrease in one case and increase in the other.

The number of combinations with repeats permitted of n things taken r at a time is the number of solutions $(x_1, x_2, \ldots, x_n)$ in nonnegative integers x_i of

$$r = x_1 + x_2 + \cdots + x_n, \tag{1.1.5}$$

where x_i is the number of times the ith object is included in the combination. This suggests another, but similar, evaluation of the number of combinations. Put $y_i = x_i + 1$, $i = 1, \ldots, n$. Then (1.1.5) becomes

$$n + r = y_1 + y_2 + \cdots + y_n, \tag{1.1.6}$$

and the number of solutions $(y_1, \ldots, y_n)$ of (1.1.6) in positive integers y_i is clearly the same as the number of solutions of (1.1.5) in nonnegative integers. If we take $(n + r)$ dots and place $(n - 1)$ marks in the $(n + r - 1)$ spaces between the dots, we may take y_1 as the number in the first set of dots, y_2 as the number in the second set, and so on. Thus, again we see that the number of solutions of (1.1.6) is $\binom{n+r-1}{n-1}$, and this is in turn the number of nonnegative solutions of (1.1.5), and so it is the number of combinations of n things r at a time with repeats permitted.

As a further application of this method we may find the number of combinations of $1, 2, \ldots, n$ taken r at a time without including repeats or consecutive numbers. Let us list $1, 2, \ldots, n$ in order and put a mark after each number selected. If there are x_1 numbers before the first mark, x_2 between the first and second mark, and finally x_{r+1} after the last mark, then these determine the choice and

$$n = x_1 + x_2 + \cdots + x_{r+1}, \tag{1.1.7}$$

where $x_1 \geq 1$, $x_2 \geq 2, \ldots, x_r \geq 2$, and $x_{r+1} \geq 0$. We now write

$$n - r + 2 = x_1 + (x_2 - 1) + \cdots + (x_r - 1) + (x_{r+1} + 1), \tag{1.1.8}$$

giving a representation of $n - r + 2$ as a sum of $(r + 1)$ positive integers, and this number is $\binom{n-r+1}{r}$, the number of ways of putting r marks in $(n - r + 1)$ spaces.

There are an enormous number of identities involving binomial coefficients, a few of which follow:

$$\sum_{k=0}^{n} \binom{n}{k} = 2^n, \tag{1.1.9a}$$

$$\sum_{k=0}^{n} (-1)^k \binom{n}{k} = 0, \tag{1.1.9b}$$

$$\sum_{k=1}^{n} k(-1)^k \binom{n}{k} = 0, \qquad n \geq 1, \tag{1.1.9c}$$

$$\sum_{k=r}^{n} (-1)^k \binom{k}{r} \binom{n}{k} = 0, \qquad n \geq r. \tag{1.1.9d}$$

These may be derived from the relation

$$(1 + x)^n = \sum_{k=0}^{n} \binom{n}{k} x^k.$$

To obtain (1.1.9a) and (1.1.9b), put $x = 1$ and $x = -1$, respectively. For (1.1.9c), differentiate with respect to x and then put $x = -1$. For (1.1.9d), differentiate r times with respect to x, divide by $r!$, and put $x = -1$.

If we have permutations $a_1 a_2 \cdots a_n$ of n objects, of which b_1 are of one kind, b_2 of a second kind, and b_i of an ith kind for i running to r, where naturally $b_1 + b_2 + \cdots + b_r = n$, we may first replace the b_i objects of the ith kind by distinct objects in every case, and then we have $n!$ permutations. But by identifying the like objects, we have counted each permutation $b_1!\, b_2! \cdots b_r!$ times. Hence, the number of permutations is

$$\frac{n!}{b_1!\, b_2! \cdots b_r!}, \qquad b_1 + b_2 + \cdots + b_r = n, \tag{1.1.10}$$

the familiar multinomial coefficient.

1.2 Applications to Probability

In a given situation let us suppose that there are n possible outcomes, which we label $x_1, x_2, \ldots, x_n$, and which are mutually exclusive. We assign to the outcome x_i a number $p_i = p(x_i)$, where p_i is a real number, $p_i \geq 0$, and $p_1 + p_2 + \cdots + p_n = 1$. If an event E occurs along with the possibilities $x_{i_1}, \ldots, x_{i_m}$ and not otherwise, we define the probability of E as $p(E) = p_{i_1} + \cdots + p_{i_m}$. The assignment of the initial probabilities $p_1, p_2, \ldots, p_n$ is not a mathematical problem, but is an estimate of the relative likelihoods of the different outcomes, and in any actual case the validity of the mathematical calculation of $p(E)$ depends on the correctness of this assignment.

There are many practical situations in which it seems reasonable to consider the n outcomes as equally likely, and so we take $p_1 = p_2 \cdots = p_n = 1/n$. In this case the probability of the event E occurring along with m possible outcomes, but no others, is $p(E) = m/n$. In such a situation the calculation of $p(E)$ becomes the purely combinatorial problem of calculating m, the number of possible outcomes yielding the event E. In throwing at random a die whose six faces are numbered from 1 to 6, it seems reasonable to assume that any face is as likely to come on top as any other, if the die is of uniform density. In this case, we take $p_1 = p_2 = \cdots = p_6 = 1/6$, where p_i is the probability that the face numbered i will come up. If we are merely interested in whether or not a 6 comes up, we consider only two possible outcomes, putting $p = 1/6$ as the probability for a 6 and a $p' = 5/6$ as the probability of not getting a 6.

Let us suppose we have N urns numbered from 1 to N. We are to place at random n balls in the urns, where $n < N$. We ask the probability that each of the urns numbered 1 to n will contain exactly one ball. This probability depends on two things: (1) whether the balls are distinguishable or indistinguishable, and (2) whether there is an exclusion principle that does not allow a second ball to be placed in an urn that already contains one ball. If the n balls are distinguishable and there is no exclusion principle, there will be N^n ways of placing the n balls in the N urns. There will be $n!$ ways of placing them in the urns numbered $1, \ldots, n$, placing one in each of these urns. With these conventions, the probability is

$$p(E) = \frac{n!}{N^n}. \tag{1.2.1}$$

If the balls are distinguishable and there is an exclusion principle, the first ball may be placed in any one of N urns, the next in any one of $(N - 1)$ urns, and the ith in any one of $(N - i + 1)$ urns, whence the number of ways of placing the n balls in the N urns is ${}_NP_n = N(N - 1)\cdots(N - n + 1)$. They may be placed in urns $1, \ldots, n$ in $n!$ ways, and under these conventions, the probability is

$$p(E) = \frac{n!}{{}_NP_n} = \frac{1}{\binom{N}{n}}. \tag{1.2.2}$$

If the balls are indistinguishable and there is no exclusion principle, we are asking for the solutions of $x_1 + x_2 + \cdots + x_N = n$ in nonnegative integers x_i, where x_i is the number of balls placed in the ith urn. This, as we noted in the preceding section, is the number of combinations of N things taken n at a time with repeats permitted, and is $\binom{N+n-1}{n}$. Exactly one of these is the solution $x_1 = x_2 = \cdots = x_n = 1$, $x_{n+1} = x_{n+2} = \cdots = x_N = 0$; in this case our probability is

$$p(E) = \frac{1}{\binom{N + n - 1}{n}}. \tag{1.2.3}$$

From the physical standpoint, "indistinguishable" means that one combination is as likely as another.

If the balls are indistinguishable and there is an exclusion principle, the number of ways of placing the balls is merely the number of combinations of N things taken n at a time without repeats, and this is ${}_NC_n = \binom{N}{n}$. The choice of the first n urns is a single one of these combinations, and here the probability is

$$p(E) = \frac{1}{\binom{N}{n}}. \tag{1.2.4}$$

Note that this is the same as (1.2.2), so that with an exclusion principle, the probability is the same whether the balls are distinguishable or not.

In statistical physics we consider a collection of n particles, which may be protons, electrons, mesons, neutrons, neutrinos, or photons, each of which may be in any one of N "states," which may be energy levels. The macroscopic state of the system of the n particles is a vector $x = (x_1, x_2, \ldots, x_N)$, where x_i is the number of particles in the ith state. The probability of any single macroscopic state depends on whether or not the particles are distinguishable and whether or not the particles obey the Pauli exclusion principle, which says that no two (indistinguishable) particles may be in the same state. If the particles are considered distinguishable and do not obey the exclusion principle, the probability of any single macroscopic state is given by (1.2.1) and the particles are said to obey the Maxwell-Boltzmann statistics. If the particles are indistinguishable and do not obey the exclusion principle, the probability is given by (1.2.3), and they are said to obey the Bose-Einstein statistics. If they are indistinguishable and do obey the exclusion principle, the probability is given by (1.2.4), and the particles are said to obey the Fermi-Dirac statistics. Electrons, protons, and neutrons obey Fermi-Dirac statistics. Photons and pi-mesons obey Bose-Einstein statistics. The case (1.2.2) of distinguishable particles with an exclusion principle does not arise in physics.

At high temperatures, when the number N is large and the different microscopic states are approximately equally likely, the Fermi-Dirac and Bose-Einstein statistics are essentially the same as the classical Maxwell-Boltzmann statistics. At low temperatures, the low-energy levels are more likely than the high-energy ones, and then the preceding models must be modified accordingly.

PROBLEMS

1. Prove

$$\sum_{i=0}^{m} \binom{r}{i}\binom{s}{m-i} = \binom{r+s}{m}.$$

Hint: $(1 + x)^r(1 + x)^s = (1 + x)^{r+s}$. Give an alternate proof of this identity by considering the number of ways of choosing a committee of m people out of a group of r men and s women.

2. A flag is to be designed with 13 horizontal stripes colored red, white, or blue, subject to the condition that no stripe be of the same color as the one above it. In how many ways may this be done?

3. How many positive integers less than 10^n (in the decimal scale) have their digits in nondecreasing order?
4. In how many ways may n identical gifts be given to r children (a) under no restriction and (b) if each child must receive at least one gift?
5. A hand of five cards is selected from a deck of $4n$ cards that contains four different suits, each with n cards, $n \geq 5$, numbered $1, \ldots, n$. Rank in order of increasing frequency, depending on the value of n, the following hands: a straight flush (five consecutively numbered cards of the same suit), four of a kind (four cards having the same number), full house (three cards of one number, the other two of another number), flush (five cards of the same suit), straight (five consecutively numbered cards), three of a kind (three cards of the same number), two pair (two cards of one number, two others of a second number) and a pair (two cards of a number).

2

Inversion Formulae

2.1 The Principle of Inclusion and Exclusion. Möbius Inversion

Suppose we have N objects and a number of properties $P(1), \ldots, P(n)$. Let N_i be the number of objects with property $P(i)$ and, more generally, $N_{i_1 i_2 \cdots i_r}$ the number of objects with properties $P(i_1), P(i_2), \ldots,$ and $P(i_r)$. Then we assert that the number of objects $N(0)$ with none of the properties is given by the inversion formula

$$N(0) = N - \sum N_i + \sum_{i_1 < i_2} N_{i_1 i_2} + \cdots + (-1)^s \sum_{i_1 < i_2 \cdots < i_s} N_{i_1 i_2 \cdots i_s} + \cdots + (-1)^n N_{12 \cdots n}. \quad (2.1.1)$$

We now prove this. An object with none of the properties is counted once in the term N and does not contribute to the remaining terms. An object A with the property $P(j)$ is counted once in N and once in N_j, and so contributes 1 to the term N, -1 to the term $-\sum_i N_i$, and thus contributes $1 - 1 = 0$ to the right-hand side of (2.1.1). An object A with exactly r properties, say, $j_1, \ldots, j_r$, contributes 1 to the sum

$$\sum_{i_1 < \cdots < i_s} N_{i_1 i_2 \cdots i_s}, \qquad \text{when } s \leq r$$

for every choice of $i_1, \ldots, i_s$ from $j_1, \ldots, j_r$—that is, for $\binom{r}{s}$ choices. Hence, A contributes to the right-hand side of (2.1.1) exactly

$$1 - \binom{r}{1} + \binom{r}{2} + \cdots + (-1)^s\binom{r}{s} + \cdots + (-1)^r\binom{r}{r} = (1-1)^r = 0. \quad (2.1.2)$$

Thus, the right-hand side of (2.1.1) counts each element with no properties exactly once, and every other element zero times; hence its value is $N(0)$, as was to be proved. Use of the formula (2.1.1) is sometimes called the method of inclusion and exclusion.

In the same way we may find the number $N(r)$ of objects with exactly r properties. This is given by

$$N(r) = \sum_{i_1<\cdots<i_r} N_{i_1\ldots i_r} + \cdots + (-1)^{s-r}\binom{s}{r}\sum_{i_1<\cdots<i_s} N_{i_1\ldots i_s} + \cdots. \quad (2.1.3)$$

On the right-hand side of (2.1.3) an object with exactly r properties is counted once in the first term and is not counted in the other terms. An object with exactly t properties, where $t > r$, contributes $(-1)^{s-r}\binom{s}{r}\binom{t}{s}$ to the term

$$(-1)^{s-r}\binom{s}{r}\sum_{i_1<\cdots<i_s} N_{i_1\cdots i_s}.$$

But

$$\sum_{s=r}^{t}(-1)^{s-r}\binom{s}{r}\binom{t}{s} = 0 \quad (2.1.4)$$

from the relation (1.1.9d), and so the relation (2.1.3) is proved.

As an application of the method of inclusion and exclusion we consider the problem of derangements. How many permutations $a_1, a_2, \ldots, a_n$ of $1, 2, \ldots, n$ are there

$$\begin{matrix} 1, 2, \ldots, & i, \ldots, n \\ a_1, a_2, \ldots, & a_i, \ldots, a_n \end{matrix} \quad (2.1.5)$$

such that we have $a_i \neq i$ for every $i = 1, 2, \ldots, n$? Here we take the N objects as the $n!$ permutations $a_1, a_2, \ldots, a_n$ and the property $P(i)$ as $a_i = i$, $i = 1, \ldots, n$. Then $N_{i_1 i_2 \cdots i_r} = (n-r)!$, this being the number of permutations fixing r specified numbers. Furthermore, for $\sum N_{i_1 i_2 \cdots i_r}$ there are $\binom{n}{r}$ summands, this being the number of ways of choosing $i_1, i_2, \ldots, i_r$ from $1, 2, \ldots, n$. Applying (2.1.1) we have

$$N(0) = n! - n\cdot(n-1)! + \binom{n}{2}(n-2)! + \cdots + (-1)^r\binom{n}{r}(n-r)! + \cdots + (-1)^n\cdot 1. \quad (2.1.6)$$

We may rewrite this in the form

$$N(0) = n!\left(1 - 1 + \frac{1}{2!} - \frac{1}{3!} + \cdots + (-1)^r \cdot \frac{1}{r!} + \cdots + \frac{(-1)^n}{n!}\right). \quad (2.1.7)$$

We recognize

$$1 - 1 + \frac{1}{2!} - \frac{1}{3!} \cdots$$

as the initial terms of an infinite series whose value is e^{-1}. The infinite series is alternating, and the first omitted term is $(-1)^{n+1}/(n+1)!$. From this we see that $N(0)$ differs from $n!/e$ by less than $1/(n+1)$, and so $n!/e$ is an extremely good approximation to the number of derangements of n letters.

If we ask not only the number of derangements of $1, 2, \ldots, n$ but also the number of permutations $a_1 a_2 \cdots a_n$ of $1, 2, \ldots, n$ for which $a_i = i$ in exactly r instances for each value of $r = 0, 1, \ldots, n$, the problem is known as the "problème des rencontres." The solution is an easy extension of the problem of derangements. We may choose r numbers from $1, \ldots, n$ in $\binom{n}{r}$ ways, and having chosen these we multiply by the number of derangements of the remaining $(n - r)$ letters. This gives the number of permutations with exactly r agreements $a_i = i$ as

$$N(r) = \frac{n!}{r!}\left(1 - 1 + \frac{1}{2!} + \cdots + (-1)^{n-r} \cdot \frac{1}{(n-r)!}\right). \quad (2.1.8)$$

This could also have been found from the rule (2.1.3).

For an additional type of inversion formula we turn to an arithmetical function, the Möbius function $\mu(n)$. This is defined for positive integers n. If $n > 1$, then n has a unique factorization as a product of prime powers

$$n = p_1^{e} p_2^{e} \cdots p_r^{e_r}, \quad (2.1.9)$$

where the p's are different primes. We define $\mu(n)$ by the rules

$$\begin{aligned} &\mu(1) = 1, \\ &\mu(n) = 0, \quad \text{if any } e_i > 1 \text{ in (2.1.9)}, \\ &\mu(n) = (-1)^r, \quad \text{if } e_1 = e_2 = \cdots = e_r = 1 \text{ in (2.1.9)}. \end{aligned} \quad (2.1.10)$$

LEMMA 2.1.1.

$$\sum_{d \mid n} \mu(d) = \begin{cases} 1, & \text{if } n = 1, \\ 0, & \text{if } n > 1, \end{cases}$$

the sum being over all positive divisors d of n.

Proof: ■ If $n = 1$, then $d = 1$ is the only divisor, and $\mu(1) = 1$. If $n > 1$ and n is given by (2.1.9), write $n^* = p_1 p_2 \cdots p_r$. Then a divisor d of n that is not a divisor of n^* will have a multiple prime factor and we have $\mu(d) = 0$. Hence,

$$\sum_{d \mid n} \mu(d) = \sum_{d \mid n^*} \mu(d). \quad (2.1.11)$$

But $\sum_{d \mid n^*} \mu(d)$ is easily evaluated as

$$1 - r + \binom{r}{2} + \cdots + (-1)^k \binom{r}{k} + \cdots = (1-1)^r = 0, \quad (2.1.12)$$

since there are $\binom{r}{k}$ divisors that are the product of k distinct primes and for each of which $\mu(d) = (-1)^k$. Thus, our lemma is proved.

THEOREM 2.1.1 (MÖBIUS INVERSION FORMULA). *Let $f(n)$ and $g(n)$ be functions defined for every positive integer n satisfying*

$$f(n) = \sum_{d \mid n} g(d). \quad (2.1.13a)$$

Then we may invert this relation to express g in terms of f by the rule

$$g(n) = \sum_{d \mid n} \mu(d) f\left(\frac{n}{d}\right). \quad (2.1.13b)$$

The second relation also implies the first.

Proof: ■ We have

$$f\left(\frac{n}{d}\right) = \sum_{d' \mid n/d} g(d') \qquad \text{for every } d \mid n.$$

Hence,

$$\sum_{d \mid n} \mu(d) f\left(\frac{n}{d}\right) = \sum_{d \mid n} \mu(d) \cdot \sum_{d' \mid n/d} g(d'). \quad (2.1.14)$$

Let us write $n = dd'n_1$. Then, for a fixed d', d ranges over the divisors of n/d'. Hence,

$$\sum_{d \mid n} \mu(d) \cdot \sum_{d' \mid (n/d)} g(d') = \sum_{d' \mid n} g(d') \sum_{d \mid (n/d')} \mu(d) = g(n), \quad (2.1.15)$$

since the sum

$$\sum_{d \mid (n/d')} \mu(d) = 0$$

by the lemma, except for $d' = n$. Thus, the right-hand side of (2.1.14) simplifies to $g(n)$ and our theorem is proved. Similarly, given (2.1.13b), we may substitute in the right-hand side of (2.1.13a) and find that it simplifies to $f(n)$, proving (2.1.13a).

Möbius inversion may be used to enumerate circular partitions. If letters $a_1, a_2, \ldots, a_n$ are arranged in a circle with a_1 following a_n, then any one of the linear sequences $a_2, a_3, \ldots, a_n, a_1$; $a_3, \ldots, a_n, a_1, a_2$; $\ldots$; $a_n, a_1, \ldots, a_{n-1}$ may be thought of as determining the same circular sequence. But not all n linear sequences corresponding to the same circular sequence need be different. If, for a divisor d of n, the sequence $a_1, a_2, \ldots, a_n$ consists of a sequence of d

letters $a_1, a_2, \ldots, a_d$ repeated n/d times, the linear sequences repeat after the first d. With each circular sequence of length n we may associate a unique minimum period d such that the circular sequence consists of n/d repetitions of a sequence of d letters. Furthermore, each circular sequence of length d and period d where $d \mid n$ may be repeated n/d times to give a circular sequence of length n and period d. Each of these sequences corresponds to exactly d different linear words of length n. If there are r different letters, there are r^n linear permutations $a_1a_2 \cdots a_n$. If $M(d)$ is the number of circular sequences of length and period d, then $dM(d)$ is the number of linear sequences of length n corresponding to them. This gives us the equation

$$\sum_{d \mid n} dM(d) = r^n. \tag{2.1.16}$$

If we take $f(x) = r^x$ and $g(x) = xM(x)$, we may apply Möbius inversion to (2.1.16) and obtain

$$nM(n) = \sum_{d \mid n} \mu(d) r^{n/d}, \tag{2.1.17}$$

whence

$$M(n) = \frac{1}{n} \sum_{d \mid n} \mu(d) r^{n/d}. \tag{2.1.18}$$

This gives the number of circular permutations of length and period n. If we wish the total number of circular permutations of length n, this number is $T(n)$:

$$T(n) = \sum_{d \mid n} M(d). \tag{2.1.19}$$

If we wish the total number of circular permutations of n objects in which the number of objects of each kind is specified (say, b_i of the ith kind, $i = 1, \ldots, r$, where $b_1 + b_2 + \cdots + b_r = n$), we recall that the number of linear permutations is the multinomial coefficient

$$\frac{n!}{b_1! \cdots b_r!}, \qquad b_1 + b_2 + \cdots + b_r = n. \tag{2.1.20}$$

Here a circular permutation of this kind of length n and period d will have d a divisor of all $b_1, b_2, \ldots, b_r$, or, what amounts to the same thing, d a divisor of $(b_1, \ldots, b_r)$, the greatest common divisor of $b_1, \ldots, b_r$. Thus, if $M(b_1, \ldots, b_r)$, $b_1 + b_2 + \cdots + b_r = n$, is the number of circular permutations of length and period n with b_i objects of type i, $i = 1, \ldots, r$, the same argument given above yields

$$M(b_1, \ldots, b_r) = \frac{1}{n} \sum_{d \mid (b_1, \ldots, b_r)} \mu(d) \frac{(n/d)!}{(b_1/d)! \cdots (b_r/d)!}. \tag{2.1.21}$$

The "problème des ménages" is as follows: A hostess wishes to place n couples at a circular table so that men and women are in alternate places, but so that no husband will sit on either side of his wife. In how many ways may this be done? It is easy to see that it cannot be done with fewer than three couples, but for three or more couples it may be done.

Let us first place the ladies at alternate places, designating them by numbers $1, 2, \ldots, n$ in circular order. Let the place to the left of the ith lady and on the right of the $(i+1)$st be numbered i, giving number n to the place between the nth lady and the first. Then the first husband can sit anywhere except in the nth or first place, and the ith husband anywhere except the $(i-1)$st or the ith. If husband number a_i sits in place i, then $a_1a_2\cdots a_n$ is a permutation of $1, 2, \ldots, n$ and in the array

$$\begin{matrix} 1 & 2 & \cdots & n-1 & n \\ 2 & 3 & \cdots & n & 1 \\ a_1 & a_2 & \cdots & a_{n-1} & a_n \end{matrix} \tag{2.1.22}$$

we see our condition is precisely that the permutation $a_1a_2\cdots a_n$ must be discordant with the first two rows. We are thus in a problem of inclusion and exclusion with properties $P(1)$: $a_1 = 1, 2; \ldots ; P(i)$: $a_i = i, i+1; \ldots ;$ $P(n)$: $a_n = n, 1$. If $P(i)$ is true for r values $a_{i_1}, \ldots, a_{i_r}$, there will be $(n-r)!$ ways of completing the permutation. Thus, we must first calculate the number of ways of having $P(i)$ true for r values, or, as we shall say, the number of ways of having r hits. The number of ways of having one hit is $2n$, and by the circular symmetry of (2.1.22) we may suppose this to be either the n in the $(n-1)$st column or the n in the nth column. We now list the remaining numbers, writing the columns one after another, giving either

$$\begin{aligned} &1, 1, 2, 2, 3, \ldots, n-2, n-2, n-1, \\ \text{or} \quad &1, 2, 2, 3, 3, \ldots, n-2, n-1, n-1. \end{aligned} \tag{2.1.23}$$

Our remaining choices of $(r-1)$ numbers are restricted by saying that in the arrays in (2.1.23) we may not choose two consecutive values because consecutive choices amount either to taking the same number twice or to taking both elements in a column of (2.1.22). This is the number of ways of choosing $(r-1)$ objects, no two consecutive, from a row of $2n-3$. This has been evaluated in Chapter 1 and is $\binom{2n-r-1}{n-1}$. This number is to be multiplied by $2n$ for the first choice, but the same set of r values could be obtained by regarding any one of the r values as the first, and so we must divide by r. Thus, our number is

$$\frac{2n}{r}\binom{2n-r-1}{r-1} = \frac{2n}{2n-r}\binom{2n-r}{r}. \tag{2.1.24}$$

We may now apply formula (2.1.1) and find as our answer that the number of solutions U_n of permutations discordant both with $1, 2, \ldots, n$ and $2, 3, \ldots, n, 1$

is given by

$$U_n = n! - 2n\cdot(n-1)! + \cdots + (-1)^r \frac{2n}{2n-r}\binom{2n-r}{r}(n-r)! + \cdots + (-1)^n\cdot 2. \quad (2.1.25)$$

From this relation we may derive the recursion

$$(n-2)U_n = n(n-2)U_{n-1} + nU_{n-2} + 4(-1)^{n+1}. \quad (2.1.26)$$

This recursion can be proved without much difficulty. For $r = 0$ and 1, the terms in $(n-2)U_n$ and $n(n-2)U_{n-1}$ are equal. For $r = 2, \ldots, n-1$, we have the identity involving the rth term of U_n and U_{n-1} and the $(r-2)$nd term of U_{n-2}:

$$(n-2)\frac{2n}{2n-r}\binom{2n-r}{r}(n-r)! = n(n-2)\frac{2(n-1)}{2n-r-2}\binom{2n-r-2}{r}(n-r-1)! + n\cdot\frac{2(n-2)}{2n-r-2}\binom{2n-r-2}{r-2}(n-r)! \quad (2.1.27)$$

Finally, the term with $r = n$ in $(n-2)U_n$ and that with $r = n-2$ in nU_{n-2} combine, so that

$$(n-2)(-1)^n\cdot 2 = n(-1)^{n-2}\cdot 2 + 4(-1)^{n+1}. \quad (2.1.28)$$

This proves the validity of the recursion (2.1.26).

2.2 Partially Ordered Sets and Their Möbius Functions

A *partially ordered set* P is a system $P\{\ldots, x, y, \ldots\}$ of elements with an ordering relation $x \geq y$ (read "x includes y") that holds for certain pairs of elements and an equality $x = y$, such that the following axioms hold:

PO 1. *$x \geq x$ for every x of P.*
PO 2. *If $x \geq y$ and $y \geq z$, then $x \geq z$.*
PO 3. *If $x \geq y$ and $y \geq x$, then $x = y$.*

A *simply ordered set* or *chain* also satisfies:

PO 4. *If x, y are elements of P, then either $x \geq y$ or $y \geq x$.*

We write $y \leq x$ as an alternate form of $x \geq y$, and $x > y$ (or $y < x$) if $x \geq y$ (or $y \leq x$) and $x \neq y$.

Partial ordering is a very general concept. Two particular cases of interest are:

1. The elements of P are all the subsets of a finite set T, where we write 0 for the void subset and 1 for the set T itself, and $y \leq x$ means that y is a subset of x.
2. The elements of P are the positive integers and $y \leq x$ means that y divides x.

It is easy to check the validity of the axioms in both instances.

If T is a subset of a partially ordered set P, then an element x of P such that $x \leq t$ for every t of T is called a *lower bound* of T. If z is a lower bound of T such that $x \leq z$ for every lower-bound x of T, then z is called a *greatest lower bound* of T. From PO 3 it follows that if T has a greatest lower bound, it is necessarily unique. Similarly, if $x \geq t$ for every t of T, x is called an *upper bound* of T, and if z is an upper bound of T such that $x \geq z$ for every upper bound x, then z is called a *least upper bound* of T and again is clearly unique if it exists. If P itself has a greatest lower bound, this is called its *zero element*, and if it has a least upper bound this is called the *all element* (or sometimes the unit element). An *interval* $[x, y]$ where $x \leq y$ is the set of elements w such that $x \leq w \leq y$. If x and y are the only elements in the interval $[x, y]$, we say that *y covers x*. A partially ordered set P is said to be *locally finite* if the number of elements in every interval $[x, y]$ is finite.

The Möbius function and Möbius inversion were defined for functions over locally finite partially ordered sets originally by L. Weisner [1] and P. Hall [2]. This idea has recently been greatly expanded by G. C. Rota [1]. A brief treatment is given here, based on Rota's work.

We consider a class of real-valued functions $f(x, y)$ defined for $x, y \in P$, a locally finite partially ordered set. We require that $f(x, y) = 0$ if $x \nleq y$. The sum of two such functions, as well as multiplication by scalars, is defined as usual. The product $h = fg$ is defined as follows:

$$h(x, y) = \sum_{x \leq z \leq y} f(x, z)g(z, y), \qquad x, y \text{ fixed.} \tag{2.2.1}$$

This product is well defined, since the sum on the right is finite, P being locally finite. Under the operations of sum, scalar product, and the product rule $h = fg$ of (2.2.1), the functions $f(x, y)$ define the *incidence algebra $A(P)$* of P. It is easy to verify that the multiplication defined for $A(P)$ is associative and distributive and that $A(P)$ has an identity, the Kronecker delta function $\delta(x, y) = 1$, $\delta(x, y) = 0$ if $x \neq y$.

LEMMA 2.2.1. *A function $f(x, y)$ of $A(P)$ has both a left and a right inverse if and only if $f(x, x) \neq 0$ for every x of P.*

Proof: ■ In (2.2.1) take $h(x, y) = \delta(x, y)$. Now, given f, we wish to solve for g. Since this requires $1 = \delta(x, x) = f(x, x)g(x, x)$ for every x, the condition $f(x, x) \neq 0$ for every x of P is clearly necessary. Thus, suppose $f(x, x) \neq 0$ for every x. Then $g(x, x) = f(x, x)^{-1}$ for every x. To evaluate $g(x, y)$ with $x < y$, we may assume inductively that we have already found $g(z, y)$ for every z satisfying $x < z \leq y$. Then

$$h(x, y) = \delta(x, y) = 0 = \sum_{x \leq z \leq y} f(x, z)g(z, y),$$

whence

$$-f(x, x)g(x, y) = \sum_{x < z \leq y} f(x, z)g(z, y), \tag{2.2.2}$$

and we may find $g(x, y)$, since $f(x, x) \neq 0$ and all terms of the finite sum on the right are known. Thus, f has a right inverse. Similarly applying our induction to terms $x \leq z < y$, we may use (2.2.1), interchanging the roles of f and g to show that f has a left inverse. But if $fg_1 = 1 = \delta(x, y)$ and $g_2 f = 1$, then by a familiar argument, $g_2 = g_2 1 = g_2(fg_1) = (g_2 f)g_1 = 1g_1 = g_1$ and the left and right inverses are the same.

DEFINITION. *Let P be a locally finite partially ordered set and $A(P)$ its incidence algebra. The zeta function $\zeta(x, y)$ of $A(P)$ is that function for which $\zeta(x, y) = 1$ for $x \leq y$, $\zeta(x, y) = 0$ otherwise. The Möbius function $\mu(x, y)$ of $A(P)$ is the inverse of the zeta function.*

Since $\zeta(x, x) = 1 \neq 0$ for every x, by our lemma $\zeta(x, y)$ has an inverse function $\mu(x, y)$, which is both a right and left inverse of it. Hence, we have

$$\mu(x, x) = 1, \qquad \text{for every } x \text{ of } P. \tag{2.2.3}$$

For $x < y$, we have

$$\mu(x, y) = -\sum_{x \leq z < y} \mu(x, z), \qquad x < y \text{ fixed}. \tag{2.2.4}$$

$$\mu(x, y) = -\sum_{x < z \leq y} \mu(z, y), \qquad x < y \text{ fixed}. \tag{2.2.5}$$

Here (2.2.4) expresses the Möbius function as the left inverse of zeta, and (2.2.5) expresses it as the right inverse.

The Möbius inversion theorem follows.

THEOREM 2.2.1. *Let P be a locally finite partially ordered set with a zero element 0. Let $f(x)$ be given for all x of P and let $g(x)$ be determined from $f(x)$ by the rule*

$$g(x) = \sum_{y \leq x} f(y), \qquad \textit{all } x \textit{ of } P. \tag{2.2.6}$$

Then, if $\mu(y, z)$ is the Möbius function of P, we have

$$f(x) = \sum_{y \leq x} g(y)\mu(y, x), \qquad \textit{all } x \textit{ of } P. \tag{2.2.7}$$

Proof: ■ Since every interval $[0, x]$ is finite, the sums in (2.2.6) and (2.2.7) are well defined. For a fixed x, consider the sum

$$S = \sum_{y \leq x} g(y)\mu(y, x) = \sum_{y \leq x}\left(\sum_{z \leq y} f(z)\right)\mu(y, z), \tag{2.2.8}$$

where we have substituted from (2.2.6). Now interchange the order of summation to get

$$\begin{aligned} S &= \sum_{z \leq y} f(z) \sum_{y \leq x} \mu(y, z) = \sum_{z} f(z)\zeta(z, y) \sum_{y \leq x} \mu(y, x) \\ &= \sum_{z} f(z) \sum_{z \leq y \leq x} \zeta(z, y)\mu(y, x) = \sum_{z} f(z)\delta(z, x) = f(x). \end{aligned} \tag{2.2.9}$$

Since $S = f(x)$, we have proved (2.2.7), the conclusion of our theorem.

Let us now determine the Möbius function for the two special cases mentioned at the beginning of this section.

In case 1, P is the partially ordered set of all subsets of a finite set T, ordered by inclusion. Here we assert that for $x \leq y$,

$$\mu(x, y) = (-1)^{n(y)-n(x)}, \tag{2.2.10}$$

where $n(x)$, $n(y)$ are respectively the number of elements of T in x and in y. The assertion is certainly true when $n(y) - n(x) = 0$ or 1. By induction assume (2.2.10) to be true for $n(y) - n(x) \leq r - 1$ and consider a case with $n(y) - n(x) = r$. Then (2.2.4) becomes

$$\mu(x, y) = -1 + \binom{r}{1} - \binom{r}{2} + \cdots - \binom{r}{j}(-1)^j + \cdots - \binom{r}{r-1}(-1)^{r-1}, \tag{2.2.11}$$

since there are $\binom{r}{j}$ z's with $x \leq z < y$ with $n(z) - n(x) = j$, namely, the subsets of T obtained by adjoining to x and j of the r elements of y not in x. Comparison of (2.2.11) with the binomial expansion of $(1 - 1)^r = 0$ gives $\mu(x, y) = (-1)^r$, as was to be shown.

Let T be the integers $1, 2, \ldots, n$ and let properties $P(1), P(2), \ldots, P(n)$ be associated with these integers. Let K be a set of N elements, each of which has the properties $P(i)$, $i \in x$ for some subset x of T. Let $f(x)$ be the number of elements of K having exactly the properties $P(i)$, $i \notin x$, x a subset of T. Then, if we put

$$g(x) = \sum_{y \leq x} f(y), \tag{2.2.12}$$

the function $g(x)$ is the number of elements of K having all the properties $P(i)$ for $i \notin x$ and possibly others. Here, for $x = T$, the inversion (2.2.7) gives us

$$f(T) = g(T) - \sum_{n(y)=n-1} g(y) + \cdots + (-1)^j \sum_{n(y)=n-j} g(y) + \cdots + (-1)^n \sum_{n(y)=0} g(y). \tag{2.2.13}$$

But here $f(T) = N(0)$ is the number of elements having none of the properties $g(T) = N$, since this counts all elements having properties of the void set and possibly others. If $n(y) = n - j$, then $g(y)$ counts all elements having the j properties not in y and possibly others. But this shows that (2.2.13) is the principle of inclusion and exclusion of (2.1.1).

In case 2, P is the partially ordered set of the positive integers, where $x \leq y$ means that x divides y.

Here in a segment $[x, y]$, if $x \leq z \leq y$, then $z = xd$, where $d \mid (y/x)$, and so this segment corresponds to the divisors of y/x. Note that the integer 1 is the zero element of P. Comparison of (2.2.4) with the lemma preceding Theorem 2.1.1 shows that $\mu(x, y) = \mu(y/x)$ in this case. Thus, Theorem 2.1.1 on Möbius inversion is the special case of Theorem 2.2.1 for P, the positive integers partially ordered by division.

PROBLEMS

1. Let A be the $n \times n$ matrix with zeros down the main diagonal and 1's elsewhere. The determinant of A is $(-1)^{n-1}(n-1)$. (This will be shown in Section 10.2.) Of the $n!$ terms in the expansion of the determinant of A, how many are $+1$, -1, 0, respectively?
2. Given the array

$$\begin{array}{cccccccccc} 0 & 1 & 2 & 3 & 4 & 5 & 6 & 7 & 8 & 9 \\ 1 & 2 & 3 & 4 & 0 & 6 & 7 & 8 & 9 & 5 \\ a_0 & a_1 & a_2 & a_3 & a_4 & a_5 & a_6 & a_7 & a_8 & a_9, \end{array}$$

   in how many ways can we choose $a_0, \ldots, a_9$ as a permutation of $0, \ldots, 9$ so that no column of the array will have a repeated number?
3. Show that U_n in (2.1.25) is approximately $n!/e^2$.
4. The function $\Lambda(n)$ is defined for positive integers n by the rule

$$\sum_{d \mid n} \Lambda(d) = \log n.$$

   Prove that $\Lambda(n) = \log p$ if $n = p^e$, p a prime, and $\Lambda(n) = 0$ otherwise.
5. Find the Möbius functions of the two partially ordered sets P_1, P_2 with five elements 0, a, b, c, 1, where: (a) in P_1, $0 \leq a \leq 1$, $0 \leq b \leq 1$, $0 \leq c \leq 1$, and there are no further inclusions; (b) in P_2, $0 \leq a \leq 1$, $0 \leq b \leq c \leq 1$, and there are no further inclusions.

3

Generating Functions and Recursions

3.1 Rules and Properties

If $u_0, u_1, u_2, \ldots, u_n, \ldots$ is a sequence of numbers, we may associate with this sequence a *generating function* $g(x)$ by the rule

$$g(x) = u_0 + u_1x + u_2x^2 + \cdots + u_nx^n + \cdots. \qquad (3.1.1)$$

If this series has a circle of convergence with a radius $R > 0$, then it may happen that the properties of the function $g(x)$ enable us to evaluate the coefficients u_n (or at least give estimates of their order of magnitude) or perhaps find other information of value. If $h(x)$ is the generating function of the sequence $v_0, v_1, v_2, \ldots, v_n, \ldots$, then

$$h(x) = v_0 + v_1x + v_2x^2 + \cdots + v_nx^n + \cdots. \qquad (3.1.2)$$

If we add (3.1) multiplied by c, and (3.2) multiplied by d, we have

$$cg(x) + dh(x) = (cu_0 + dv_0) + (cu_1 + dv_1)x + \cdots + (cu_n + dv_n)x^n + \cdots, \qquad (3.1.3)$$

and if we multiply, we have

$$g(x)h(x) = w_0 + w_1x + w_2x^2 + \cdots + w_nx^n + \cdots, \qquad (3.1.4)$$

where for every $n = 1, 2, 3, \ldots,$

$$w_n = u_0v_n + u_1v_{n-1} + \cdots + u_{n-1}v_1 + u_nv_0. \tag{3.1.5}$$

Even if the series for $g(x)$ and $h(x)$ are not convergent, we may regard (3.1.3), (3.1.4), and (3.1.5) as defining formal operations on formal series. In these terms we easily verify that the addition, multiplication by scalars, and series multiplication satisfy the associative, commutative, and distributive laws. Furthermore, if $u_0 \neq 0$ and if we take $v_0 = u_0^{-1}$, we may use (3.1.5) to determine $v_1, v_2, \ldots$ recursively to make $g(x)h(x) = 1$.

Instead of the generating function $g(x)$ associated with $u_0, u_1, \ldots, u_n, \ldots,$ we may associate an *exponential generating function*, $G(x)$, by the rule

$$G(x) = u_0 + u_1x + \frac{u_2x^2}{2!} + \cdots + \frac{u_nx^n}{n!} + \cdots. \tag{3.1.6}$$

Similarly, let $H(x)$ be associated with $v_0, v_1, \ldots$ by

$$H(x) = v_0 + v_1x + \frac{v_2x^2}{2!} + \cdots + \frac{v_nx^n}{n!} + \cdots. \tag{3.1.7}$$

Then, for the product $G(x)H(x) = K(x)$,

$$G(x)H(x) = K(x) = w_0 + w_1x + \frac{w_2x^2}{2!} + \cdots + \frac{w_nx^n}{n!} + \cdots, \tag{3.1.8}$$

where

$$w_n = u_0v_n + \binom{n}{1}u_1v_{n-1} + \cdots + \binom{n}{r}u_rv_{n-r} + \cdots + \binom{n}{n}u_nv_0, \tag{3.1.9}$$

or symbolically,

$$w^n = (u + v)^n. \tag{3.1.10}$$

In the symbolical formula (3.1.10), it is to be understood that after expansion of $(u + v)^n$ by the binomial formula, all exponents are to be replaced by subscripts.

Suppose that a sequence $u_0, u_1, u_2, \ldots, u_n, \ldots$ satisfies the recurrence of rth order:

$$u_{n+r} = a_1u_{n+r-1} + a_2u_{n+r-2} + \cdots + a_ru_n, \qquad n = 0, 1, 2, \ldots, \tag{3.1.11}$$

where the a_i, $i = 1, \ldots, r$ are constants. Then, if $g(x)$ is the generating function for the sequence $\{u_n\}$, and if we take $k(x)$ as the polynomial

$$k(x) = 1 - a_1x - a_2x^2 - \cdots - a_rx^r, \tag{3.1.12}$$

we find that

$$g(x)k(x) = c_0 + c_1x + c_2x^2 + \cdots + c_{r-1}x^{r-1} = C(x), \tag{3.1.13}$$

where $C(x)$ is a polynomial of degree at most $r - 1$, since if c_{n+r} is the coefficient of x^{n+r}, $n \geq 0$, in the product $g(x)k(x)$, we find

$$c_{n+r} = u_{n+r} - a_1u_{n+r-1} \cdots - a_ru_n = 0, \tag{3.1.14}$$

using the recursion (3.1.11). Thus, for a sequence $\{u_n\}$ satisfying the linear recurrence (3.1.11), the generating function $g(x)$ is a rational function

$$g(x) = \frac{C(x)}{k(x)}. \tag{3.1.15}$$

With the linear recurrence (3.1.11) we associate the *characteristic polynomial* $f(x)$ given by

$$f(x) = x^r - a_1 x^{r-1} - \cdots - a_r. \tag{3.1.16}$$

Without loss of generality we shall assume that $a_r \neq 0$, since if $a_r = 0$, the recurrence is not truly of order r but is of lower order. Let the factorization of $f(x)$ into linear factors be

$$f(x) = (x - \alpha_1)^{e_1}(x - \alpha_2)^{e_2} \cdots (x - \alpha_s)^{e_s}, \qquad e_1 + e_2 + \cdots + e_s = r, \tag{3.1.17}$$

where $\alpha_1, \ldots, \alpha_s$ are the (possibly) complex roots of $f(x)$. Comparing $f(x)$ of (3.1.16) and $k(x)$ of (3.1.12), we see that

$$k(x) = x^r f\left(\frac{1}{x}\right), \tag{3.1.18}$$

and corresponding to the factorization (3.1.17) of $f(x)$, we have the factorization of $k(x)$:

$$k(x) = (1 - \alpha_1 x)^{e_1} \cdots (1 - \alpha_s x)^{e_s}, \qquad e_1 + e_2 + \cdots + e_s = r. \tag{3.1.19}$$

We may express the rational function $g(x) = C(x)/k(x)$ in terms of partial fractions

$$g(x) = \frac{C(x)}{k(x)} = \sum_{i=1}^{s} \sum_{k=1}^{e_i} \frac{\beta_{ik}}{(1 - \alpha_i x)^k}, \tag{3.1.20}$$

where the β's are appropriate constants.

Thus, (3.1.20) expresses the generating function as a sum of functions of the form

$$\frac{\beta}{(1 - \alpha x)^k} = \beta(1 - \alpha x)^{-k}. \tag{3.1.21}$$

We may easily expand (3.1.21) by the binomial formula to find

$$\beta(1 - \alpha x)^{-k} = \beta\left(1 + (-k)(-\alpha x) + \cdots + \frac{(-k)\cdots(-k - n + 1)(-\alpha x)^n}{n!} + \cdots\right). \tag{3.1.22}$$

In this, the coefficient of x^n is

$$\frac{\beta(n + k - 1)\cdots(k)}{n!}\alpha^n = \beta\binom{n + k - 1}{n}\alpha^n = \beta\binom{n + k - 1}{k - 1}\alpha^n. \tag{3.1.23}$$

We note that

$$\sum_{k=1}^{e_i} \beta_{ik}\binom{n+k-1}{k-1}\alpha_i^n = P_i(n)\alpha_i^n, \tag{3.1.24}$$

where $P_i(n)$ is a polynomial of degree at most $e_i - 1$ in n, and that any polynomial $P_i(n)$ can be obtained by using an appropriate choice of constants β_{ik}. Substituting back in (3.1.20), we have

$$\begin{aligned} g(x) &= \sum_{n=0}^{\infty} u_n x^n \\ &= \sum_{n=0}^{\infty} \sum_{i=1}^{s} P_i(n)\alpha_i^n x^n, \end{aligned} \tag{3.1.25}$$

and comparing coefficients of x^n, we have

$$u_n = \sum_{i=1}^{s} P_i(n)\alpha_i^n, \tag{3.1.26}$$

where $P_i(n)$ is degree at most $e_i - 1$.

We shall state this result as a theorem.

THEOREM 3.1.1. *Suppose a sequence* $u_0, u_1, u_2, \ldots, u_n, \ldots$ *satisfies the linear recurrence with constant coefficients*

$$u_{n+r} = a_1 u_{n+r-1} + \cdots + a_r u_n, \qquad n \geq 0.$$

Let us call $f(x) = x^r - a_1 x^{r-1} - \cdots - a_r$ *the characteristic polynomial of this recurrence and let*

$$f(x) = (x-\alpha_1)^{e_1} \cdots (x-\alpha_s)^{e_s}, \qquad e_1 + e_2 + \cdots + e_s = r$$

be the factorization of $f(x)$ *as a product of linear factors. Then*

$$u_n = \sum_{i=1}^{s} P_i(n)\alpha_i^n$$

for all n, *where* $P_i(n)$ *is a polynomial of degree at most* $e_i - 1$ *in* n. *The coefficients of the polynomials* $P_i(n)$ *are determined by the initial values* $u_0, u_1, \ldots, u_{r-1}$ *of the sequence* $\{u_n\}$.

3.2 Combinatorial Problems

Let us consider a combinatorial problem whose solution depends on a linear recurrence. Let u_t, $t \geq 2$, be the number of ways of finding a permutation $a_1 a_2 \ldots, a_t$ of $1, 2, \ldots, t$ such that for each i, a_i is in the ith column of the array:

$$\begin{matrix} & 1 & 2 & \cdots & t-3 & t-2 & t-1 \\ 1 & 2 & 3 & \cdots & t-2 & t-1 & t \\ 2 & 3 & 4 & \cdots & t-1 & t & \end{matrix} \tag{3.2.1}$$

Here we find directly that $u_2 = 2$, $u_3 = 3$, $u_4 = 5$. The number t must be used in either the tth or $(t - 1)$st column. Thus, our two choices are:

$$\begin{array}{ccccccc} & 1 & 2 & \cdots & t-3 & t-2 & \textcircled{t} \\ 1 & 2 & 3 & \cdots & t-2 & t-1 & \\ 2 & 3 & 4 & \cdots & t-1 & & \end{array} \tag{3.2.2}$$

or

$$\begin{array}{ccccccc} & 1 & 2 & \cdots & t-3 & \textcircled{t} & \textcircled{t-1} \\ 1 & 2 & 3 & \cdots & t-2 & & \\ 2 & 3 & 4 & \cdots & & & \end{array} \tag{3.2.3}$$

In both cases chosen numbers have been circled. In (3.2.2) we have omitted the t from the $(t - 1)$st column, since it cannot be used there. In (3.2.3), having chosen t from the $(t - 1)$st column, we must choose $t - 1$ from the tth column and then omit $t - 1$ from the $(t - 2)$nd column. The number of choices in (3.2.2) for $1, 2, \ldots, t - 1$ is u_{t-1}, and the number of choices in (3.2.3) is u_{t-2}. Hence, as these combine to give all choices in (3.2.1), we have

$$u_t = u_{t-1} + u_{t-2}, \tag{3.2.4}$$

a linear recurrence of second order for u_t. Although the combinatorial problem is not meaningful for $t = 0$ or 1, the values $u_0 = 1$, $u_1 = 1$ are consistent with the recurrence (3.2.4) and the succession of values $u_0 = 1$, $u_1 = 1$, $u_2 = 2$, $u_3 = 3$, $u_4 = 5, \ldots$. The characteristic polynomial of (3.2.4) is

$$f(x) = x^2 - x - 1 = (x - \alpha_1)(x - \alpha_2), \tag{3.2.5}$$

where

$$\alpha_1 = \frac{1 + \sqrt{5}}{2}, \quad \alpha_2 = \frac{1 - \sqrt{5}}{2}. \tag{3.2.6}$$

We easily find from Theorem 3.1.1 and our initial values that

$$u_n = \frac{1}{\sqrt{5}} \cdot (\alpha_1^{n+1} - \alpha_2^{n+1}). \tag{3.2.7}$$

Another, more natural, combinatorial problem can be reduced to the evaluation of u_t just made. What is the number z_n, $n \geq 3$, of permutations $a_1 a_2 \cdots a_n$ of $1, 2, \ldots, n$ such that a_i is in the ith column of the following array?

$$\begin{array}{cccccccc} 1 & 2 & 3 & \cdots & n-3 & n-2 & n-1 & n \\ 2 & 3 & 4 & \cdots & n-2 & n-1 & n & 1 \\ 3 & 4 & 5 & \cdots & n-1 & n & 1 & 2 \end{array} \tag{3.2.8}$$

The complete set of choices can be subdivided according to the choice of column in which n is selected, and if this is not the $(n - 1)$st column, the choice of $n - 1$ or 1 is in the $(n - 1)$st column. These choices may be indicated by the circled values shown in the last three columns as follows:

$$
\begin{array}{llll}
\text{(a)} & n-2 & \textcircled{$n-1$} & \textcircled{n} \\
& n-1 & n & 1 \\
& n & 1 & 2
\end{array}
\qquad
\begin{array}{llll}
\text{(b)} & n-2 & n-1 & \textcircled{n} \\
& n-1 & n & 1 \\
& n & \textcircled{1} & 2
\end{array}
\tag{3.2.9}
$$

$$
\begin{array}{llll}
\text{(c)} & n-2 & n-1 & n \\
& n-1 & \textcircled{n} & 1 \\
& n & 1 & 2
\end{array}
\qquad
\begin{array}{llll}
\text{(d)} & n-2 & n-1 & n \\
& n-1 & n & 1 \\
& \textcircled{n} & \textcircled{1} & 2
\end{array}
\qquad
\begin{array}{llll}
\text{(e)} & n-2 & \textcircled{$n-1$} & n \\
& n-1 & n & 1 \\
& \textcircled{n} & 1 & 2
\end{array}
$$

In case (a) there is exactly one choice, namely, the top row of (3.2.8), for only a single 1 remains to be chosen, and this in the first column. Having chosen this 1, only the 2 in the second column remains, and similarly the choices of $3, 4, \ldots, n-2$ in the first row are forced. In case (b) we are to choose from

$$
\begin{array}{ccccccc}
 & 2 & 3 & \cdots & n-3 & n-2 \\
2 & 3 & 4 & \cdots & n-2 & n-1 \\
3 & 4 & 5 & \cdots & n-1 &
\end{array}
\tag{3.2.10}
$$

and this number is u_{n-2}. In (c) the choices are from the array for u_{n-1}. In (d) there is exactly one choice, namely, the third row, since only one $n-1$ remains to be chosen and that is in the $(n-3)$rd column; similarly, the $n-2$, $n-3, \ldots, 2$ in the third row must be chosen. In (e) the choices are from the array for u_{n-2}. Hence, the total number of choices for (3.2.8), z_n, is given by

$$
\begin{aligned}
z_n &= 1 + u_{n-2} + u_{n-1} + 1 + u_{n-2} \\
&= u_n + u_{n-2} + 2 \\
&= \alpha_1^n + \alpha_2^n + 2,
\end{aligned}
\tag{3.2.11}
$$

where

$$
\alpha_1 = \frac{1+\sqrt{5}}{2}, \qquad \alpha_2 = \frac{1-\sqrt{5}}{2}
$$

as before.

The number u_n of derangements of $1, 2, \ldots, n$ evaluated in (2.1.6) may also be found recursively. Consider a derangement

$$
\begin{pmatrix} 1 & 2 & \cdots & n \\ a_1 & a_2 & \cdots & a_n \end{pmatrix}.
\tag{3.2.12}
$$

If $a_1 = j$, we consider the partial permutation

$$
\begin{pmatrix} 2 & \cdots & j & \cdots & n \\ a_2 & \cdots & a_j & \cdots & a_n \end{pmatrix}.
\tag{3.2.13}
$$

Two cases are to be considered here. First are cases with $a_j \neq 1$, and second those with $a_j = 1$. These cases are collectively exhaustive and mutually exclusive. In the first case we have

$$\begin{pmatrix} 2 & \cdots & i & \cdots & j & \cdots & n \\ a_2 & \cdots & 1 & \cdots & a_j & \cdots & a_n \end{pmatrix}, \qquad i \neq j \tag{3.2.14}$$

which may be associated with the derangement of $2, \ldots, n$:

$$\begin{pmatrix} 2 & \cdots & i & \cdots & j & \cdots & n \\ a_2 & \cdots & j & \cdots & a_j & \cdots & a_n \end{pmatrix}. \tag{3.2.15}$$

Conversely, every derangement of $2, \ldots, n$ such as (3.2.15) leads to $(n - 1)$ derangements of $1, 2, \ldots, n$ of this first kind by taking j as $2, \ldots, n$ in turn. In the second case, with $a_j = 1$, the partial permutation of (3.2.13) takes the form

$$\begin{pmatrix} 2 & \cdots & j & \cdots & n \\ a_2 & \cdots & 1 & \cdots & a_n \end{pmatrix}. \tag{3.2.16}$$

If we delete the $\binom{j}{1}$, this is a derangement of $2, \ldots, j-1, j+1, \ldots, n$; conversely, for each $j = 2, \ldots, n$ such derangements of $(n - 2)$ numbers lead to derangements of $1, \ldots, n$ of the second type. These two cases combine to give us the recursion

$$u_n = (n-1)u_{n-1} + (n-1)u_{n-2}. \tag{3.2.17}$$

We may easily verify that this recursion yields the same numbers as (2.1.6).

A sequence $x_1x_2 \cdots x_n$ may be combined in this order by a binary non-associative product in a number of ways. What is this number u_n? For $n = 3$, 4 we have the possibilities

$$\begin{aligned} &x_1(x_2x_3), && (x_1x_2)x_3; \\ &x_1(x_2(x_3x_4)), && x\,((x_2x_3)x_4); \\ &(x_1x_2)(x_3x_3); && \\ &(x_1(x_2x_3))x_4, && ((x\;x_2)x_3)x_4. \end{aligned} \tag{3.2.18}$$

Thus, $u_3 = 2$, $u_4 = 5$. We also have $u_2 = 1$ and will take $u_1 = 1$ as a convention. The last product will be some composite of the first r letters multiplied by some composite of the last $n - r$, of the form $(a_1 \cdots a_r)(a_{r+1} \cdots a_n)$. The first r can be combined in u_r ways (here the convention $u_1 = 1$ fits) and the last $n - r$ in u_{n-r} ways. Thus

$$u_n = u_1u_{n-1} + u_2u_{n-2} + \cdots + u_{n-1}u_1, \qquad n \geq 2. \tag{3.2.19}$$

Let us write the generating function $f(x)$ as

$$f(x) = u_1x + u_2x^2 + \cdots + u_nx^n + \cdots, \tag{3.2.20}$$

postponing the consideration of its convergence. The recursion (3.2.19) is equivalent formally to the relation

$$(f(x))^2 = -x + f(x) \tag{3.2.21}$$

We note that $u_1 = 1$ and that the recursion (3.2.19) holds only for $n \geq 2$ and must account for this by the $-x$ on the right-hand side of (3.2.21). Solving (3.2.21) for $f(x)$ as a quadratic equation, we have

$$f(x) = \frac{1 - \sqrt{1 - 4x}}{2}. \tag{3.2.22}$$

Here we take the minus sign, since the series for $f(x)$ has no constant term. Expanding (3.2.22) as a power series, we find the coefficient of x^n to be v_n, where

$$v_n = \frac{(\frac{1}{2})(-\frac{1}{2}) \cdots ((3 - 2n)/2)(-4)^n(-\frac{1}{2})}{n!}. \tag{3.2.23}$$

This simplifies to

$$v_n = \frac{(2n - 2)!}{n!\,(n - 1)!}. \tag{3.2.24}$$

We may now observe that the series for $f(x)$ as given by (3.2.22) must converge for $|x| < \frac{1}{4}$, and for these values the equation (3.2.21)—and hence the recursion (3.2.19)—with v_n in place of u_n must hold. But as $u_1 = v_1 = 1$, we have $u_n = v_n$ for all $n \geq 1$, and so our solution is

$$u_n = \frac{(2n - 2)!}{n!\,(n - 1)!} \tag{3.2.25}$$

for all $n \geq 2$. We observe that an attempt to prove the convergence of (3.2.20) on the basis of (3.2.19) alone is exceedingly difficult.

PROBLEMS

1. The Fibonacci numbers are the sequence of numbers $u_0, u_1, \ldots, u_n, \ldots$, with $u_0 = 0$, $u_1 = 1$, and $u_{n+2} = u_{n+1} + u_n$. Show that every positive integer N has a unique representation

$$N = \sum_{i=1}^{\infty} a_i u_i$$

with $a_i = 0$ or 1 and $a_i a_{i+1} = 0$, $i \geq 1$.

2. Numbers $s(n, r)$ and $S(n, r)$, called the Stirling numbers of the first and second kind, respectively, are defined by the rules

$$(x)_n = x(x - 1) \cdots (x - n + 1) = \sum_{r=0}^{n} s(n, r)x^r, \qquad n > 0$$

and

$$x^n = \sum_{r=0}^{n} S(n, r)(x)_r, \qquad n > 0.$$

Here we take $x^0 = (x)_0 = 1$. Show that

$$\sum_r S(n, r)s(r, m) = \delta_{nm},$$

the Kronecker delta where $\delta_{nn} = 1$, $\delta_{nm} = 0$ if $n \neq m$. From this, show that of the two relations

(a) $$a_n = \sum_r s(n, r)b_r, \qquad n = 1, 2, \ldots,$$

(b) $$b_n = \sum_r S(n, r)a_r, \qquad n = 1, 2, \ldots,$$

each relation implies the other.

3. Use the relation $(x)_{m+1} = (x - m)(x)_m$ to derive the recursions for the Stirling numbers of the first and second kind:

$$s(n + 1, r) = s(n, r - 1) - ns(n, r),$$

$$S(n + 1, r) = S(n, r - 1) + rS(n, r).$$

4. Let $P_n = \sum_{r=0}^{n} (n)_r$, this being the total number of permutations of n distinct objects.
   (a) Show that P_n satisfies the recurrence $P_n = nP_{n-1} + 1$, $n \geq 1$, $P_0 = 1$.
   (b) Show that $P_n = n! \sum_{r=0}^{n} 1/r!$ and conclude that for $n \geq 1$, P_n is the nearest integer to $n!e$.
   (c) Show that $\sum_{n=0}^{\infty} P_n x^n/n! = e^x/(1 - x)$.
5. A Dirichlet generating function $A(s)$ for a sequence of numbers $a_1, a_2, \ldots$ is a formal series

$$A(s) = \sum_{n=1}^{\infty} \frac{a_n}{n^s}.$$

If

$$B(s) = \sum_{n=1}^{\infty} \frac{b_n}{n^s},$$

we define

$$A(s) + B(s) = C(s) = \sum_{n=1}^{\infty} \frac{c_n}{n^s},$$

where $c_n = a_n + b_n$ and

$$A(s)B(s) = V(s) = \sum_{n=1}^{\infty} \frac{v_n}{n^s},$$

where

$$v_n = \sum_{d \mid n} \frac{a_d b_n}{d}.$$

(a) Prove that this product rule is commutative and associative.
(b) Show that if $i_1 = 1, 0 = i_2 = i_3 = \cdots$, the series $I(s)$ is a unit for this multiplication.
(c) Show that $a_1 \neq 0$, the series $A(s)$ has an inverse $B(s)$ satisfying

$$A(s)B(s) = I(s).$$

6. Define the zeta function as

$$\zeta(s) = \sum_{n=1}^{\infty} \frac{1}{n^s}.$$

Show that the inverse of $\zeta(s)$ is the series

$$\zeta(s)^{-1} = \sum_{n=1}^{\infty} \frac{\mu(n)}{n^s},$$

where the function $\mu(n)$ is the Möbius function defined in (2.1.10). If $a_n = g(n)$, $b_n = f(n)$ for all n, and $B(s) = A(s)\zeta(s)$, then $A(s) = B(s)\zeta(s)^{-1}$. Show that this corresponds to the Möbius inversion formula of Theorem 2.1.1.

4

Partitions

4.1 Partitions. Identities and Arithmetic Properties

A partition of a positive integer n is a representation of n as a sum of positive integers

$$n = x_1 + x_2 + \cdots + x_k, \qquad x_i > 0, \qquad i = 1, \ldots, k. \tag{4.1.1}$$

We can easily find the number of ordered partitions of n into k parts, this being the number of ways of putting $(k - 1)$ markers in the $(n - 1)$ spaces between n dots. This number is $\binom{n-1}{k-1}$. If we do not specify the number k of parts, we may or may not place a marker in each space, and this may be done in 2^{n-1} ways.

The theory of unordered partitions is far more complicated and presents a number of interesting problems. Let us designate by $p_k(n)$ the number of (unordered) partitions of n into k parts. An unordered partition of n into k parts may be put into a standard form by listing the parts in decreasing order. Thus, $p_k(n)$ is the number of solutions in positive integers x_i of

$$n = x_1 + x_2 + \cdots + x_k, \qquad x_1 \geq x_2 \geq \cdots \geq x_k \geq 1. \tag{4.1.2}$$

It is not difficult to calculate $p_k(n)$ for small values of k and all values of n, but with increasing k the calculations become more tedious. From (4.1.2) we may write

$$n - k = (x_1 - 1) + (x_2 - 1) + \cdots + (x_k - 1), \tag{4.1.3}$$

where $x_1 \geq x_2 \geq \cdots \geq x_k \geq 1$. Putting $y_i = x_i - 1$, $i = 1, \ldots, k$, we have

$$n - k = y_1 + y_2 + \cdots + y_k, \qquad y_1 \geq y_2 \geq \cdots \geq y_k \geq 0. \tag{4.1.4}$$

But here, if $y_s > 0$ but $y_{s+1} = \cdots = y_k = 0$, (4.1.4) gives a partition of $n - k$ into s parts. Thus, the number of solutions of (4.1.4) is the number of partitions of $n - k$ into at most k parts; in other words, the number is

$$p_k(n-k) + p_{k-1}(n-k) + \cdots + p_1(n-k).$$

Comparing this with (4.1.2), we have the recursion

$$p_k(n) = p_k(n-k) + p_{k-1}(n-k) + \cdots + p_1(n-k). \tag{4.1.5}$$

This recursion determines $p_k(n)$ subject to the initial conditions $p_k(n) = 0$ for $n < k$ and $p_k(k) = 1$, since the only way of writing k as a sum of k positive integers is to write it as a sum of k 1's. We can use (4.1.5) to construct a table of $p_k(n)$, as in Table 4.1.

We may also use the recursion (4.1.5) to calculate $p_k(n)$ for small values of k and all n. Thus, trivially, $p_1(n) = 1$ for all $n \geq 1$. Now writing

$$p_2(n) = p_2(n-2) + p_1(n-2) = p_2(n-2) + 1,$$

we find, since $p_2(1) = 0$ and $p_2(2) = 1$, that $p_2(n) = n/2$ for n even, and $p_2(n) = (n-1)/2$ for n odd; $p_3(n)$ can be expressed as a polynomial of second degree in n, depending on the value of n modulo 6. The values are as follows:

$$\begin{aligned}
p_1(n) &= 1, \\
p_2(n) &= \frac{n}{2}, & n &\equiv 0(2), \\
&= \frac{n-1}{2}, & n &\equiv 1(2), \\
p_3(n) &= \frac{n^2}{12}, & n &\equiv 0(6), \\
&= \frac{n^2}{12} - \frac{1}{12}, & n &\equiv 1(6), \\
&= \frac{n^2}{12} - \frac{1}{3}, & n &\equiv 2(6), \\
&= \frac{n^2}{12} + \frac{1}{4}, & n &\equiv 3(6), \\
&= \frac{n^2}{12} - \frac{1}{3}, & n &\equiv 4(6), \\
&= \frac{n^2}{12} - \frac{1}{12}, & n &\equiv 5(6).
\end{aligned} \tag{4.1.6}$$

It can be shown without much difficulty that $p_k(n)$ depends on the value of n modulo $k!$ and that, for $n \equiv n_0 (\text{mod } k!)$, $p_k(n)$ is a polynomial of degree $k-1$ whose leading term is $n^{k-1}/(k-1)!\,k!$. This is true for $k = 1, 2, 3$ from (4.1.6). We proceed by induction. If $n \equiv n_0 (\text{mod } k!)$, then $n - k \equiv n_0 - k (\text{mod } k!)$,

TABLE 4.1 Values of $p_k(n)$

k \ n	1	2	3	4	5	6	7
1	1	1	1	1	1	1	1
2	0	1	1	2	2	3	3
3	0	0	1	1	2	3	4
4	0	0	0	1	1	2	3
5	0	0	0	0	1	1	2
6	0	0	0	0	0	1	1
7	0	0	0	0	0	0	1

whence in (4.1.5) each of $p_{k-1}(n-k) + \cdots + p_1(n-k)$ is a polynomial in n and the sum is a polynomial with leading term $n^{k-2}/(k-2)!\,(k-1)!$. But then

$$p_k(n) - p_k(n-k) = \frac{n^{k-2}}{(k-2)!\,(k-1)!} + R_{k-3}(n), \tag{4.1.7}$$

where $R_{k-3}(n)$ is a polynomial of degree at most $k-3$ in n. In the same way there is a rule

$$p_k(n-ik) - p_k(n-(i+1)k) = \frac{n^{k-2}}{(k-2)!\,(k-1)!} + R^i_{k-3}(n) \tag{4.1.8}$$

for $i = 1, 2, \ldots, (k-1)! - 1$. Adding all these, we find

$$p_k(n) - p_k(n-k!) = \frac{n^{k-2}}{(k-2)!} + S_{k-3}(n), \tag{4.1.9}$$

where $S_{k-3}(n)$ is some polynomial of degree at most $k-3$ in n. The equation (4.1.9) now determines all values of $p_k(n)$ with $n \equiv n_0 (\text{mod } k!)$ in terms of the least value $p_k(n_0)$. But it is not difficult to see that the solution of (4.1.9) is of the form

$$p_k(n) = \frac{n^{k-1}}{(k-1)!\,k!} + R_{k-2}(n), \qquad n \equiv n_0(k!), \tag{4.1.10}$$

where $R_{k-2}(n)$ is a polynomial in n of degree at most $k-2$. Thus (4.1.10) has been proved by induction.

We recall that the number of ordered partitions of n into k parts was $\binom{n-1}{k-1}$, a polynomial of degree $k-1$ in n with leading term $n^{k-1}/(k-1)!$. If the k parts are distinct, each unordered partition yields $k!$ ordered partitions. Here

$(1/k!)\binom{n-1}{k-1}$ is a polynomial of degree $k - 1$ in n, whose leading term is $n^{k-1}/(k-1)!\,k!$; comparison with (4.1.10) establishes the fact that in some sense for large n, most partitions of n into k parts have all k parts different.

With a partition of n we may associate a diagram using a row of dots for each part, putting the largest at the top and in succession the decreasing parts. Thus, we represent $15 = 5 + 5 + 3 + 2$ by the diagram

```
· · · · ·
· · · · ·
· · ·
· ·
```

We start all the rows in the same column, as shown. We may use a diagram to read by columns rather than by rows. In this instance the partition is $15 = 4 + 4 + 3 + 2 + 2$. Two partitions related in this way are called *conjugate*. Clearly, the relationship of being conjugate is symmetric.

THEOREM 4.1.1. *The number of partitions of an integer n into k parts is the same as the number of partitions of n into parts the greatest of which is k.*

Proof: ■ The conjugate of a partition with k parts is a partition whose greatest part is k, and conversely.

We now turn our attention to the unrestricted partitions of n, designating their number as $p(n)$. Arranging the parts in decreasing order, since there are at most n parts, $p(n)$ is the number of solutions in integers of

$$n = x_1 + x_2 + \cdots + x_n, \qquad x_1 \geq x_2 \geq \cdots \geq x_n \geq 0. \tag{4.1.11}$$

We may also describe a partition of n by the number of 1's 2's, ..., n's in the sum. Thus, $p(n)$ is also the number of solutions in integers of

$$n = 1y_1 + 2y_2 + \cdots + ny_n, \qquad y_i \geq 0, i = 1, \ldots, n. \tag{4.1.12}$$

THEOREM 4.1.2. *The generating function of $p(n)$,*

$$f(x) = 1 + p(1)x + p(2)x^2 + \cdots + p(n)x^n + \cdots$$

is the function

$$P(x) = \prod_{i=1}^{\infty} (1 - x^i)^{-1}.$$

Proof: ■ $(1 - x^i)^{-1} = 1 + x^i + x^{2i} + \cdots + x^{ri} + \cdots$. Each representation of n in (4.1.12) thus corresponds to a term x^n in the product

$$P(x) = \prod_{i=1}^{\infty} (1 - x^i)^{-1},$$

taking x^{jy_j} from the term $(1 - x^j)^{-1}$; conversely, every x^n in the expansion of

$f(x)$ is of this form. Hence,

$$P(x) = \sum_{n=0}^{\infty} p(n)x^n,$$

if we adopt the convention $p(0) = 1$. The representations

$$\begin{array}{lll} 1 = 1, & 4 = 4, & 5 = 5, \\ 2 = 2, & 4 = 3 + 1, & 5 = 4 + 1, \\ 2 = 1 + 1, & 4 = 2 + 2, & 5 = 3 + 2, \\ 3 = 3, & 4 = 2 + 1 + 1, & 5 = 3 + 1 + 1, \\ 3 = 2 + 1, & 4 = 1 + 1 + 1 + 1, & 5 = 2 + 2 + 1, \\ 3 = 1 + 1 + 1, & & 5 = 2 + 1 + 1 + 1, \\ & & 5 = 1 + 1 + 1 + 1 + 1, \end{array} \tag{4.1.13}$$

give us directly the values

$$p(1) = 1, \quad p(2) = 2, \quad p(3) = 3, \quad p(4) = 5, \quad p(5) = 7. \tag{4.1.14}$$

The inverse of $P(x)$, namely, $\phi(x) = \prod_{i=1}^{\infty}(1 - x^i)$ is also a generating function of a series

$$\phi(x) = \prod_{i=1}^{\infty} (1 - x^i) = \sum_{n=0}^{\infty} c_n x^n, \qquad c_0 = 1, \tag{4.1.15}$$

where the coefficient c_n has a combinatorial interpretation:

$$c_n = p_e(n) - p_o(n), \tag{4.1.16}$$

where $p_e(n)$ is the number of partitions of n into an even number of distinct parts and $p_o(n)$ is the number of partitions into an odd number of distinct parts. We may use this interpretation to evaluate c_n.

THEOREM 4.1.3 (EULER IDENTITY). We have

$$\prod_{i=1}^{\infty} (1 - x^i) = 1 + \sum_{k=1}^{\infty}(-1)^k(x^{(3k^2-k)/2} + x^{(3k^2+k)/2}).$$

Proof: ■ The following proof, due to Franklin [1], depends on the combinatorial nature of the coefficient c_n as given in (4.1.16). We take a diagram representing a partition of n into any number of unequal parts in descending order. We call the lowest line the base b of the diagram. From C, the extreme upper right-hand point, we draw the longest possible line of slope 1 through points of our diagram; it might of course contain only one point. We call this line the slope s.

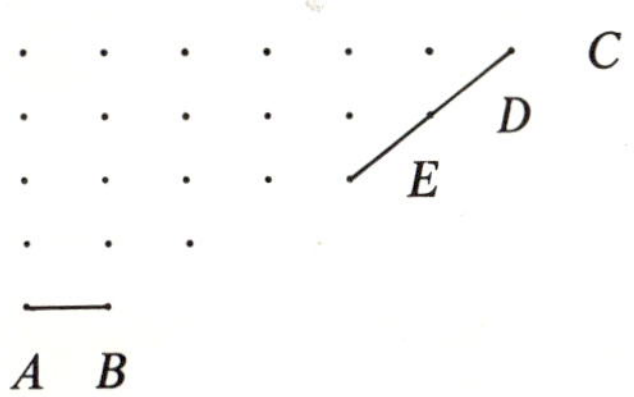

In the accompanying diagram, the base b is the line AB and CDE is the slope s. We write $b = 2$, $s = 3$, in this case meaning that b contains two points and s contains three points. In general, we may use b, s also to represent the number of points in the base and slope, respectively. We now define two operations on the diagrams, which we designate by α and β.

1. For α: If $b \leq s$ and if the base and slope do not have a common point, remove the base and adjoin its points to the first lines to form a new slope. If the base and slope do intersect and $b \leq s - 1$, this operation is still possible.
2. For β: If $b > s$, remove the slope and adjoin this as a new base. This is possible if the base and slope do not intersect. If the base and slope do intersect, this is still possible if $b \geq s + 2$.

Since α requires $b \leq s$ and β requires $b > s$, at most one of α, β is possible for any diagram D. Further, we easily verify that if α is possible, then when α is applied to D we get a diagram D' for which β is possible, and β applied to D' gives D. Similarly, if β can be applied to D to give D'', then α can be applied to D'' to give D. Thus, these operations set up a one-to-one correspondence between partitions with an even number of distinct parts and partitions with an odd number of parts whenever the operations may be applied. When $b \leq s$, we may apply α except when the base and slope interact and $b = s = k$. In this case the partition is

$$n = k + (k + 1) + \cdots + (2k - 1) = \frac{3k^2 - k}{2}. \tag{4.1.17}$$

If $b > s$, we may apply β except when the base and slope intersect and $b - 1 = s = k$. Here the partition is

$$n = (k + 1) + (k + 2) + \cdots + 2k = \frac{3k^2 + k}{2}. \tag{4.1.18}$$

Thus, $p_e(n) = p_o(n)$ and $c_n = 0$ except when n is of the form $(3k^2 \pm k)/2$; in this case there is one partition of k parts left over, given by (4.1.17) or (4.1.18), and its contribution to c_n is $(-1)^k$, whence $c_n = (-1)^k$ for $n = (3k^2 \pm k)/2$. Thus, we have shown that

$$\begin{aligned} c_n &= 0, && \text{if } n \text{ is not of the form } (3k^2 \pm k)/2, \\ c_n &= (-1)^k, && \text{if } n = (3k^2 \pm k)/2. \end{aligned} \tag{4.1.19}$$

Thus, (4.1.19) evaluates the coefficients c_n in (4.1.15) and Theorem 4.1.3 is now proved.

Theorems 4.1.2 and 4.1.3 give the identity

$$\left[1+\sum_{k=1}^{\infty}(-1)^k(x^{(3k^2-k)/2}+x^{(3k^2+k)/2})\right]\cdot\left[\sum_{n=0}^{\infty}p(n)x^n\right]=1, \tag{4.1.20}$$

where we use the convention $p(0) = 1$. This may be used to give a recurrence for $p(n)$ by calculating the coefficient of x^n, $n \geq 1$, on the left of (4.1.20) and putting this equal to zero. This yields

$$p(n) = p(n-1) + p(n-2) - p(n-5) - p(n-7) + \cdots$$
$$+ (-1)^{k-1}p\left(n - \frac{3k^2-k}{2}\right) + (-1)^{k-1}p\left(n - \frac{3k^2+k}{2}\right) + \cdots, \tag{4.1.21}$$

including only arguments ≥ 0 on the right and recalling that we use the convention $p(0) = 1$. In this way we may calculate the values of $p(n)$ as shown in Table 4.2.

TABLE 4.2 Values of $p(n)$

n	$p(n)$	n	$p(n)$	n	$p(n)$	n	$p(n)$
1	1	26	2436	51	239943	76	9289091
2	2	27	3010	52	281589	77	10619863
3	3	28	3718	53	329931	78	12132164
4	5	29	4565	54	386155	79	13848650
5	7	30	5604	55	451276	80	15796476
6	11	31	6842	56	526823	81	18004327
7	15	32	8349	57	614154	82	20506255
8	22	33	10143	58	715220	83	23338469
9	30	34	12310	59	831820	84	26543660
10	42	35	14883	60	966467	85	30167357
11	56	36	17977	61	1121505	86	34262962
12	77	37	21637	62	1300156	87	38887673
13	101	38	26015	63	1505499	88	44108109
14	135	39	31185	64	1741630	89	49995925
15	176	40	37338	65	2012558	90	56634173
16	231	41	44583	66	2323520	91	64112359
17	297	42	53174	67	2679689	92	72533807
18	385	43	63261	68	3087735	93	82010177
19	490	44	75175	69	3554345	94	92669720
20	627	45	89134	70	4087968	95	104651419
21	792	46	105558	71	4697205	96	118114304
22	1002	47	124754	72	5392783	97	133230930
23	1255	48	147273	73	6185689	98	150198136
24	1575	49	173525	74	7089500	99	169229875
25	1958	50	204226	75	8118264	100	190569292

Theorem 4.1.3 and other results may be derived from the more general identity in

THEOREM 4.1.4 (GAUSS-JACOBI IDENTITY). *The following identity holds for all* $z \neq 0$ *and all* q *with* $|q| < 1$.

$$\prod_{k=1}^{\infty}(1-q^{2k})(1-q^{2k-1}z^2)(1-q^{2k-1}z^{-2}) = \sum_{k=-\infty}^{+\infty}(-1)^k z^{2k} q^{k^2}. \tag{4.1.22}$$

Proof: ∎ Let us write

$$\phi(z) = \prod_{k=1}^{n}(1-q^{2k-1}z^2)(1-q^{2k-1}z^{-2}). \tag{4.1.23}$$

Then we can verify at once that

$$\phi(qz)(-qz^2 + q^{2n}) = \phi(z)(1-q^{2n+1}z^2). \tag{4.1.24}$$

It is also true that

$$\phi(z) = A_0 - A_1(z^2 + z^{-2}) + A_2(z^4 + z^{-4}) + \cdots + (-1)^n A_n(z^{2n} + z^{-2n}), \tag{4.1.25}$$

where the A's are polynomials in q and

$$A_n = q^{1+3+\cdots+2n-1} = q^{n^2}. \tag{4.1.26}$$

Substituting from (4.1.25) into (4.1.24) and equating the coefficients of z^{-2k+2}, we find

$$A_k = A_{k-1}q^{2k-1}\frac{1-q^{3n-2k+2}}{1-q^{2n+2k}}. \tag{4.1.27}$$

From (4.1.26) and (4.1.27) we calculate A_k and find

$$A_k = \frac{q^{k^2}}{(1-q^2)(1-q^4)\cdots(1-q^{2n})}\prod_{2s=2n-2k+2}^{2n}(1-q^{2s})\prod_{2s=2n+2k+2}^{4n}(1-q^{2s}), \tag{4.1.28}$$

which tends to

$$\frac{q^{k^2}}{(1-q^2)(1-q^4)\cdots} \tag{4.1.29}$$

as $n \to \infty$. Here (4.1.28) tells us that

$$\prod_{k=1}^{n}(1-q^{2k})(1-q^{2k-1}z^2)(1-q^{2k-1}z^{-2}) = \sum_{k=-n}^{+n}(-1)^k z^{2k} q^{k^2} B_k, \tag{4.1.30}$$

where B_k is given by

$$B_k = \prod_{2s=2n-2k+2}^{2n}(1-q^{2s})\prod_{2s=2n+2k+2}^{4n}(1-q^{2s}). \tag{4.1.31}$$

Here, for each k, $\lim_{n\to\infty} B_k = 1$, and with a little care about the passage to the limit, this establishes the identity of the theorem.

From the Gauss-Jacobi identity we can easily derive Theorem 4.1.3 by replacing q, z, respectively, by $q^{3/2}$, $q^{1/4}$, obtaining

$$\prod_{k=1}^{\infty}(1-q^{3k})(1-q^{3k-1})(1-q^{3k-2}) = \sum_{-\infty<k<+\infty}(-1)^k q^{(3k^2+k)/2}, \quad (4.1.32)$$

which is a modified form of Theorem 4.1.3. We wish to obtain an additional formula of Jacobi.

THEOREM 4.1.5 (JACOBI IDENTITY). We have

$$\prod_{k=1}^{\infty}(1-q^k)^3 = 1+\sum_{k=1}^{\infty}(-1)^k(2k+1)q^{-(k(k+1))/2}. \quad (4.1.33)$$

Proof: ■ Although the product on the left can be associated with partitions into parts, no one of which may be used more than three times, no combinatorial proof of this identity is known to the writer.

In (4.1.22) replace q, z by $q^{1/2}$, $q^{+1/4}$, and make $\epsilon \to 0$. This gives

$$\prod_{k=1}^{\infty}(1-q^k)(1-q^{k+2\epsilon})(1-q^{k-1-2\epsilon}) = \sum_{-\infty<k<+\infty}(-1)^k q^{(k+2\epsilon k)/2} q^{k^2/2}. \quad (4.1.34)$$

The factor $1-q^{-2\epsilon}$ on the left can be written

$$\begin{aligned} 1-e^{-2\epsilon\log q} &= 1-(1-2\epsilon\log q+\cdots) \\ &= 2\epsilon\log q+\cdots \end{aligned} \quad (4.1.35)$$

where subsequent terms involve higher powers of ϵ.

On the right we use

$$q^{2\epsilon k} = 1+(2k\log q)\epsilon+\cdots, \quad (4.1.36)$$

where again additional terms involve higher powers of ϵ. In the sum

$$\sum_{-\infty<k<+\infty}(-1)^k(1+2k\epsilon\log q+\cdots)q^{(k^2+k)/2} \quad (4.1.37)$$

the terms free of ϵ vanish, those for which $k=r$ and $k=-r-1$ canceling, and (4.1.34) now becomes

$$\begin{aligned} (2\epsilon\log q+\cdots)\prod_{k=1}^{\infty}(1-q^k)(1-q^{k+2\epsilon})\prod_{k=2}^{\infty}(1-q^{k-1-2\epsilon}) \\ = \sum_{-\infty<k<+\infty}(-1)^k(2k\epsilon\log q+\cdots)q^{(k^2+k)/2}. \end{aligned} \quad (4.1.38)$$

If we now divide by $2\epsilon\log q$ and then make $\epsilon\to 0$ we have

$$\prod_{k=1}^{\infty}(1-q^k)^3 = \sum_{-\infty<k<\infty}(-1)^k\cdot kq^{(k^2+k)/2}. \quad (4.1.39)$$

This proves the formula of Jacobi in our theorem.

Ramanujan, the self-taught Indian genius who was discovered by G. H. Hardy, was the first to discover some arithmetical curiosities of the partition numbers $p(n)$. These are

$$p(5m + 4) \equiv 0 (\text{mod } 5), \tag{4.1.40a}$$

$$p(7m + 5) \equiv 0 (\text{mod } 7), \tag{4.1.40b}$$

$$p(11m + 6) \equiv 0 (\text{mod } 11). \tag{4.1.40c}$$

We shall prove the first two of these, as did Ramanujan, using the Euler and Jacobi identities. Multiplying, we have

$$\begin{aligned} &q\{(1 - q)(1 - q^2)\cdots\}^4 \\ &\quad = q\{(1 - q)(1 - q^2)\cdots\}\{(1 - q)(1 - q^2)\cdots\}^3 \\ &\quad = q(1 - q - q^2 + q^5 + q^7 + \cdots)(1 - 3q + 5a^3 - 7q^6 \cdots), \end{aligned} \tag{4.1.41}$$

using the Euler and Jacobi identities.

We write this as

$$q\{(1 - q)(1 - q^2)\cdots\}^4 = \sum_r \sum_s (-1)^{r+s}(2s + 1)q^{E(r,s)}, \tag{4.1.42}$$

where

$$E(r, s) = 1 + \tfrac{1}{2}r(3r + 1) + \tfrac{1}{2}s(s + 1), \tag{4.1.43}$$

and r runs from $-\infty$ to $+\infty$, while s runs from 0 to ∞. We ask under what circumstances $E(r, s)$ is a multiple of 5. This requires that

$$2(r + 1)^2 + (2s + 1)^2 = 8E(r, s) - 10r^2 - 5 \tag{4.1.44}$$

shall also be a multiple of 5. But

$$2(r + 1)^2 \equiv 0, 2, 3 (\text{mod } 5), \quad (2s + 1)^2 \equiv 0, 1, 4 (\text{mod } 5). \tag{4.1.45}$$

Hence, $E(r, s) \equiv 0 (\text{mod } 5)$ only when

$$r + 1 \equiv 2s + 1 \equiv 0 (\text{mod } 5). \tag{4.1.46}$$

Therefore in (4.1.42) the exponent $E(r, s)$ is a multiple of 5 only when the coefficient $(2s + 1)$ is also a multiple of 5. Hence, in

$$q\{(1 - q)(1 - q^2)\cdots\}^4$$

the coefficient of q^{5m+5} is a multiple of 5. In the expansion of $(1 - q)^{-5}$ by the binomial theorem, all coefficients are divisible by 5 except $1, q^5, q^{10}, \ldots,$ which have residue 1(mod 5). We express this by writing

$$\frac{1}{(1 - q)^5} \equiv \frac{1}{1 - q^5} (\text{mod } 5) \tag{4.1.47}$$

or

$$\frac{1 - q^5}{(1 - q)^5} \equiv 1 (\text{mod } 5), \tag{4.1.48}$$

where these congruences are taken to mean that all coefficients are congruent modulo 5. Hence, the coefficient of q^{5m+5} in

$$q\,\frac{(1-q^5)(1-q^{10})\cdots}{(1-q)(1-q^2)\cdots}$$

$$\equiv q\,\frac{\{(1-q)(1-q^2)\cdots\}^4(1-q^5)(1-q^{10})\cdots}{\{(1-q)(1-q^2)\cdots\}^5} \qquad (4.1.49)$$

is a multiple of 5, and so therefore is that in

$$\frac{q}{(1-q)(1-q^2)\cdots}, \qquad (4.1.50)$$

and this coefficient is $p(5m+4)$. This proves (4.1.40a). In the same way we may use the square of Jacobi's identity to prove (4.1.40b). No such simple proof of (4.1.40c) is known.

The generating function $P(x) = \sum_{n=1}^{\infty} p(n)x^n$ is associated with arithmetic in another way, specifically with the divisor function $\sigma(n)$ defined by

$$\sigma(n) = \sum_{d\,|\,n} d. \qquad (4.1.51)$$

We start from the basic generating function

$$P(x) = \prod_{k=1}^{\infty} (1-x^k)^{-1}. \qquad (4.1.52)$$

This gives

$$\log(P(x)) = \sum_{k=1}^{\infty} -\log(1-x^k). \qquad (4.1.53)$$

Now take the derivative of (4.1.50) with respect to x and multiply the result by x. This yields

$$\frac{xP'(x)}{P(x)} = \sum_{k=1}^{\infty} \frac{kx^k}{1-x^k}. \qquad (4.1.54)$$

The kth term of the right-hand side, when expanded, contributes k to the coefficient of every power x^{mk}. Hence, on the right, the coefficient of x^n is $\sigma(n)$, the sum of all divisors of n. Thus,

$$\frac{xP'(x)}{P(x)} = \sum_{n=1}^{\infty} \sigma(n)x^n. \qquad (4.1.55)$$

4.2 Asymptotic Properties of $p(n)$

If we ask "How large is $p(n)$ for large n?" we are dealing with a question that properly belongs to the analytic theory of numbers. This is beyond the scope of this book, but the work and the results are so fascinating that a brief sketch of this will be given.

The divisor function $\sigma(n)$ is very irregular in its behavior, but its average value is well behaved. If we define $s(n)$ by

$$s(n) = \sum_{k=1}^{n} \sigma(k), \tag{4.2.1}$$

then an easy calculation shows that

$$s(n) = \sum_{k=1}^{n} \frac{1}{2}\left\{\left[\frac{n}{h}\right]^2 + \left[\frac{n}{h}\right]\right\}. \tag{4.2.2}$$

where the bracket function $[x]$ stands for the greatest integer not exceeding x. Since the series

$$\sum_{h=1}^{\infty} \frac{1}{h^2} \tag{4.2.3}$$

is convergent and has the sum $\pi^2/6$, we may get a reasonably good estimate of $s(n)$, namely

$$\frac{\pi^2 n^2}{12} - \frac{n}{2}\log n - \frac{3n}{2} < s(n) < \frac{\pi^2 n^2}{12} + \frac{n}{2}\log n + \frac{n}{2}. \tag{4.2.4}$$

From the identity

$$(1 - x)^{-1} \sum_{n=1}^{\infty} \sigma(n) x^n = \sum_{n=1}^{\infty} s(n) x^n \tag{4.2.5}$$

we may put (4.1.51) into the form

$$\frac{xP'(x)}{P(x)} = (1 - x) \sum_{n=1}^{\infty} s(n) x^n. \tag{4.2.6}$$

With a certain amount of relatively elementary calculations we may use the estimate (4.2.4) of $s(n)$ in (4.2.6) to conclude that, asymptotically,

$$\log p(n) \sim A\sqrt{n}, \qquad A = \pi\sqrt{\tfrac{2}{3}}, \tag{4.2.7}$$

where by $f(n) \sim g(n)$ we mean

$$\lim_{n\to\infty} \frac{f(n)}{g(n)} = 1.$$

The derivation of (4.2.7) from (4.2.4) and (4.2.6) is a lengthy process, and since the result is a relatively weak one, it is not presented here. The asymptotic value of $p(n)$ is in fact given by

$$p(n) \sim \frac{1}{4n\sqrt{3}} e^{A\sqrt{n}}. \tag{4.2.8}$$

The series $P(x)$ is convergent within the unit circle, but has the unit circle as a natural boundary. By Cauchy's theorem we have

$$p(n) = \frac{1}{2\pi i}\int_C \frac{P(x)}{x^{n+1}}\,dx, \tag{4.2.9}$$

where C is a contour around the origin. Hardy and Ramanujan [1] based their study of the asymptotic value of $p(n)$ on (4.2.9) and made use of the fact that there are a number of remarkable functional equations that they satisfy. In particular, if complex numbers x, x', with $|x|, |x'| < 1$, are related by

$$\log\frac{1}{x}\log\frac{1}{x} = 4\pi^2 \quad \text{or} \quad x' = \exp\left\{\frac{-4\pi^2}{\log(1/x)}\right\}, \tag{4.2.10}$$

we have the identity

$$P(x) = \frac{x^{1/24}}{\sqrt{2\pi}}\sqrt{\log\frac{1}{x}}\exp\left\{\frac{\pi^2}{6\log(1/x)}\right\}P(x'). \tag{4.2.11}$$

If x is real and near to 1, then x' is extravagantly small and $P(x')$ is practically 1, so that (4.2.11) expresses $P(x)$ effectively in terms of elementary functions. There are similar formulae associated with the roots of unity, so that if x is near $x_{p,q} = \exp(2\pi i(p/q))$ for relatively prime integers p and q, we can approximate $P(x)$ with great accuracy by elementary functions.

Taking C in (4.2.9) to be a circle with just the right radius, which turns out to be $1 - 1/n$, and substituting from (4.2.11) into (4.2.9), replacing $P(x')$ by 1 (with an error that turns out to be of order $e^{h\sqrt{n}}$, where $H < A = \pi\sqrt{\frac{2}{3}}$), Hardy and Ramanujan found the formula

$$p(n) = \frac{1}{2\pi\sqrt{2}}\frac{d}{dn}\left(\frac{e^{A\lambda_n}}{\lambda_n}\right) + O(e^{H\sqrt{n}}), \tag{4.2.12}$$

where

$$\lambda_n = \sqrt{n - \tfrac{1}{24}}. \tag{4.2.13}$$

In (4.2.12) the O notation is a special instance of the general notation $f(n) = O(g(n))$, meaning that a constant K exists such that

$$\limsup_{n\to\infty}\left|\frac{f(n)}{g(n)}\right| \le K.$$

Here (4.2.12) includes (4.2.8), but is much more precise.

The formula (4.2.12) can be considerably improved. As x approaches $x_{p,q}$ along a radius, $P(x)$ behaves roughly like $\exp\{\pi^2/6q^2(1 - |x|\}$. Thus, in the integral (4.2.9), the "rational singularities" $x_{p,q}$ are the heaviest singularities, and it is to be expected that their contributions to the integral (4.2.9) will outweigh the rest. This led Hardy and Ramanujan to look for a formula of the form

$$p(n) = P_1(n) + P_2(n) + \cdots + P_Q(n) + R(n), \tag{4.2.14}$$

where $P_1(n)$ is the dominant term in (4.2.12) and $P_q(n)$, $q = 2, \ldots, Q$ are similar in form, and $P_q(n)$ combines the contributions near the rational points of denominator q; and $R(n)$ is a remainder term. Since Major MacMahon had calculated $p(n)$ for values of n up to 200, it was natural to test the formulas for $n = 200$. Much to the surprise of Hardy and Ramanujan they found that eight terms of their formula gave $p(200)$ with an error of only 0.004. Here MacMahon's numerical analysis, as we would call it nowadays, led Hardy and Ramanujan to push their theory further, and they were able to show the existence of constants α, M such that

$$p(n) = \sum_{q<\alpha\sqrt{n}} P_q(n) + R(n), \tag{4.2.15}$$

where

$$|R(n)| < Mn^{-1/4}. \tag{4.2.16}$$

Thus, a formula was obtained that would give $p(n)$ exactly, once α and M had been found. In trying to simplify the work of Hardy and Ramanujan, H. Rademacher [1] made some slight changes and found a convergent series for $p(n)$. Putting

$$\psi_q(n) = \frac{q^{1/2}}{\pi\sqrt{2}} \cdot \frac{d}{dn}\left\{\frac{\sinh(A\lambda_n/q)}{\lambda_n}\right\}, \qquad A = \pi\sqrt{\tfrac{2}{3}}, \tag{4.2.17}$$

and

$$L_q(n) = \sum_p w_{p,q} \exp\left(\frac{-2np\pi i}{q}\right), \tag{4.2.18}$$

where $w_{p,q}$ is a certain twenty-fourth root of unity, Rademacher proved that

$$p(n) = \sum_{q=1}^{\infty} L_q(n)\psi_q(n) \tag{4.2.19}$$

and that the remainder after Q terms is less than

$$CQ^{-1/2} + D\left(\frac{Q}{n}\right)^{1/2} \sinh\left(\frac{An^{1/2}}{Q}\right), \tag{4.2.20}$$

where C and D are constants for which he found definite values. For Q of order $\sqrt{n}$ this is similar to the results of (4.2.15) and (4.2.16). But in one important respect, Rademacher's result is an improvement, since it was shown by D. H. Lehmer [1] that the infinite series of Hardy and Ramanujan does not converge.

PROBLEMS

1. Calculate $p(101)$, $p(102)$, $p(103)$, $p(104)$, and $p(105)$. Check these values, using the congruences of (4.1.40).
2. Show that $p^*(n) = p(n) - p(n-1)$ for $n \geq 1$ is the number of partitions of n into parts greater than 1. From the properties of $p^*(n)$ show that $p(n+2) - 2p(n+1) + p(n) \geq 0$ for $n \geq 0$.
3. Prove that the number of partitions of any positive integer n into distinct parts is the same as the number of partitions of n into odd parts by showing that

$$(1+x)(1+x^2)\cdots(1+x^i)\cdots = \frac{1}{(1-x)(1-x^3)\cdots(1-x^{2i-1})\cdots}$$

4. Prove by a combinatorial argument that

$$(1+x)(1+x^3)\cdots(1+x^{2n-1})\cdots = 1 + \frac{x}{1-x^2} + \frac{x^4}{(1-x^2)(1-x^4)} + \cdots + \frac{x^{n^2}}{(1-x^2)\cdots(1-x^{2n})} + \cdots.$$

5. Show that

$$f_1(n, k) = \frac{1}{k!}\binom{n-1}{k-1} \leq p_k(n)$$

for all $n \geq 1$ and $k \geq 1$. Find some explicit function $f_2(n, k)$ such that $p_k(n) \leq f_2(n, k)$ for all $n \geq 1$ and $k \geq 1$, and so that for fixed k, the ratio $f_2(n, k)/f_1(n, k)$ approaches 1 as n goes to infinity.
6. Express $p_4(n)$ by polynomials in n depending on the residue of n modulo 12.

5

Distinct Representatives

5.1 The Theorems of P. Hall and D. König

A problem of the type we shall discuss here is: Given the following five sets,

$$
\begin{aligned}
S_1 &= (1, 2, 3),\\
S_2 &= (1, 2, 4),\\
S_3 &= (1, 2, 5),\\
S_4 &= (3, 4, 5, 6),\\
S_5 &= (3, 4, 5, 6),
\end{aligned}
\tag{5.1.1}
$$

we wish to choose numbers x_1, x_2, x_3, x_4, x_5 such that $x_i \in S_i, i = 1, \ldots, 5$ and such that the x's are all distinct. One such choice is $x_1 = 1, x_2 = 2, x_3 = 5$, $x_4 = 3, x_5 = 4$. But if our sets were instead

$$
\begin{aligned}
T_1 &= (1, 2),\\
T_2 &= (1, 2),\\
T_3 &= (1, 2),\\
T_4 &= (3, 4, 5, 6),\\
T_5 &= (3, 4, 5, 6),
\end{aligned}
\tag{5.1.2}
$$

no such choice would be possible, since clearly we could choose three distinct numbers from T_1, T_2, T_3, because they contain only the two numbers between them, 1 and 2. We ask under what circumstances subsets S_i, $i = 1, \ldots, n$ of a set S possess distinct representatives x_i, $i = 1, \ldots, n$—that is, $x_i \in S_i$, $x_i \neq x_j$—if $i \neq j$? Note that we do not require that subsets S_i and S_j with $i \neq j$ be distinct as subsets of S. A clearly necessary condition, as seen in (5.1.2), for the existence of distinct representatives is that any k sets S_i contain between them at least k distinct elements.

We shall abbreviate "system of distinct representatives" to SDR when convenient. It is a remarkable fact, first proved by Philip Hall in 1935 [1], that this obviously necessary condition for the existence of distinct representatives is also sufficient.

THEOREM 5.1.1 (PHILIP HALL). *Let I be a finite set of indices, $I = \{1, 2, \ldots, n\}$. For each $i \in I$ let S_i be a subset of a set S. A necessary and sufficient condition for the existence of distinct representatives x_i, $i = 1, \ldots, n$, $x_i \in S_i$, $x_i \neq x_j$, when $i \neq j$ is condition C: For every $k = 1, \ldots, n$ and choice of k distinct indices $i_1, \ldots, i_k$, the subsets $S_{i_1}, \ldots, S_{i_k}$ contain between them at least k distinct elements.*

Proof: ■ We have already remarked on the necessity of condition C for the existence of distinct representatives and must prove its sufficiency. We note that if each S_i contains only a single element x_i, then condition C asserts that $x_1, \ldots, x_n$ are distinct, and so are distinct representatives. Our basic operation will be the deletion of certain elements from certain of the subsets S_i, leaving subsets $\bar{S}_i \subseteq S_i$, $i = 1, \ldots, n$ so that condition C remains valid. If, after a succession of deletions preserving condition C, the remaining sets $\bar{S}_i$ each contain only a single element x_i, then $x_1, \ldots, x_n$ are distinct representatives, and our theorem is proved.

As a first trivial deletion that leaves condition C valid, we may remove all but n elements from any set S_i that contains more than n elements. Let us call a set of r subsets $S_{i_1}, \ldots, S_{i_r}$ a *block* and designate it as $B_{r,s}$, where s is the number of distinct elements in the r subsets. Then condition C is equivalent to the assertion that $s \geq r$ for any block $B_{r,s}$. If $s = r$, we call such a block $B_{r,r}$ a critical block. By convention we regard the void block as a critical block $B_{0,0}$. We may also define the union and intersection of blocks. Suppose that $A_1, \ldots, A_m$, $C_{m+1}, \ldots, C_r$ are the subsets S_i in a block $B_{r,s}$ and that $A_1, \ldots, A_m$, $D_{m+1}, \ldots, D_t$ are the subsets in a block $B_{t,v}$, where $A_1, \ldots, A_m$ are the subsets in both blocks (recall that subsets S_i, S_j are distinguished by their indices and not by the elements they contain). Then we define the intersection $B_{r,s} \cap B_{t,v}$ as the block $B_{u,w}$ whose subsets are $A_1, \ldots, A_m$ and the union $B_{r,s} \cup B_{t,v}$ as the block whose subsets are $A_1, \ldots, A_m$, $C_{m+1}, \ldots, C_r$, $D_{m+1}, \ldots, D_t$, this being a block $B_{y,z}$ with $y = r + t - u$.

LEMMA 5.1.1. *The union* $B_{r,r} \cup B_{t,t}$ *and intersection* $B_{r,r} \cap B_{t,t}$ *of critical blocks are again critical blocks, assuming condition C to be valid.*

Proof: ■ Let $B_{r,r} \cap B_{t,t} = B_{u,v}$ and $B_{r,r} \cup B_{t,t} = B_{y,z}$. The z elements of the union will be the $r + t$ elements of $B_{r,r}$ and $B_{t,t}$ reduced by the number of elements in both blocks, and this number includes at least the v elements in the intersection. Thus, $z \leq r + t - v$. Also, $v \geq u$ and $z \geq y$ by condition C. As $y + u = r + t$, we have $r + t - v \geq z \geq y = r + t - u \geq r + t - v$. Hence, we have equality throughout and $z = y$, $u = v$, and the lemma is proved.

LEMMA 5.1.2. *If* $B_{k,k}$ *is a critical block, the deletion of elements of* $B_{k,k}$ *from sets not belonging to* $B_{k,k}$ *leaves condition C valid.*

Proof: ■ Let $B_{r,s}$ be an arbitrary block. We must show that if $(B_{r,s})' = B'_{r,s'}$ is the block after deletion, then we have $s' \geq r$. Let $B_{r,s} \cap B_{k,k} = B_{u,v}$ and $B_{r,s} \cup B_{k,k} = B_{y,z}$. Here, let $B_{r,s}$ be the sets $A_1, \ldots, A_m, C_{m+1}, \ldots, C_r$ and $B_{k,k}$ be the sets $A_1, \ldots, A_m, D_{m+1}, \ldots, D_k$, where $A_1, \ldots, A_m$ are the sets in common. Then $B_{u,v}$ is $A_1, \ldots, A_m$, and $B_{y,z}$ is

$$A_1, \ldots, A_m, \quad C_{m+1}, \ldots, C_r, \quad D_{m+1}, \ldots, D_k.$$

The deleted block $(B_{r,s})' = B'_{r,s'}$ is

$$A_1, \ldots, A_m, \quad C'_{m+1}, \ldots, C'_r.$$

But $C_{m+1}, \ldots, C_r$ as blocks of the union $B_{y,z}$ contain $z - k$ elements not in $B_{k,k}$. Thus, $s' = v + z - k$, being the elements of the intersection $B_{u,v}$ together with the $z - k$ elements of $C'_{m+1}, \ldots, C'_r$. As $y = r + k - u$ and $z \geq y$, $v \geq u$, we have $s' = v + z - k \geq u + y - k = r$. Thus, $s' \geq r$ and so, after deletion, condition C still holds.

We may now prove our theorem, using induction on the number n of sets, this being trivial when $n = 1$. First suppose that in the system $U\{S_1, \ldots, S_n\}$, there is a critical block $B_{k,k}$ not the whole system—that is, $1 \leq k \leq n$. Deleting the elements of $B_{k,k}$ from the remaining sets, U consists of $B_{k,k}$ and a block $B_{n-k,v}$, which have no elements in common. By Lemma 5.1.2 condition C remains valid, and we may assume by induction that $B_{k,k}$ and $B'_{n-k,v}$ both have SDR's, and being disjoint, they form together an SDR for U. Next suppose that in the system $U(S_1, \ldots, S_n)$ there is no critical block except possibly the entire system. Now select an arbitrary element of any set as its representative and delete this element from all other sets. In this circumstance a block $B_{r,s}$, with $r < n$, becomes a block $B'_{r,s'}$, where $s' = s$ or $s - 1$. But by assumption $B_{r,s}$ was not critical and so $s \geq r + 1$, whence $s' \geq r$ and condition C holds for the remaining $(n - 1)$ blocks; thus, by induction, they have an SDR, which together with the representative selected for the one set forms an SDR for the entire system. This completes the proof of our theorem.

COROLLARY TO THEOREM 5.1.1. *If n sets $S_1, \ldots, S_n$ have an SDR and if the smallest of these sets contains t objects, then if $t \geq n$, there are at least $t(t-1)\cdots(t-n+1)$ different SDR's and if $t < n$, there are at least $t!$ different SDR's.*

This corollary comes from a closer inspection of the proof. There must be at least one of the sets from which any object may be chosen as a representative, for if there are no critical blocks, this is true of any set. But if there are critical blocks, then this is true of a set in a minimal critical block. One set, say S_1, can have any element chosen as its representative. This choice may be made in at least t ways. Let us now delete the representative chosen for S_1 from $S_2, \ldots, S_n$, giving sets $S_2', \ldots, S_n'$ that possess an SDR and in which the smallest set contains at least $t-1$ objects. Continuing in this way we have at least $t(t-1)\cdots(t-n+1)$ choices in all if $t \geq n$ and at least $t!$ if $t < n$.

The preceding proof is not Philip Hall's original proof, but is due to the writer. It lends itself to an extension of Theorem 5.1.1 to the case in which the system U contains infinitely many sets S_i, though every set S_i is finite. If we have infinitely many sets and also allow the sets to be infinite, it is not clear which condition is appropriate for guaranteeing the existence of an SDR. For example, if we have sets $S_0 = \{1, 2, 3, \ldots\}$ and $S_i = \{i\}$, $i = 1, 2, \ldots$, then the system does not have an SDR because we have no freedom of choice for the S_i, $i = 1, 2, \ldots$, and this leaves no element free to be a representative of S_0. And yet, for any cardinal k, finite or infinite, k of the sets given above contain between them at least k elements. Thus the requirement in the following theorem that the sets be finite is not superfluous.

THEOREM 5.1.2. *Suppose that for each i of a system of indices I we are given a finite subset S_i of a set S. The system $U = \{S_i\}$, $i \in I$ has a system of distinct representatives if and only if the following condition holds: condition C—for every finite k and any choice of k distinct indices $i_1, i_2, \ldots, i_k$, the subsets $S_{i_1}, S_{i_2}, \ldots, S_{i_k}$ contain between them at least k distinct elements.*

Proof: ■ We may establish a partial ordering on deletions, writing $D_1 \subseteq D_2$ for deletions D_1 and D_2 if every element deleted from a set by D_1 is also deleted by D_2. We are interested in deletions that preserve condition C. If all deletions $D_1 \subseteq D_2 \subseteq \ldots \subseteq D_i \subseteq \ldots$ in an ascending chain preserve condition C, let D be the deletion that consists of deleting an element b from a set S in all instances in which this happens for any D_i in the ascending chain. Then we assert that D also preserves condition C, for in any block $B_{r,s}$ of U (since both r and s are finite), only a finite number of deletions in the chain affect $B_{r,s}$ and so there is a last deletion D_n affecting it. But under D_n, $(B_{r,s})' = B'_{r,s'}$ still satisfies condition C, $s' \geq r$ by hypothesis. Hence, D preserves condition C. By Zorn's lemma there will then be for all deletions preserving condition C a maximal one, D. We shall show that under a maximal deletion D preserving condition C, every deleted set S_i' has only a single element, and these elements therefore form an SDR for the original system U.

If there is an element a_1 that does not belong to any critical block, let us delete a_1 from every set S_i containing a_1. For this type of deletion a block $B_{r,s}$ is replaced by $B'_{r,s'}$, where $s' = s$ if a_1 is not an element of any set of $B_{r,s}$ and $s' = s - 1$ if a_1 is an element of $B_{r,s}$. But in this case we assumed $s > r$ whence $s' \geq s - 1 \geq r$, and condition C holds after such a deletion. If there is a critical block $B_{k,k}$, we may apply Lemma 5.1.2, delete the elements of $B_{k,k}$ from all other sets, and still preserve the validity of condition C. By Theorem 5.1.1, a critical block $B_{k,k}$, being finite, possesses an SDR when condition C holds, and after these elements are deleted from all remaining sets, the remaining sets satisfy condition C. Hence, for a maximal deletion D preserving condition C, every element is in a critical block and every critical block has only one element in each set. Thus, for a maximal deletion D preserving condition C, every set consists of a single element and these elements form an SDR for the system U. The proof of our theorem is now finished.

Given n subsets $S_1, \ldots, S_n$ of a set S, it is not in general practical to test the validity of condition C, since a direct verification involves examining $2^n - 1$ choices of subsets. We give an algorithm here that requires no preliminary verifications and terminates either with the desired set of distinct representatives or exhibits k sets that do not contain as many as k distinct elements between them.

Choose an arbitrary element a_1 from S_1 and, as long as we can, choose an arbitrary a_i from S_i different from the previously chosen a's. This will yield distinct representatives if the process can be continued to S_n. This procedure will fail if we reach a set S_r, all of whose elements $b_1, \ldots, b_t$ have been used as representatives. We now list $b_1, \ldots, b_t$, regarded as being ordered to form a first list T_1. The second list, T_2, will consist of T_1 followed by the elements of the set $S(b_1)$ whose representative is b_1, excluding elements of T_1, say, $b_{t+1}, \ldots, b_s$. This further set of elements may, of course, be void. In general, having constructed a list T_i, we form the list T_{i+1} by following T_i by the elements (those not already listed) of the set $S(b_{i+1})$ whose representative is b_{i+1}. If at any stage we list an element b_u that has not been used as a representative, then b_u is not in the list T_1 but is in some list T_{u_2}, being included as an element of $S(b_{u_2})$ not in T_{u_2-1}. Here, $u_2 < u$, and if b_{u_2} is not in T_1, then b_{u_2} is in a set $S(b_{u_3})$ with $u_3 < u_2$. Continuing, we have

$$b_u \in S(b_{u_2}), \quad b_{u_2} \in S(b_{u_3}), \ldots, \quad b_{u_{s-1}} \in S(b_{u_s}), \qquad \text{and} \qquad b_{u_s} \in T_1.$$

Thus, we may use b_u as the representative of $S(b_{u_2})$ and, generally, b_{u_i} as the representative of $S(b_{u_{i+1}})$, freeing b_{u_s} to be a representative of S_r. We still have distinct representatives for $S_1, \ldots, S_{r-1}$, but now we also have a representative for S_r. Repeated application of this procedure may yield a complete SDR. The foregoing procedure will fail if we reach a final list T_m with elements

$b_1, \ldots, b_{k-1}$ in which every element b_i has been used as a representative of a set $S(b_i)$, $i = 1, \ldots, k-1$, and if every element of all these sets is included in the list. But then these $(k-1)$ sets together with S_r, for which no representative has been found, form k sets, and since these between them contain only $(k-1)$ elements, condition C is violated. This algorithm could easily be programmed on a digital computer.

A consequence of this algorithm, not immediately obvious from Theorem 5.1.1, is

THEOREM 5.1.3. *Let $U\{S_1, \ldots, S_i, \ldots, S_n\}$ be an indexed set of subsets of a set S. If $S_1, \ldots, S_r$ have distinct representatives $a_1, \ldots, a_r$ and if U has an SDR, then U has an SDR including $a_1, \ldots, a_r$ as representatives, though not necessarily as representatives of $S_1, \ldots, S_r$.*

Note that in (5.1.1) we have 3, 4, 5 as representatives of S_1, S_2, S_3, but in an SDR these cannot all be used for these sets.

A theorem of König [1] on matrices is essentially equivalent to Philip Hall's theorem. Let us use the term "line" of a matrix to mean either a row or a column.

THEOREM 5.1.4 (KÖNIG). *If the elements of a rectangular matrix are zeros and* 1*'s, the minimum number of lines that contain all of the* 1*'s is equal to the maximum number of* 1*'s that can be chosen with no two on a line.*

Proof: ■ Let $A = (a_{ij})$ be an $n \times t$ matrix of zeros and 1's. Let m be the minimum number of lines containing all the 1's, and M be the maximum number of 1's, no two on a line. Then, trivially, $m \geq M$, since no line can pass through two of the 1's in M. We may use the Philip Hall theorem to prove the other inequality $M \geq m$. Suppose the minimum covering by m lines consists of r rows and s columns, where $r + s = m$. We may rearrange rows and columns so that these are the first r rows and first s columns, since permutations of rows and columns clearly does not affect the values of M and m. With the first r rows, $R_1, \ldots, R_r$, we associate sets $S_1, S_2, \ldots, S_r$, the set S_i, $i = 1, \ldots, r$ consisting of those j's for which $a_{ij} = 1$ and also $j > s$. In other words, S_i is the set of column numbers (excluding the first s columns) of the entries that are 1 in the ith row.

We assert that the sets S_i satisfy condition C, for if k of these sets contained at most $(k-1)$ entries, these k rows could be replaced by the appropriate $(k-1)$ or fewer columns, and all the 1's would still be contained in this new choice of rows and columns. By the minimality of m, this cannot be done. Hence, the sets S_i satisfy condition C and, by Philip Hall's theorem, they have r distinct representatives, these being 1's in the first r rows, no two in the same line and none in the first s columns. By the same argument we may choose s 1's, no two on a line, from the first s columns, but none in the first r rows.

These $(r + s = m)$ number of 1's chosen have the property that no two lie on a line. Hence, $M \geq m$. Combining this with our earlier observation $m \geq M$, we have $m = M$, and our theorem is proved.

Conversely, it is easy to prove the Philip Hall theorem from the König theorem. If the sets are $S_1, \ldots, S_n$ and the elements are $x_1, \ldots, x_m$, form a matrix $A = (a_{ij})$, where $a_{ij} = 1$ if x_j is in S_i and $a_{ij} = 0$ if it is not. If the 1's in A can be contained in r rows and s columns with $r + s < n$, then for the $(k = n - r)$ rows not in the r-covered rows, there are 1's only in the $(s < n - r = k)$ columns, and condition C is violated for these k sets. But if a minimal covering of lines has $r + s = n$, then the König theorem yields n 1's, no two on a line, and the corresponding elements form distinct representatives for the n sets.

We may apply the theory of distinct representatives to the study of Latin rectangles and Latin squares. We recall that an array

$$\begin{matrix} a_{11} & a_{12} & \cdots & a_{1n} \\ a_{21} & a_{22} & \cdots & a_{2n} \\ a_{r1} & a_{r2} & \cdots & a_{rn} \end{matrix} \tag{5.1.3}$$

is an $r \times n$ Latin rectangle if each row contains the numbers $1, 2, \ldots, n$ in some order and if each column does not contain any digit repeated. In general, $r \leq n$, and if $r = n$, the array is a Latin square and each number $1, 2, \ldots, n$ occurs exactly once in each row and column. A natural question to ask, if $r < n$, is: Can we add another row to the array (5.1.3) to form an $(r + 1) \times n$ Latin rectangle and, if so, in how many ways? If we have only a single row, the problem of the number of ways of adding a second row is the problem of derangements, and we have already seen that this number is the integer nearest to $n!/e$. If we have the two rows

$$\begin{matrix} 1 & 2 & 3 & \cdots & n-1 & n \\ 2 & 3 & 4 & \cdots & n & 1, \end{matrix} \tag{5.1.4}$$

the number of ways of adding a third row is the "problème des ménages" discussed in Chapter 2, and this number is approximately $n!/e^2$. It has been shown by Erdös and Kaplansky [1] that if r is small compared with n, specifically if $r < \sqrt[3]{n}$, then the number of ways of adding a row is approximately $n!/e^r$. The following theorem gives a lower bound for all values of r, which is almost surely too low if $r \leq n - 3$.

THEOREM 5.1.5. *The number of ways of adding a row to an $r \times n$ Latin rectangle to give an $(r + 1) \times n$ Latin rectangle is at least $(n - r)!$*

Proof: ■ Let S_i, $i = 1, \ldots, n$ be the set of numbers that do not appear in the ith column of the given $r \times n$ Latin rectangle R. Then a set of distinct representatives of the sets S_i will be a row that may be added to R to form an

$(r + 1) \times n$ Latin rectangle, since it will contain all numbers 1 to n and will not repeat any digits in the columns. Conversely, a row that may be added to R to form an $(r + 1) \times n$ Latin rectangle will be an SDR for the sets S_i.

Our problem now is to show that the sets S_i have an SDR and to find how many SDR's there are. The set S_i consists of the $(n - r)$ numbers not in the ith column of R. Each number $1, 2, \ldots, n$ appears r times in R, and hence $(n - r)$ times in the sets $S_1, \ldots, S_n$ taken together. A choice of k of the sets $S_1, \ldots, S_r$ will contain $k(n - r)$ numbers, counting multiplicities. But since no one of these numbers occurs more than $(n - r)$ times, there must be at least k different numbers in the k sets. Hence, condition C is satisfied. Since every one of the sets S_i contains $(n - r)$ elements, the corollary to Theorem 5.1.1 tells us that there are at least $(n - r)!$ SDR's and our theorem is proved.

A further application of the theorem on distinct representations is

THEOREM 5.1.6 (SIMULTANEOUS REPRESENTATIVES). *If a set S is divided into a finite number n of subsets in two ways, $S = A_1 + A_2 + \cdots + A_n = B_1 + B_2 + \cdots + B_n$, and if no k of the A's are contained in fewer than k of the B's for each $k = 1, 2, \ldots, n$, there will exist elements $x_1, \ldots, x_n$ that are simultaneously representatives of the A's and the B's.*

Proof: ■ For each A_i define a set S_i, where S_i is the set of indices j such that the intersection $A_i \cap B_j$ is not void. The condition of the theorem is precisely condition C for the sets S_i. If $j_1, j_2, \ldots, j_n$ are distinct representatives for the sets S_i, choose an object x_i in the intersection $A_i \cap B_{j_i}$. Then $x_1, \ldots, x_n$ are simultaneously representatives of the A's and the B's. The condition of the theorem is trivially necessary and has been shown to be sufficient for the existence of simultaneous representatives. By considering complements we see that the condition is really symmetric in the A's and the B's.

We note that this theorem is also valid if n is infinite, provided the subsets are finite, since then Theorem 5.1.2 applies.

This theorem has an interesting application to groups.

THEOREM 5.1.7. *If H is a finite subgroup of a group G, there exists a set of elements that are simultaneously representatives for the right cosets of H and the left cosets of H.*

Proof: ■ Both right cosets x_iH and left cosets H_{y_j} have the same number of elements as does the subgroup H. Thus, the condition of Theorem 5.1.6 holds trivially and the conclusion of Theorem 5.1.7 follows. Simultaneous right- and left-coset representatives exist for a subgroup in a variety of other circumstances. This problem has been investigated at some length by Ore [1].

THEOREM 5.1.8. *In a vector space of infinite dimension any two bases have the same cardinal number.*

Proof: ■ Let x_i, $i \in I$, and y_j, $j \in J$ be two indexed bases for a vector space V over a field F. If we express each x_i as a linear combination of y's, we have a set S_i of y's associated with x_i, namely, those y's with nonzero coefficients in the expression for x_i in terms of the y's. The sets S_i, $i \in I$ are finite and must satisfy condition C, for if k sets S_i contained less than a k number of y's together, the corresponding x's would be linearly dependent. Hence, we can choose distinct y's from the S_i's, and so the cardinal number of y's is at least that of the x's. Similarly, the cardinal number of x's is at least that of the y's, and the two cardinals must be equal.

We may apply the theorem on distinct representatives to obtain information on matrices. The following theorem is an important application.

THEOREM 5.1.9. *Let $A = (a_{ij})$, $i = 1, \ldots, n$, $j = 1, \ldots, n$ be an $n \times n$ matrix where the a_{ij} are nonnegative real numbers such that every row and every column has the same sum. Then A is a sum of nonnegative multiples of permutation matrices.*

Proof: ■ A permutation matrix P is a square matrix of zeros and 1's with a single 1 in each row and in each column. We are to prove that if $A = (a_{ij})$, where

$$\left.\begin{aligned} \sum_{i=1}^{n} a_{ij} &= t, \qquad j = 1, \ldots, n \\ \sum_{j=1}^{n} a_{ij} &= t, \qquad i = 1, \ldots, n \end{aligned}\right\} a_{ij} \geq 0,$$

then

$$A = u_1P_1 + u_2P_2 + \cdots + u_sP_s, \qquad u_i \geq 0,$$

where each of $P_1, P_2, \ldots, P_s$ is a permutation matrix. We proceed by induction on the number w of nonzero entries a_{ij}. If $A \neq 0$, then $w \geq n$, and if $w = n$, then we easily see that $A = tP$, where P is a permutation matrix. For each $i = 1, \ldots, n$, let S_i be the set of j's such that $a_{ij} > 0$. We assert that the sets S_i satisfy condition C, for suppose to the contrary that k sets S_i contain between them at most $(k - 1)$ indices j. In this case the positive entries in k rows lie in at most $(k - 1)$ columns. But if we add these entries by rows, we obtain kt, whereas when we add them by columns, the sum is at most $(k - 1)t$. This is a conflict and so condition C must be valid. Then let $j_1, \ldots, j_n$ be distinct representatives of $S_1, \ldots, S_n$. This means that $a_{ij_i} > 0$ for $i = 1, \ldots, n$, and as $j_1, \ldots, j_n$ are distinct, they are a permutation of $1, \ldots, n$. Here the matrix $P_1 = (c_{ij})$, where $c_{ij_i} = 1$, $c_{ij} = 0$, $j \neq j_i$ is a permutation matrix. If u_1 is the minimum of a_{ij_i}, $i = 1, \ldots, n$, then $A_1 = A - u_1P_1$ is a matrix with nonnegative entries in which every row and column has a sum $t - u_1$. Furthermore, by the choice of u_1, A_1 has fewer nonzero entries, and so by induction there are permutation matrices $P_2, \ldots, P_s$ and nonnegative numbers $u_2, \ldots, u_s$ such that $A_1 = u_2P_2 + \cdots + u_sP_s$, whence $A = u_1P_1 + u_2P_2 + \cdots + u_sP_s$, as we were to prove. This completes the proof of our theorem.

PROBLEMS

1. With the columns of the following arrays as sets, show that there are respectively 31 and 24 SDR's.

$$\begin{matrix} 1 & 2 & 3 & 4 & 5 & 6 & 7 \\ 2 & 3 & 4 & 5 & 6 & 7 & 1 \\ 3 & 4 & 5 & 6 & 7 & 1 & 2 \end{matrix} \qquad \begin{matrix} 1 & 2 & 3 & 4 & 5 & 6 & 7 \\ 2 & 3 & 4 & 5 & 6 & 7 & 1 \\ 4 & 5 & 6 & 7 & 1 & 2 & 3 \end{matrix}$$

2. Sets $S_1, S_2, \ldots, S_n$ contain respectively $2, 3, \ldots, n + 1$ elements. Show that there are at least 2^n SDR's. Exhibit such sets in which 2^n is the exact number of SDR's.
3. Let $S_1, \ldots, S_n$ be n sets that have an SDR. Suppose that $a_1, \ldots, a_t$, $t < n$, are an SDR for $S_1, \ldots, S_t$. Prove that $S_1, \ldots, S_n$ have an SDR including the elements $a_1, \ldots, a_t$, but not necessarily as representatives of $S_1, \ldots, S_t$. Give an example of sets $S_1, \ldots, S_n$ and elements $a_1, \ldots, a_t$ that comprise an SDR for $S_1, \ldots, S_t$, but which cannot be representatives of $S_1, \ldots, S_t$ in any SDR of $S_1, \ldots, S_n$.
4. Let A be an $n \times n$ matrix of nonnegative real numbers such that every row and column has the same sum. Prove that A can be written as a sum of nonnegative multiples of at most $(n^2 - 2n + 2)$ permutation matrices. (This is in fact the best possible limit.)

6

Ramsey's Theorem

6.1 Statement of the Theorem

Suppose we have six points joined in pairs by arcs colored either red or blue. We are asked to show that there are three points such that the arcs of the triangle they form are of the same color. We may demonstrate this in the following way. Take one point P_0 and consider the five arcs from P_0 to the remaining five points. Three of these arcs must be of the same color (say, red); let these be numbered P_0P_1, P_0P_2, P_0P_3. If any one of the arcs P_1P_2, P_1P_3, or P_2P_3 is red, this arc together with the two arcs joining its ends to P_0 form a red triangle. Otherwise, all three of P_1P_2, P_1P_3, and P_2P_3 are blue, and $P_1P_2P_3$ is a blue triangle. This solves our little problem, and it is easy to see that six is the right number of points to guarantee a triangle of one color. The following list gives an example of a set of five points, with red and blue arcs joining the pairs so that there is no triangle of one color:

P_1P_2:	red	P_2P_4:	red
P_1P_3:	red	P_2P_5:	blue
P_1P_4:	blue	P_3P_4:	blue
P_1P_5:	blue	P_3P_5:	red
P_2P_3:	blue	P_4P_5:	red

The problem may be greatly generalized and the solution given above can be also generalized. The generalization we give here is known as Ramsey's theorem.

THEOREM 6.1.1 (RAMSEY'S THEOREM) [1]. *Let S be a set containing N elements and suppose that the family T of all subsets of S containing exactly r elements is divided into two mutually exclusive families α and β. Let $p \geq r, q \geq r, r \geq 1$. Then, if $N \geq n(p, q, r)$, a number depending solely on the integers p, q, r and not on the set S, it will be true that there is either a subset A of p elements, all of whose r subsets are in the family α, or there is a subset B of q elements, all of whose r subsets are in the family β.*

Proof: ■ We note that the problem solved above shows that $n(3, 3, 2)$ is at most 6 and the example shows that $n(3, 3, 2)$ is greater than 5. Thus, $n(3, 3, 2) =$ 6 exactly.

We now proceed to the proof of the theorem. We shall use complete induction on p, q, r, assuming the theorem to be true for $r - 1$ and any values $p^* \geq r - 1$, $q^* \geq r - 1$, and for r and triples p', q, r with $p' < p$ and triples p, q', r with $q' < q$. For the initial values of our induction we must evaluate $n(p, q, 1)$, $n(r, q, r)$, and $n(p, r, r)$. We assert that $n(p, q, 1) = p + q - 1$, for we may take $(p - 1)$ α's and $(q - 1)$ β's, whence $n(p, q, 1) > p + q - 2$. But with $N \geq p + q - 1$ elements, if there are not p α's, there are at most $p - 1$, and then there are at least $N - (p - 1) \geq q$ β's. If $p = r$ and there are any r sets of the family α, any one of them will satisfy our choice. If not, every r set is a β, and with q or more elements we have q elements, all of whose r sets are β's. Thus, $n(r, q, r) = q$. Similarly, $n(p, r, r) = p$. Write $p_1 = n(p - 1, q, r)$, $q_1 = n(p, q - 1, r)$. Then we assert that $n(p, q, r) \leq n(p_1, q_1, r - 1) + 1$.

Let S be a set with $(N \geq n(p_1, q_1, r - 1) + 1)$ elements. Choose a particular element a_0 of S. We now define α' and β' sets of $(r - 1)$ elements in the set $S' = S - a_0$ by calling a set of $(r - 1)$ elements of S' an α' set, if together with a_0 they form an α set of S, and a β' set if with a_0 they form a β set of S. Since S' has at least $(n(p_1, q_1, r - 1))$ elements, by our induction S' contains either p_1 elements, all of whose $(r - 1)$ subsets are α' sets, or q_1 elements, all of whose $(r - 1)$ subsets are β' sets. In the first case, since $p_1 = n(p - 1, q, r)$, if the p_1 elements contain a subset of q elements, all of whose r subsets are β sets, these q elements satisfy our requirement. Otherwise, $p - 1$ of the p_1 elements have all their r subsets as α sets, and these $(p - 1)$ elements together with a_0 form p elements of S, all of whose r subsets are α sets—and again our requirement is satisfied. A similar argument applies if S' contains $(q_1 = n(p, q - 1, r))$ elements, all of whose $(r - 1)$ subsets are β' sets. Thus, we have shown that $n(p, q, r) \leq n(p_1, q_1, r - 1) + 1$ with $p_1 = n(p - 1, q, r)$, $q_1 = n(p, q - 1, r)$, and our theorem is proved.

6.2 Application of Ramsey's Theorem

As an application of Ramsey's theorem, Erdös and Szekeres [1] have proved the following.

THEOREM 6.2.1. *For a given integer n, there is an integer $N = N(n)$ such that any N points in a plane, no three on a line, will contain n points forming a convex n-gon.*

Proof: ■ We shall study convex bodies in more detail in Chapter 8. A body is *convex* if any line segment joining two points of it lies entirely in the body. The *convex hull* of any set of points is the smallest convex body containing all points. For a finite set of points in a plane, not all on a line, the convex hull is a polygon containing all the points, either on its boundary or in its interior.

LEMMA 6.2.1. *Of five points in a plane, no three on a line, four are the vertices of a convex quadrilateral.*

Proof: ■ If the convex hull of the five points is a quadrilateral or a pentagon, the proof is immediate. If the convex hull is a triangle A, B, C, and the other two points D, E, are inside, then the line DE extended will intersect two of the sides of the triangle but not the third, say, BC. Then B, C, D, E form the vertices of a convex quadrilateral.

LEMMA 6.2.2. *If all the quadrilaterals formed from n points, no three on a line, are convex, then the n points are the vertices of a convex n-gon.*

Proof: ■ This is trivial for $n = 4$ and we proceed by induction on n. Suppose $n \geq 5$ and that all the quadrilaterals formed from $A_1, \ldots, A_n$ are convex. Then, by induction, $A_1, \ldots, A_{n-1}$ are the vertices of a convex $(n - 1)$-gon C_{n-1} and we may suppose this to be their order around the perimeter. Suppose first that A_n is inside C_{n-1}. Then A_n must be inside one of the triangles $A_1A_2A_3$, $A_1A_3A_4, \ldots, A_1A_iA_{i+1}, \ldots, A_1A_{n-2}A_{n-1}$. If A_n is inside the triangle $A_1A_iA_{i+1}$, then the four points A_1, A_i, A_{i+1}, A_n do not form a convex quadrilateral, contrary to our hypothesis. (Note that since no three points are on a line, A_n cannot lie on a side of any of these triangles.) Hence, A_n must lie outside the polygon C_{n-1}. Let us join A_n to each of $A_1, \ldots, A_{n-1}$ by lines. Since A_n is outside C_{n-1}, there will be two extreme lines (say, A_nA_i and A_nA_j) forming an angle containing the convex polygon C_{n-1}. Unless A_i and A_j are consecutive vertices, the triangle $A_nA_iA_j$ will contain one more point, A_k, inside it so that the four points A_n, A_i, A_j, A_k do not form a convex quadrilateral, contrary to our hypothesis. Hence, A_i and A_j are consecutive, and by putting A_n between them, we now have a convex n-gon, and our lemma is proved.

With these geometrical lemmas established, Theorem 6.1.2 is an almost immediate application of Ramsey's theorem, for N points in a plane, no three on a line, divide the quadrilaterals into a family α of convex quadrilaterals and a family β of concave quadrilaterals. Then Ramsey's theorem says that if $N \geq N(n, 5, 4)$, there will either be an n-gon, all of whose quadrilaterals are convex, or a pentagon, all of whose quadrilaterals are concave. By Lemma

6.2.1, the second alternative cannot arise, and so the first must hold. Thus, by Lemma 6.2.2, these n points form the vertices of a convex n-gon and our theorem is proved.

PROBLEMS

1. Given six points joined in pairs by either a blue or a red arc. Calling a triangle chromatic if its three arcs are of the same color, show that there must be at least two chromatic triangles.
2. Using the result of Problem 1, show that, given seven points joined in pairs by either a blue or a red arc, there must be at least three chromatic triangles.
3. Prove $n(k, m, 2) = n(m, k, 2)$ when $m \geq 2, k \geq 2$.
4. Prove for $k \geq 2, m \geq 2$,

$$n(k, m, 2) \geq \binom{k + m - 2}{k - 1} = \binom{k + m - 2}{m - 1}.$$

5. State and prove the natural generalization of Ramsey's theorem for a set S whose r subsets are divided into m exclusive families $\alpha_1, \alpha_2, \ldots, \alpha_m$.
6. Prove $n(3, 4, 2) \leq 9$, where α is red, β blue. Show that the conclusion holds if there are four red arcs at a vertex or six blue arcs. Show that it is impossible to have three red arcs and five blue arcs at every vertex because the numbers 3, 5, and 9 are odd.
7. Prove $n(3, 4, 2) > 8$, and so $n(3, 4, 2) = 9$, by taking vertices as the residues modulo 8, 0, 1, . . . , 7 and making the color of the arc joining i and j depend only on the difference $i - j \pmod 8$. Note that we must assign the same color to $d \equiv i - j$ and $-d \equiv j - i$.
8. Show that $n(4, 4, 2) > 17$ by taking vertices as the residues 0, 1, . . . , 16(mod 17) and denoting as blue the arc joining i and j if $i - j \equiv \pm 1$, ± 2, ± 4, $\pm 8 \pmod{17}$, and red otherwise. Hence, conclude $n(4, 4, 2) = 18$.

7

Some Extremal Problems

7.1 The Assignment Problem

The theorem on distinct representatives is useful in solving certain problems that, at least superficially, do not have any relation to distinct representatives. One of these is the *assignment problem.* Suppose that we have n positions to be filled by n men, and scores a_{ij} for the ith man in the jth position. Any assignment is specified by a permutation

$$\pi = \begin{pmatrix} 1, 2, \ldots, n \\ j_1, j_2, \ldots, j_n \end{pmatrix},$$

in which the ith man is assigned to position $j_i = \pi(i)$. An optimal assignment is one that maximizes $\sum_i a_{i\pi(i)}$. There are of course $n!$ permutations to be tried. In practice, if n is of even moderate size, this is a prohibitive number of trials. The following theorem has two virtues, the first being to find the solution in a relatively small number of steps, and the second being that the method makes immediate the verification that the solution found is indeed an optimal solution.

THEOREM 7.1.1. *Let $A = (a_{ij})$ be an $n \times n$ matrix of real numbers. Then the maximum of the sum*

$$\sum_{i=1}^{n} a_{i\pi(i)}$$

over all permutations π is equal to the minimum of

$$\sum_{i=1}^{n} u_i + \sum_{j=1}^{n} v_j$$

for all numbers u_i, $i = 1, \ldots, n$ and v_j, $j = 1, \ldots, n$ such that $u_i + v_j \geq a_{ij}$ in all cases. This common value is attained when $u_i + v_{\pi(i)} = a_{i\pi(i)}$, $i = 1, \ldots, n$, and these values solve the assignment problem.

Proof: ■ Given the matrix $A = (a_{ij})$, we can certainly find numbers u_i, v_j satisfying $u_i + v_j \geq a_{ij}$ in all cases by taking all the v's equal to zero and each u_i as the maximum of a_{ij}, $j = 1, \ldots, n$. For any permutation π, we have $u_i + v_{\pi(i)} \geq a_{i\pi(i)}$, whence summing on i we have

$$\sum_{i=1}^{n} u_i + \sum_{j=1}^{n} v_j \geq \sum_{i=1}^{n} a_{i\pi(i)}$$

From this the minimum m of

$$\sum_{i=1}^{n} u_i + \sum_{j=1}^{n} v_j$$

exists, and the maximum M of

$$\sum_{i=1}^{n} a_{i\pi(i)}$$

over all permutations must satisfy $m \geq M$. We must prove that $m = M$.

Initially, let us assume that the scores a_{ij} are integers. For any choice of u's and v's satisfying the inequalities $u_i + v_j \geq a_{ij}$ for all i, j, we may keep the v's fixed and reduce the u's if necessary so that for each i there is at least one j such that $u_i + v_j = a_{ij}$. For a given choice of u's and v's and for each $i = 1, \ldots, n$, let S_i be the set of j's such that $u_i + v_j = a_{ij}$. If the sets S_i have an SDR, j_1, $j_2, \ldots, j_n$, then $u_i + v_{j_i} = a_{ij_i}$ and the permutation $\pi(i) = j_i$, $i = 1, \ldots, n$ yields a solution of the assignment problem. If there is no SDR, then condition C (Chapter 5) is violated, and we may find k sets S_i, $i \in K$, where K is a set of k indices such that there are at most $(k - 1)$ j's in the sets S_i, $i \in K$ taken together. (We note that the algorithm for the SDR's will find sets violating condition C, if it is violated.) Designate by J the subset of indices such that $j \in J$ if and only if $j \in S_i$ for some $i \in K$. The violation of condition C means that J contains at most $(k - 1)$ indices j. We now replace the u's and v's by new values, putting

$$\begin{aligned} u_i^* &= u_i - 1, && \text{if } i \in K, \\ u_i^* &= u_i, && \text{if } i \notin K, \\ v_j^* &= v_j + 1, && \text{if } j \in J, \\ v_j^* &= v_j, && \text{if } j \notin J. \end{aligned} \tag{7.1.1}$$

Let there be $t < k$ indices in J. Then the new values u_i^* and v_j^* satisfy

$$\sum_{i=1}^{n} u_i^* + \sum_{j=1}^{n} v_j^* = \sum_{i=1}^{n} u_i + \sum_{j=1}^{n} v_j - k + t, \tag{7.1.2}$$

and as $t < k$, the sum of the u's and v's has been reduced by $k - t$. We also assert that

$$u_i^* + v_j^* \geq a_{ij}, \qquad \text{all } i, j. \tag{7.1.3}$$

Here, from (7.1.1) it follows that (7.1.3) is surely satisfied unless $i \in K$. If $j \in J$, then

$$u_i^* + v_j^* = (u_i - 1) + (v_j + 1) = u_i + v_j \geq a_{ij}. \tag{7.1.4}$$

But if $j \notin J$, then from the definition of the S_i,

$$u_i + v_j > a_{ij} \qquad \text{or} \qquad u_i + v_j \geq a_{ij} + 1, \tag{7.1.5}$$

whence

$$u_i^* + v_j^* = u_i + v_j - 1 \geq a_{ij}. \tag{7.1.6}$$

If condition C is satisfied with this new choice of u's and v's, we are led to a solution of the theorem and the assignment problem. If not, we continue as before and find still newer values for u_i and v_j for which $\sum u_i + \sum v_j$ has been reduced; but the condition $u_i + v_j \geq a_{ij}$ for all i, j still holds. Since our initial choice of u's and v's was integral and since we have reduced $\sum u_i + \sum v_j$ by a positive integer, at least one at each stage, the process must terminate, and at termination we have proved our theorem and found a solution for the assignment problem.

We note that if for a particular choice of u's and v's we have an SDR $j_1, j_2, \ldots, j_n$ of the sets S_i, of course we then have a solution of the assignment problem with $a_{1j_1} + a_{2j_2} + \cdots + a_{nj_n} = \sum u_i + \sum v_j$. If, however, $f_1, f_2, \ldots, f_n$ is another permutation, we have $u_i + v_{f_i} \geq a_{if_i}$ and have equality only if $f_i \in S_i$. Hence,

$$a_{1f_1} + a_{2f_2} + \cdots + a_{nf_n} \leq \sum u_i + \sum v_j$$

and we have equality only if $f_1, f_2, \ldots, f_n$ are an SDR of the sets S_i. In other words, a choice of u's and v's that gives one optimal assignment will lead to all optimal assignments by finding all SDR's of the sets S_i.

This process may easily be modified to prove the theorem and solve the assignment problem for nonintegral values of the a_{ij}. Let us first replace the a_{ij} by integers $\bar{a}_{ij}$, where $\bar{a}_{ij} \geq a_{ij} > \bar{a}_{ij} - 1$. Then, as in the preceding paragraph, permutation $j_1, \ldots, j_n$ and integers u_i, v_j may be found such that

$$\bar{a}_{1j_1} + \bar{a}_{2j_2} + \cdots + \bar{a}_{nj_n} = \sum u_i + \sum v_j = m_1 \tag{7.1.7}$$

and

$$u_i + v_j \geq \bar{a}_{ij} \qquad \text{in every case.} \tag{7.1.8}$$

We note that from the choice of the $\bar{a}_{ij}$, we have

$$a_{1j_1} + a_{2j_2} + \cdots + a_{nj_n} \geq m_1 - n \tag{7.1.9}$$

and that for any permutation $f_1, \ldots, f_n$,

$$a_{1f_1} + a_{2f_2} + \cdots + a_{nf_n} \leq \bar{a}_{1f_1} + \cdots + \bar{a}_{nf_n} \leq m_1 . \tag{7.1.10}$$

Now choose a denominator N and replace a_{ij} by $g_N(a_{ij})$, where $g_N(a_{ij})$ is rational with denominator N and satisfies

$$g_N(a_{ij}) \geq a_{ij} > g_N(a_{ij}) - \frac{1}{N} . \tag{7.1.11}$$

If we alter the rule (7.1.1) so that

$$\begin{aligned} u_i^* &= u_i - \frac{1}{N}, &\quad &\text{if } i \in K, \\ u_i^* &= u_i, & &\text{if } i \notin K, \\ v_j^* &= v_j + \frac{1}{N}, & &\text{if } j \in J, \\ v_j^* &= v_j, & &\text{if } j \notin J, \end{aligned} \tag{7.1.12}$$

the proof goes through as before and we find a permutation $j_1, \ldots, j_n$ and rational numbers u_i, v_j of denominator N such that

$$g_N(a_{1j_1}) + \cdots + g_N(a_{nj_n}) = \sum u_i + \sum v_j = m_N \tag{7.1.13}$$

and

$$u_i + v_j \geq g_N(a_{ij}) \geq a_{ij} \qquad \text{in all cases.} \tag{7.1.14}$$

We also have

$$a_{ij_1} + \cdots + a_{nj_n} \geq m_N - \frac{n}{N}, \tag{7.1.15}$$

and for any permutation $f_1, \ldots, f_n$,

$$a_{1f_1} + \cdots + a_{nf_n} \leq m_N . \tag{7.1.16}$$

Clearly, if the a_{ij} are rational, this leads to a solution directly, but if not, by taking larger and larger denominators N, the relations (7.1.15) and (7.1.16) show that in the limit we reach a solution of the assignment problem. Also let us take a succession of N's (say, 2, 4, 8, . . .), each a multiple of the preceding, and take the last values of u's and v's for one value of N as the first for the next value of N. We note that in our process at each step, the u's that are changed decrease, and the v's that are changed increase, but that neither changes more than $1/N$. Hence, in the limit,

$$\sum u_i + \sum v_j = m = M = a_{1j_1} + \cdots + a_{nj_n}. \qquad (7.1.17)$$

This proves all parts of the theorem.

7.2 Dilworth's Theorem

The following theorem of Dilworth [1] gives an extremal property of partially ordered sets, which is similar to the König theorem (Theorem 5.1.4) on zero-one matrices.

A *partially ordered* set P is a set of elements S with an ordering relation $a \subseteq b$, a is *contained* in b, holding for certain pairs a, b of elements of S such that

PO 1. (*Reflexive law.*) $a \subseteq a$.
PO 2. (*Transitive law.*) *If* $a \subseteq b$ *and* $b \subseteq c$, *then* $a \subseteq c$.
PO 3. (*Equality law.*) *If* $a \subseteq b$ *and* $b \subseteq a$, *then* $a = b$.

The partially ordered set P may satisfy one more law:

PO 4. *For any* $a, b, \in S$, *either* $a \subseteq b$ *or* $b \subseteq a$.

If PO 4 is satisfied, we call P a *simply ordered set.* A subset of P that is simply ordered is called a *chain.* Two elements of a partially ordered set are said to be *comparable* if they form a chain; otherwise, they are said to be *incomparable.*

THEOREM 7.2.1 (DILWORTH). *Given a partially ordered set P. The minimal number of disjoint chains which together contain all elements of P is equal to the maximal number of elements in a subset of P whose elements are pairwise incomparable, if this number is finite.*

Proof: ∎ Let m be the minimal number of disjoint chains containing all elements of P, and let M be the maximal number of elements in a set S whose elements are pairwise incomparable. If $x_1, \ldots, x_M$ are mutually incomparable, no chain can contain two of these elements, and so we have trivially

$$m \geq M. \qquad (7.2.1)$$

We must prove the inequality in the opposite direction. Thus, our theorem is reduced to the following lemma.

LEMMA 7.2.1: *If P is a partially ordered set and if any* $(k + 1)$ *elements, k finite, contain a comparable pair, then there exists in P a set of at most k disjoint chains, which together contain all elements of P.*

We shall first prove the lemma for the case in which P is finite. We shall use induction on the number n of elements in P. If there are t disjoint chains C_1, $C_2, \ldots, C_t$ that together contain all elements of P, we say that P is the sum of $C_1, \ldots, C_t$ and write $P = C_1 + C_2 + \cdots + C_t$. If a subset S of elements of P does not contain a comparable pair, we shall say that these elements are independent. In this terminology our lemma says that if k is the maximal number of independent elements in P, then P is the sum of k chains. Let us delete an arbitrary element b from P. Then we have in $P^* = P - b$ a partially ordered set in which the maximal number of independent elements is $k - 1$ or k. By our induction, P^* is the sum of $(k - 1)$ or k chains. If this number is $k - 1$, these $(k - 1)$ chains together with b alone as one more chain have P as their sum, and our lemma is proved. Hence, we suppose that P^* is the sum of k chains:

$$P - b = P^* = C_1 + C_2 + \cdots + C_k. \tag{7.2.2}$$

We now consider the relation of the chains $C_1, C_2, \ldots, C_k$ to b in P. For each $i = 1, 2, \ldots, k$, let U_i be the subset of C_i containing b, L_i, the subset of C_i contained in b, and N_i the subset of C_i not comparable with b. Let us write

$$\begin{aligned} U &= U_1 + U_2 + \cdots + U_k, \\ L &= L_1 + L_2 + \cdots + L_k, \\ N &= N_1 + N_2 + \cdots + N_k. \end{aligned} \tag{7.2.3}$$

Here,

$$C_i = U_i + N_i + L_i, \qquad i = 1, \ldots, k.$$

We note that $L_i \subseteq b \subseteq U_j$ for all i and j. Thus, if for some i, N_i is empty, then $L_i + b + U_i = C_i'$ is a chain and $C_1 + \cdots + C_i' + \cdots + C_k = P$, and our theorem is proved. Hence, we may assume that no N_i is empty.

We now show that for some m, the maximal number of independent elements in $N + U - U_m$ is less than k, for suppose to the contrary that for every $j = 1, 2, \ldots, k$, $N + U - U_j$ contains a set S_j of k independent elements. As S_j contains k independent elements, it contains exactly one from each of $C_1, C_2, \ldots, C_k$, and as S_j contains no element of U_j, it follows that S_j contains exactly one element from N_j. Then, in the set sum $S = S_1 + S_2 + \cdots + S_j + \cdots + S_k$ (the S_j's need not be disjoint), let s_i, $i = 1, \ldots, k$ be the minimal element of C_i contained in S. Here, s_i is in N_i, since S_i contains an element of N_i and all elements of N_i are contained in those of U_i. Suppose, if possible, that for some $i \neq j$, $s_i \supseteq s_j$. Then s_j arose in a set S_r of k independent elements, and let t_i be the element of S_r coming from C_i. By the minimality of s_i, we have $t_i \supseteq s_i$. But then $t_i \supseteq s_i \supseteq s_j$, contrary to the independence of the elements t_i and s_j of S_r. Hence, for no $i \neq j$ can we have $s_i \supseteq s_j$, and so the elements $s_1, s_2, \ldots, s_k$ are independent. But as s_i is in N_i for $i = 1, \ldots, k$, the elements $s_1, s_2, \ldots, s_k, b$ are $(k + 1)$ independent elements of P. But we were given that P did not contain $(k + 1)$ independent elements, and so we must reject the assumption that each of $N + U - U_j$, $j = 1, \ldots, k$ contains k independent elements. Hence, for some m, $N + U - U_m$ contains less than k independent elements. In the same way, for some w, $N + L - L_w$ contains less than k independent elements.

Let T be an independent subset of

$$(N + U - U_m) + (N + L - L_w) = P^* - U_m - L_w .$$

T cannot contain an element x of $U - U_m$ and also an element y of $L - L_w$, since $x \supseteq b \supseteq y$, contrary to the independence of T. But then T is contained in either $N + U - U_m$ or $N + L - L_w$, and T has at most $(k - 1)$ elements. By our induction, $P^* - U_m - L_w$ is the sum of at most $(k - 1)$ chains $P^* - U_m - L_w = C_1^* + C_2^* \cdots + C_{k-1}^*$. But $C_k^* = U_m + b + L_w$ is a chain, and then we have

$$P = C_1^* + C_2^* + \cdots + C_k^* . \tag{7.2.4}$$

This proves the lemma and theorem when P is finite.

When P is infinite, we proceed by induction on k, the conclusion being trivial if $k = 1$. We assume that the lemma holds for all partially ordered sets having at most $(k - 1)$ independent elements. A subset C of P is said to be *strongly dependent* if for every finite subset S of P there is a representation of S as a sum of at most k disjoint chains, one of which contains all elements of C belonging to S. Clearly, C must be a chain. The property of being strongly dependent is a property of finite character, as defined by Tukey [1, p. 7]. Hence, by the maximal principle, there is a maximal strongly dependent set C_1 in P. Suppose, if possible, that $P - C_1$ contains k independent elements $a_1, a_2, \ldots, a_k$. By the maximality of C_1, no one of $C_1 + a_1$, $C_1 + a_2, \ldots$, $C_1 + a_k$ is strongly dependent. Hence, for each $i = 1, \ldots, k$, there is a finite set S_i such that the elements of $C_1 + a_i$ in S_i cannot be in a single chain when S_i is represented as a sum of k disjoint chains. In particular, S_i must contain a_i because C_1 is strongly dependent. Now consider $S = S_1 + S_2 + \cdots S_k$. Here, S contains all $a_1, a_2, \ldots, a_k$. S is finite, and by the strong dependence of C_1, we can write S as a sum of disjoint chains, $S = K_1 + \cdots + K_k$, so that all C_1 contained in S lies in one of them, say, K_1. But since $a_1, \ldots, a_k$ are independent, one of them (say, a_r) lies in K_1. Thus, the intersection of $K_1, \ldots, K_k$ with S_r is a set of disjoint chains $T_1, \ldots, T_r$ such that $S_r = T_1 + \cdots + T_r$ and all $C_1 + a_r$ lying in S_r belongs to T_1. But this is contrary to our choice of S_r. Hence, it is not possible that $a_1, \ldots, a_k$ are independent. Thus, $P - C_1$ contains at most $(k - 1)$ independent elements. By induction, $P - C_1 = C_2 + \cdots + C_k$, and we have

$$P = C_1 + C_2 + \cdots + C_k, \tag{7.2.5}$$

proving our lemma and our theorem

PROBLEMS

1. Find a 5×5 matrix of scores for the assignment problem such that neither of the two largest scores may be used in the optimal assignment.
2. In the problem of assigning four men to four positions, the scores a_{ij}, $i, j = 1, 2, 3, 4$ are the numbers $1, 2, \ldots, 16$ in some order. Show that the total score for an optimal assignment is at least 34, and exhibit a matrix with such scores for which 34 is the optimal value.
3. A set of white mice is being studied experimentally. If there are $mn + 1$ mice, show that either (a) there is a sequence of $(m + 1)$ mice, each a descendant of the next, or (b) there are $(n + 1)$ mice, no one of which is a descendant of another.

8

Convex Spaces and Linear Programming

8.1 Convex Spaces. Convex Cones and Their Duals

Let us denote by E_n the Euclidean geometry of n dimensions (n a finite integer) over the real numbers. A point of E_n is a vector $x = (x_1, x_2, \ldots, x_n)$, the x_i being real numbers. The norm of x is $|x| = \sqrt{x_1^2 + x_2^2 + \cdots + x_n^2}$, and the distance between x and y is $|x - y|$.

A *convex space* C is a set of points in E_n such that if $x \in C$, $y \in C$, then also $ax + (1 - a)y \in C$ for $0 \leq a \leq 1$. Geometrically, this means that if two points belong to C, the line segment joining them also belongs to C.

LEMMA 8.1.1. *If $x_1, \ldots, x_m$ belongs to C, a convex space, then $x = a_1x_1 + a_2x_2 + \cdots + a_mx_m \in C$ if the a_i are real numbers,*

$$a_i \geq 0, \qquad \sum_{i=1}^{m} a_i = 1.$$

Proof: ■ This follows by induction on m, being trivial if $m = 1$, and the definition of convexity for $m = 2$. If $a_1 = 0$ or 1, induction yields the result $x \in C$. Otherwise, write $b = 1 - a_1$ and $b_i = a_i/b$, $i = 2, \ldots, m$. Then $b_i \geq 0$, $i = 2, \ldots, m$, and $\sum_{i=2}^{m} b_i = 1$, whence, by induction, $b_2x_2 + \cdots + b_mx_m = y \in C$. But then $a_1x_1 + (1 - a_1)y = a_1x_1 + by = a_1x_1 + a_2x_2 + \cdots + a_mx_m = x \in C$.

By a *neighborhood* of a point x we understand as usual an open sphere consisting of all points y such that $|x - y| < \epsilon$ for some $\epsilon > 0$. A point x is said to be the *limit of a sequence* x_m, $m = 1, 2, \ldots$, written

$$\lim_{m \to \infty} x_m = x,$$

if for each $\epsilon > 0$ there is an N depending on ϵ such that $|x_m - x| < \epsilon$ for all $m \geq N$. Note that we do permit $x_m = x$. The *closure* $\bar{S}$ of a set S consists of all points of S together with limit points of S. The *boundary* of a set S consists of points y such that every neighborhood of y contains points of S as well as points not in S.

The preceding terms apply to general point sets. We also use some terms that apply more specifically to convex spaces. An *extreme point* x of a convex space C is a point $x \in C$ such that if $x = ay + (1 - a)z$, $0 < a < 1$, $y, z \in C$, then $y = z = x$. A hyperplane $u_1x_1 + u_2x_2 + \cdots + u_nx_n = u_0$, which we write $f(x) = u_0$, is a *bounding hyperplane* of C if $f(y) > u_0$ for every $y \in \bar{C}$. The hyperplane $f(x) = u_0$ *separates* a point w from $\bar{C}$ if $f(w) < u_0$ and $f(y) > u_0$ for every $y \in \bar{C}$. A hyperplane $f(x) = u_0$ is a *supporting hyperplane* of the convex space C if $f(y) \geq u_0$ for every $y \in \bar{C}$ and if the greatest lower bound of $f(y)$ for $y \in \bar{C}$ is equal to u_0. For example, if C is in a plane and bounded by one branch of a hyperbola, then an asymptote of the hyperbola is a supporting hyperplane of C. Tangents are also supporting hyperplanes of convex bodies, but through an extreme point there may be several supporting hyperplanes. Thus, at a corner of a square, which is an extreme point of the square, there are many supporting lines.

A special kind of convex body is a *convex cone*. A convex body is a convex cone C if for every $x \in C$ and any real number $b \geq 0$, $bx \in C$. We use a slightly different definition for extreme points of a convex cone. We say that $x \in C$ is an extreme point of a convex cone C if $x = ay + (1 - a)z$, $0 < a < 1$, $y \in C$, $z \in C$ implies that $y = ux$, $z = vx$ for some real $u \geq 0$, $v \geq 0$.

Many familiar geometric objects are convex bodies. For our purposes, we shall also be interested in other convex spaces that we do not ordinarily consider geometric. The mn real entries of an $m \times n$ matrix $A = (a_{ij})$ may be listed in some order as the coordinates of a point in E_{mn}. Here the set of all these matrices form the full space E_{mn}, while the matrices with nonnegative entries form a convex cone. Similarly, the doubly stochastic matrices, that is, $n \times n$ matrices $A = (a_{ij})$ with

$$a_{ij} \geq 0, \qquad \sum_i a_{ij} = 1, \qquad j = 1, \ldots, n, \qquad \sum_j a_{ij} = 1, \qquad i = 1, \ldots, n,$$

form a convex space. The symmetric $n \times n$ matrices form a convex cone, and those symmetric matrices $A = (a_{ij})$, $a_{ij} = a_{ji}$, such that the associated quadratic form

$$Q = \sum_{i,j} a_{ij}x_ix_j$$

is semidefinite, will form a convex cone.

THEOREM 8.1.1. *If C is a convex set, the closure $\bar{C}$ of C is also a convex set.*

Proof: ■ If $x \in \bar{C}$, $y \in \bar{C}$, there exist sequences $\{x_m\}$, $\{y_m\}$ of points of C such that

$$\lim_{m\to\infty} x_m = x, \qquad \lim_{m\to\infty} y_m = y.$$

But then, for $0 \leq a \leq 1$, we also have

$$\lim_{m\to\infty} (ax_m + (1-a)y_m) = ax + (1-a)y,$$

whence every point of the line joining x and y is also a point of $\bar{C}$. This is the assertion of our theorem.

THEOREM 8.1.2. *If C is a convex set and P is a point not in $\bar{C}$, there is a hyperplane separating P from $\bar{C}$.*

Proof: ■ Let s be any particular point of C. Then points x, such that $|P - x| \leq |P - s| = b$, form a closed sphere of radius b and these points form a compact set. Hence, there is a point Q of $\bar{C}$ nearest to P, namely, such that $|P - Q|$ is minimum with $Q \in \bar{C}$. Furthermore, Q is unique, since if Q_1, Q_2 are two points of $\bar{C}$ the same distance from P, then PQ_1Q_2 is an isosceles triangle with sides PQ_1 and PQ_2 equal in length. But if $Q_1 \neq Q_2$, the midpoint of Q_1Q_2 is a point of $\bar{C}$ nearer to P than to Q_1 or Q_2.

Now let $f(x) = u_1x_1 + u_2x_2 + \cdots + u_nx_n = u_0$ be the hyperplane that is the perpendicular bisector of PQ. We shall normalize this equation, dividing by an appropriate constant so that $u_1^2 + \cdots + u_n^2 = 1$. Here, $f(P) - u_0$ and $f(Q) - u_0$ are of opposite sign, and we multiply by -1, if necessary, so that $f(P) < u_0$ and $f(Q) > u_0$. To show that the hyperplane $f(x) = u_0$ separates P from $\bar{C}$, it is sufficient to prove that $f(x) = u_0$ contains no point of $\bar{C}$. Suppose to the contrary that S is a point of $f(x) = u_0$, which is in $\bar{C}$. Then S is farther from P than from Q; in particular, S is not in midpoint of the line PQ, so that PSQ is a proper isosceles triangle with equal sides PS and SQ and with the third side PQ shorter than these. But then all three angles of the triangle PSQ are acute, and so the perpendicular from P to SQ meets SQ in an interior point T, and since both S and Q are in $\bar{C}$, T is also a point of $\bar{C}$. But PT is shorter than PQ, contrary to our choice of Q. This conflict shows that the hyperplane $f(x) = u_0$ contains no point of $\bar{C}$. As $f(Q) > u_0$ and $f(P) < u_0$, it follows that $f(y) > u_0$ for every $y \in \bar{C}$, and so $f(x) = u_0$ is a hyperplane separating P from $\bar{C}$.

The next theorem is similar, but a little more delicate.

THEOREM 8.1.3. *Let P be a point on the boundary of the closure $\bar{C}$ of a convex space C. Then there is a supporting hyperplane of $\bar{C}$ through P.*

Proof: ■ Since P is on the boundary of $\bar{C}$, every neighborhood of P contains a point not in $\bar{C}$. Hence, we may find an infinite sequence $\{z_m\}m = 1, 2, \ldots$ of points z_m not in $\bar{C}$ such that $\lim_{m\to\infty} z_m = P$. By Theorem 8.1.3, there is a hyperplane $f_m(x) = u_{m0}$ for each z_m, separating z_m from $\bar{C}$. We shall assume this to be normalized, so that with

$$f_m(x) = u_{1m}x_1 + \cdots + u_{nm}x_n,$$

we have $u_{1m}^2 + \cdots + u_{nm}^2 = 1$ and so that $f_m(y) > u_{m0}$ for every $y \in \bar{C}$. Since the coefficients of $f_m(x)$ lie on a unit sphere, they form a compact set, and so some subsequence of the points $(u_{1m}, u_{2m}, \ldots, u_{nm})$, $m = 1, \ldots, k, \ldots,$ approaches a limit $(u_1, u_2, \ldots, u_n)$. Let these m's be the subsequence $\{m_j\}$ $j = 1, 2, \ldots$. Then

$$\lim_{j\to\infty} f_{m_j}(x) = f(x) = u_1x_1 + u_2x_2 + \cdots + u_nx_n.$$

As

$$\lim_{j\to\infty} (z_{m_j}) = P,$$

we have

$$\lim_{j\to\infty} u_{m_j0} = u_0,$$

where $f(P) = u_0$. For every $y \in \bar{C}$, we had $f_{m_j}(y) > u_{m_j0}$. Hence, in the limit, as $j \to \infty$, $f(y) \geq u_0$. Thus, the hyperplane $f(x) = u_0$ is a supporting plane for C through P, since $f(y) \geq u_0$ for every $y \in C$ and $f(P) = u_0$, where P is either a point of C or a limit of points of C.

LEMMA 8.1.2. *If P is a boundary point of the closure C of a convex cone, a supporting hyperplane of C through P goes through the origin.*

Proof: ■ If P is the origin, there is nothing to prove. Hence, suppose $P \neq (0, \ldots, 0)$. Let $f(x) = u_0$ be the supporting hyperplane through P, where $f(P) = u_0$. Then, for any $c > 0$, cP is a point of $\bar{C}$, whence $f(cP) = cf(P) \geq u_0$, since $f(x) = u_0$ is a supporting plane. As $f(P) = u_0$, this requires $cu_0 \geq u_0$ for every $c > 0$. This is possible only if $u_0 = 0$, whence the hyperplane is $f(x) = 0$ and necessarily goes through the origin.

The *inner product* (x, y) of vectors $x = (x_1, \ldots, x_n)$ and $y = (y_1, \ldots, y_n)$ is defined by the rule

$$(x, y) = x_1y_1 + x_2y_2 + \cdots + x_ny_n. \tag{8.1.1}$$

We note the following properties of the inner product:

$$\begin{aligned} (x, y) &= (y, x), \\ (bx, y) &= (x, by) = b(x, y), \qquad b \text{ a real scalar}, \\ (x + y, z) &= (x, z) + (y, z). \end{aligned} \tag{8.1.2}$$

THEOREM 8.1.4. *If we have any nonempty set S of vectors $\{x\}$, the set T of vectors y such that $(x, y) \geq 0$ for every $x \in S$ is a closed convex cone.*

Proof: ■ This theorem is almost immediate, since we note that for vectors y_1, y_2 satisfying $(x, y_1) \geq 0$ and $(x, y_2) \geq 0$, for every $x \in S$ and for any positive numbers a, b we have $(x, ay_1 + by_2) = a(x, y_1) + b(x, y_2) \geq 0$ for every $x \in S$. Similarly, if

$$\lim_{m\to\infty} y_m = y \qquad \text{and} \qquad (x, y_m) \geq 0,$$

then $(x, y) \geq 0$ for every $x \in S$. This proves the theorem.

DEFINITION. *If C is a closed convex cone, the set of points y such that $(x, y) \geq 0$ for every $x \in C$ is called the dual space C^*, and C^* is a closed convex cone.*

We merely note from Theorem 8.1.4 that C^* is necessarily a closed convex cone.

THEOREM 8.1.5. *If C is a closed convex cone and C^* is its dual, then $(C^*)^* = C$.*

Proof: ■ This is a major theorem, but we shall see that it is almost equivalent to the separation theorem, Theorem 8.1.2. From the definition of dual spaces, we have

$$\begin{aligned} C^* &= \{y \mid (x, y) \geq 0, \quad \text{all } x \in C\}, \\ (C^*)^* &= \{z \mid (y, z) \geq 0, \quad \text{all } y \in C^*\}. \end{aligned} \tag{8.1.3}$$

From the symmetry of the inner product we see that every vector x of C satisfies $(y, x) \geq 0$ for every $y \in C^*$ and so $C \subseteq (C^*)^*$. We must prove the inclusion in the opposite direction. If z is a vector (point) of $(C^*)^*$ that is not in $\bar{C} = C$, then by Theorem 8.1.2, there is a hyperplane $f(x) = u_0$ that separates z from $\bar{C} = C$, such that $f(z) < u_0$ and $f(x) > u_0$ for every $x \in C$. Here, as the origin $\mathcal{O}$ belongs to C, we have $0 = f(\mathcal{O}) > u_0$ and u_0 is negative. If $f(x) = u_1x_1 + u_2x_2 + \cdots + u_nx_n$, it then follows that $u = (u_1, u_2, \ldots, u_n) \in C^*$; but $(u, z) < u_0 < 0$, whence z is not in $(C^*)^*$, a conflict. Hence, $(C^*)^* \subseteq C$, and we have shown that $C = (C^*)^*$.

8.2 Linear Inequalities

We now turn our attention to the theory of linear inequalities, or linear programming, as it has come to be called. If $x = (x_1, \ldots, x_m)$ is a vector, we write

$$x \geqq \mathcal{O}, \qquad \text{if } x_i \geq 0, \qquad i = 1, \ldots, m, \qquad x \geq \mathcal{O} \qquad \text{for} \qquad x \geqq \mathcal{O},$$

excluding $x = \mathcal{O}$, and $x > \mathcal{O}$ if $x_i > 0$, $i = 1, \ldots, m$. We define $x \geqq y$, $x \geq y$, $x > y$ according to the relation of $x - y$ to $\mathcal{O}$.

If $A = (a_{ij})$, $i = 1, \ldots, m$, $j = 1, \ldots, n$ is an $m \times n$ matrix and $x = (x_1, \ldots, x_m)$ is such that $x \geqq 0$, the set of all vectors xA is readily seen to be the convex cone C, determined by the rows of A. Thus, if $r_1, r_2, \ldots, r_m$ are the rows of A and $x = (x_1, \ldots, x_m) \geqq 0$, then $xA = x_1r_1 + x_2r_2 + \cdots + x_mr_m$. For a vector $w = (w_1, \ldots, w_n)$, the transpose w^t is a column vector, and the assertion $Aw^t \geqq 0$ is equivalent to the assertion that the inner products $(r_i, w) \geq 0$ for $i = 1, \ldots, m$; whence, by the linearity of the inner product $(y, w) \geq 0$ for every $y \in C$ and whence $w \in C^*$, the dual cone of C^*. We note that C is a closed convex cone. Thus, we may translate Theorem 8.1.5 into matrix form. This theorem was originally due to Farkas [1].

THEOREM 8.2.1 (FARKAS). *Let A be an $m \times n$ matrix and suppose that $y = (y_1, \ldots, y_n)$ is a vector such that $yw^t = (y, w) \geq 0$ for every vector $w = (w_1, \ldots, w_n)$ such that $Aw^t \geqq 0$. Then y is of the form $y = xA$, where $x = (x_1, \ldots, x_m) \geqq 0$.*

Proof: ■ Let C be the closed convex cone spanned by the rows of A—that is, C consists of all vectors $x_1r_1 + x_2r_2 + \cdots + x_mr_m$, where $r_1, \ldots, r_m$ are the rows of A and $x_r \geq 0$. Then the hypothesis of the theorem says that $y \in (C^*)^*$, and the conclusion asserts that $y \in C$. This follows from Theorem 8.1.5.

A stronger, nonhomogeneous form of this theorem is as follows.

THEOREM 8.2.2. *Let A be an $m \times n$ matrix. Let d be a vector $d = (d_1, \ldots, d_m)$ and k be a scalar, and suppose that for at least one vector, $w = (w_1, \ldots, w_n)$, we have*

$$Aw^t \geqq d^t. \tag{8.2.1}$$

Then a vector $y = (y_1, \ldots, y_n)$ will satisfy

$$yw^t = (y, w) \geq k, \tag{8.2.2}$$

for all w's satisfying (8.2.1) *if and only if there is a vector $x = (x_1, \ldots, x_m) \geqq 0$ such that*

$$y = xA \qquad \text{and} \qquad xd^t \geq k. \tag{8.2.3}$$

Proof: ■ If (8.2.3) holds, immediately

$$yw^t = xAw^t \geqq xd^t \geq k. \tag{8.2.4}$$

We must prove the converse. To do this, we reformulate the problem in homogeneous form in order to apply Theorem 8.2.1. If we write

$$A' = \begin{pmatrix} A & -d^t \\ 0 & 1 \end{pmatrix}, \qquad \begin{aligned} w' &= (w_1, w_2, \ldots, w_n, z_0), \\ y' &= (y_1, y_2, \ldots, y_n, -k), \end{aligned} \tag{8.2.5}$$

then (8.2.1) takes the form

$$A'w'^t \geqq 0, \qquad \text{with } z_0 = 1, \tag{8.2.6}$$

and (8.2.2) the form

$$y'w'^t \geqq 0, \qquad \text{with } z_0 = 1, \tag{8.2.7}$$

and the desired conclusion (8.2.3), has the form

$$y' = x'A', \qquad \text{with } x' = (x_1, \ldots, x_m, x_{m+1}) \geqq 0, \tag{8.2.8}$$

which states that

$$y = xA \qquad \text{and} \qquad -xd^t + x_{m+1} = -k \qquad \text{or} \qquad xd^t = k + x_{m+1} \geq k,$$

which is equivalent to (8.2.3). Apart from the restriction $z_0 = 1$, the hypotheses (8.2.6) and (8.2.7) are those of Theorem 8.2.1, and (8.2.8) is the conclusion of Theorem 8.2.1. We must relax the effect of this restriction. If $w' = (w_1, w_2, \ldots, w_n, z_0)$, with $z_0 > 0$, satisfies (8.2.6), then $w'' = (1/z_0)w'$ is of the form $w'' = (w_1'', \ldots, w_n'', 1)$, whence by assumption, w'' also satisfies (8.2.7); and so multiplying by z_0, we find that w' also satisfies (8.2.7). Thus, our assumption becomes: whenever

$$A'w'^t \geqq 0 \qquad \text{for} \qquad w' = (w_1, \ldots, w_n, z_0), \qquad \text{with } z_0 > 0, \tag{8.2.9}$$

then also

$$y'w'^t \geqq 0, \tag{8.2.10}$$

and we wish to conclude

$$y' = x'A', \qquad \text{with } x' = (x_1, \ldots, x_m, x_{m+1}) \geqq 0. \tag{8.2.11}$$

If (8.2.9) also implied (8.2.10), with $z_0 = 0$, we could apply Theorem 8.2.1 to conclude (8.2.11). Hence, the only remaining alternative is to suppose that for some $w'' = (w_1'', \ldots, w_n'', 0)$, with $z_0 = 0$, (8.2.9) holds but (8.2.10) does not. This gives

$$A'w''^t \geqq 0 \qquad \text{and} \qquad y'w''^t < 0. \tag{8.2.12}$$

Now take a particular w'_0 satisfying (8.2.6) and (8.2.7), with $z_0 = 1$, and form w' by

$$w' = w'' + x_0 w_0', \qquad \text{with } x_0 > 0. \tag{8.2.13}$$

Then this w' satisfies (8.2.9) and therefore also (8.2.10), whence

$$y'w'^t \geq 0 \qquad \text{or} \qquad y'(w'' + x_0 w_0')^t \geq 0. \tag{8.2.14}$$

But this is

$$y'w''^t \geq -x_0 y' w_0'^t. \tag{8.2.15}$$

But $y'w''^t < 0$ and $y'w_0'^t \geq 0$ and so (8.2.15) must be false for some sufficiently small positive x_0. Thus, (8.2.12) cannot arise, and so (8.2.9) also implies (8.2.10), even when $z_0 = 0$, and we may conclude that (8.2.11) holds, which is the conclusion of our theorem, now proved in full.

Linear programming is the problem of maximizing or minimizing a linear form when the variables are subject to a number of linear inequalities. The treatment given here is due to M. Flood [1]. The first analysis is in Gale, Kuhn, and Tucker [1].

Suppose we are given an $r \times s$ matrix $A = (a_{ij})$ and two vectors $b = (b_1, \ldots, b_r)$ and $c = (c_1, \ldots, c_s)$. We now consider two problems.

Problem I: ■ To find a vector $x = (x_1, \ldots, x_s)$ that minimizes

$$cx^t = (c, x) \tag{8.2.16}$$

subject to the inequalities

$$Ax^t \geqq b^t, \qquad x \geqq 0. \tag{8.2.17}$$

Problem II: ■ To find a vector $y = (y_1, \ldots, y_r)$ that maximizes

$$yb^t = (y, b) \tag{8.2.18}$$

subject to the inequalities

$$yA \leqq c, \qquad y \geqq 0. \tag{8.2.19}$$

These two problems are said to be dual to each other. If there exists a vector x satisfying (8.2.17), then Problem I is said to be feasible and x is called a feasible vector. Feasibility is described in the same way for Problem II.

THEOREM 8.2.3 (DUALITY THEOREM). *Problems I and II possess solutions if and only if both problems are feasible. Furthermore, if m is the minimum value of cx^t in Problem I and M is the maximum value of yb^t in Problem II, then $m = M$. If either problem has a solution, so does the other.*

Proof: ■ First assume that both problems are feasible. Then $x = (x_1, \ldots, x_s)$ and $y = (y_1, \ldots, y_r)$ exist with

$$Ax^t \geqq b^t, \qquad x \geqq 0, \tag{8.2.20a}$$

$$yA \leqq c, \qquad y \geqq 0. \tag{8.2.20b}$$

Then

$$cx^t \geqq yAx^t \geqq yb^t. \tag{8.2.21}$$

Hence, cx^t is bounded below by yb^t for any feasible y, and yb^t is bounded above by cx^t for any feasible x. Hence, there is a greatest lower bound m of cx^t and a least upper bound M of yb^t; furthermore,

$$m \geq M. \tag{8.2.22}$$

Our proof will depend on Theorem 8.2.2, but to apply this theorem we must replace the two inequalities of (8.2.20a) by a single one. Let us define an $(r + s)$

dimensional vector $b_0 = (b_1, \ldots, b_r, 0, \ldots, 0)$ by adding s zeros to b. Also, we form an $(r + s) \times s$ matrix A_0 from A by adding s rows, these last being the identity $s \times s$ matrix. Then

$$A_0 x^t = \begin{pmatrix} A \\ I_s \end{pmatrix} x^t \geqq b_0^t = \begin{pmatrix} b_1 \\ b_2 \\ \cdot \\ \cdot \\ \cdot \\ b_r \\ 0 \\ \cdot \\ \cdot \\ \cdot \\ 0 \end{pmatrix} \tag{8.2.23}$$

This is precisely equivalent to the two inequalities $Ax^t \geqq b, x \geqq 0$ of (8.2.20a). Here, with m the greatest lower bound of cx^t, we have

$$cx^t \geq m \tag{8.2.24}$$

for all x satisfying (8.2.23). We are now in a position to apply Theorem 8.2.2, and we conclude that there exists a vector y_0 such that

$$y_0 = (u_1, \ldots, u_r, v_1, \ldots, v_s), \quad y_0 \geqq 0, \quad c = y_0 A_0, \quad y_0 b_0^t \geq m. \tag{8.2.25}$$

If we now write

$$u = (u_1, \ldots, u_r), \qquad v = (v_1, \ldots, v_s), \tag{8.2.26}$$

the conclusion (8.2.25) may be put in the form

$$c = uA + v, \qquad u \geqq 0, \qquad v \geqq 0, \qquad ub^t \geq m. \tag{8.2.27}$$

But from (8.2.22) and (8.2.27) we have, with $y = u$,

$$\begin{aligned} uA &\leqq c, \\ ub^t &\geq m \geq M \geq ub^t. \end{aligned} \tag{8.2.28}$$

Thus, also,

$$ub^t = m = M, \tag{8.2.29}$$

and $y = u$ is a solution of Problem II.

Similarly, if we write

$$B_0 = (A, -I_r), \tag{8.2.30}$$

adjoining r columns to A that form $-I_r$, I_r, being the identity $r \times r$ matrix, then the two inequalities $yA \leqq c$ and $y \geqq 0$ of (8.2.19) take the form

$$(-B_0^t)y^t \geqq -c_0^t, \tag{8.2.31}$$

where $c_0 = (c_1, \dots c_s, 0, \dots, 0)$ is an $(s + r)$ dimensional vector obtained by adjoining r zeros to c. Here we have as a consequence of (8.2.21) that

$$-by^t \geq -M \tag{8.2.32}$$

for all vectors y satisfying (8.2.31). We may apply Theorem 8.2.2 to conclude the existence of a vector $t = (w_1, \dots, w_s, z_1, \dots, z_r)$ such that

$$-b = t(-B_0^t), \qquad t \geqq 0, \qquad t(-c_0^t) \geq -M. \tag{8.2.33}$$

If we write

$$w = (w_1, \dots, w_s), \qquad z = (z_1, \dots, z_r), \tag{8.2.34}$$

then (8.2.33) becomes

$$b = wA^t - z, \qquad w \geqq 0, \qquad z \geqq 0, \qquad wc^t \leq M. \tag{8.2.35}$$

But then, with $x = w$, we have $wA^t \geqq b$:

$$Aw^t \geqq b^t, \qquad cw^t \leq M \leq m \leq cw^t, \qquad w \geqq 0. \tag{8.2.36}$$

Thus, $x = w$ solves Problem I, and again we have shown that $M = m$.

We have now shown that if both problems are feasible, they both have solutions and that the minimum m for Problem I is equal to the maximum M for Problem II. We further note that the existence of m in (8.2.24) implies that $y = u$ in (8.2.26) is a solution of Problem II, and that the existence of M in (8.2.32) implies that $x = w$ in (8.2.34) is a solution of Problem I. Thus, if either problem has a solution, so does the other, and now all statements of the theorem are proved.

On a line segment in n-dimensional space, a linear function of the coordinates is either constant or takes on its maximum at one end of the segment and its minimum at the other end. Thus, a linear function that is bounded above (or below) in a convex space will take on its maximum (or minimum) at an extreme point. Clearly, permutation matrices are extreme points of the space of doubly stochastic matrices, and Theorem 5.1.9 shows that they are the only extreme points. Hence, the maximum or minimum of a linear function over the space of doubly stochastic matrices is attained for a permutation matrix. We may use this observation to attack the assignment problem, using the duality theorem. This will show that the numbers u_i and v_j, which appeared somewhat mysteriously in Theorem 7.1.1, arise very naturally from the problem dual to the assignment problem.

Let $A = (a_{i,j})$, $i, j = 1, \dots, n$ be the matrix of assignment scores treated in Theorem 7.1.1. Let us define n^2 b's by the rule $b_{i+(j-1)n} = a_{i,j}$. Let us also introduce variables $z_{i,j}$, $i, j = 1, \ldots, n$ and single subscript variables $y_1, \ldots, y_{n^2}$, where $y_{i+(j-1)n} = z_{i,j}$. Now define an $n^2 \times 4n$ matrix K and a vector c:

$$K = \begin{vmatrix} J_1 & -J_1 & I_n & -I_n \\ J_2 & -J_2 & I_n & -I_n \\ \cdot & \cdot & \cdot & \cdot \\ \cdot & \cdot & \cdot & \cdot \\ \cdot & \cdot & \cdot & \cdot \\ J_n & -J_n & I_n & -I_n \end{vmatrix}, \qquad c = (i_n, -i_n, i_n, -i_n), \tag{8.2.37}$$

where I_n is the $n \times n$ identity matrix and J_i is the $n \times n$ matrix with 1's in the ith column and zeros elsewhere, and i_n is the row vector of n 1's. Take $y = (y_1, \ldots, y_{n^2})$. Then

$$yK \leqq c, \qquad y \geqq 0, \tag{8.2.38}$$

says that the matrix $Z = (z_{i,j})$ is doubly stochastic, for it says that the entries of Z are nonnegative, that the row and column sums of Z are at most $+1$, and that the negatives of the row and column sums are at most -1. Hence, with

$$b = (b_1, \ldots, b_{n^2}), \qquad b_{i+(j-1)n} = a_{i,j}, \tag{8.2.39}$$

the problem of maximizing yb^t is precisely the problem of maximizing

$$\sum_{i,j} a_{i,j} z_{i,j},$$

where $Z = (z_{i,j})$, is doubly stochastic. We know the problem has a solution that must be attained when Z is a permutation matrix, since the permutations are the extreme points of the space of doubly stochastic matrices. Thus, the assignment problem is precisely the problem of maximizing yb^t subject to the inequalities (8.2.38). This is Problem II of our duality theorem. Problem I is to find the minimum of cx^t, where $x = (x_1, \ldots, x_{4n})$ subject to the inequalities

$$Kx^t \geqq b^t, \qquad x \geqq 0. \tag{8.2.40}$$

If we write $u_i = x_i - x_{n+i}$ for $i = 1, \ldots, n$, and $v_j = x_{2n+j} - x_{3n+j}$ for $j = 1, \ldots, n$, then (8.2.40) becomes

$$u_i + v_j \geq a_{i,j}, \qquad i, j = 1, \ldots, n, \tag{8.2.41}$$

and the duality theorem asserts that the minimum m of

$$cx^t = \sum_i u_i + \sum_j v_j$$

subject to (8.2.41) is the maximum M of

$$\sum_{i,j} z_{ij} a_{i,j},$$

where $Z = (z_{ij})$ is a permutation matrix, which was the chief part of the proof of Theorem 7.1.1.

8.3 Linear Programming. The Simplex Method

Problems I and II of the preceding section deal with maximizing (or minimizing) a linear form in real variables $x_1, \ldots, x_r$, subject to certain linear inequalities and the additional condition that $x_i \geq 0$, $i = 1, 2, \ldots, r$. In this section we consider the more general problem in which the variables are not restricted to be nonnegative. And we also give a method due to G. Dantzig [1], called the "simplex method," which yields a constructive solution of the problem when a solution exists. We give our general problem as Problem III and also define Problem IV, which is dual to Problem III just as Problems I and II were dual.

Problem III: ■ To find a vector $z = (z_1, \ldots, z_m)$ that maximizes the linear function

$$b_1 z_1 + b_2 z_2 + \cdots + b_m z_m = u \tag{8.3.1}$$

subject to the inequalities

$$\begin{aligned} a_{11} z_1 + a_{21} z_2 + \cdots + a_{m1} z_m + c_1 &\geq 0, \\ a_{12} z_1 + a_{22} z_2 + \cdots + a_{m2} z_m + c_2 &\geq 0, \\ \cdots\cdots\cdots\cdots\cdots\cdots\cdots\cdots \\ a_{1n} z_1 + a_{2n} z_2 + \cdots + a_{mn} z_m + c_n &\geq 0. \end{aligned} \tag{8.3.2}$$

Problem IV: ■ To find a vector $y = (y_1, \ldots, y_n)$, $y \geq 0$, that minimizes the function

$$c_1 y_1 + c_2 y_2 + \cdots + c_n y_n = v \tag{8.3.3}$$

and satisfies the equations

$$\begin{aligned} a_{11} y_1 + a_{12} y_2 + \cdots + a_{1n} y_n + b_1 &= 0, \\ a_{21} y_1 + a_{22} y_2 + \cdots + a_{2n} y_n + b_2 &= 0, \\ \cdots\cdots\cdots\cdots\cdots\cdots\cdots\cdots \\ a_{m1} y_1 + a_{m2} y_2 + \cdots + a_{mn} y_n + b_m &= 0. \end{aligned} \tag{8.3.4}$$

We say that Problem III is feasible if there is a vector $z = (z_1, \ldots, z_m)$ satisfying the inequalities (8.3.2) and that Problem IV is feasible if there is a vector $y = (y_1, \ldots, y_n)$, $y \geq 0$, satisfying the equations (8.3.4).

THEOREM 8.3.1. *Problems III and IV possess solutions if and only if both problems are feasible. For solutions of the problems, we have $u = v$. If either problem has a solution, then so does the other.*

Proof: ■ For our proof we alter the problems in such a way that the duality theorem applies. Write

$$z = (z_1, \dots, z_m), \qquad y = (y_1, \dots, y_n), \qquad b = (b_1, \dots, b_m),$$
$$c = (c_1, \dots, c_n), \qquad A = [a_{ij}], \qquad i = 1, \dots, m, \qquad j = 1, \dots, n.$$

Also define

$$x = (x_1, \dots, x_m, x_{m+1}, \dots, x_{2m}),$$

where

$$x_{m+i} - x_i = z_i, \qquad i = 1, \dots, m \qquad \text{and} \qquad x \geq 0,$$

Also put

$$x^1 = (x_1, \dots, x_m), \qquad x^2 = (x_{m+1}, \dots, x_{2m}),$$

so that $x^2 - x^1 = z$ and $d = (-b_1, \dots, -b_m, b_1, \dots, b_m) = [-b, b]$. Finally define the $2m \times n$ matrix A_1 by

$$A_1 = \begin{bmatrix} A \\ -A \end{bmatrix} \tag{8.3.5}$$

We now define dual problems in terms of these vectors and matrices.

Problem I: ■ To find $y = (y_1, \dots, y_n)$, that minimizes

$$cy^t = m \tag{8.3.6}$$

subject to

$$y \geq 0, \qquad A_1 y^t \geq d^t. \tag{8.3.7}$$

Problem II: ■ To find $x = (x_1, \dots, x_m, x_{m+1}, \dots, x_{2m})$, that maximizes

$$xd^t = M \tag{8.3.8}$$

subject to

$$x \geq 0, \qquad xA_1 \leq c. \tag{8.3.9}$$

The duality theorem is applicable to these two problems. Here, (8.3.7) takes the form

$$y \geq 0, \qquad \begin{bmatrix} Ay^t \\ -Ay^t \end{bmatrix} \geq \begin{bmatrix} -b^t \\ b^t \end{bmatrix}, \tag{8.3.10}$$

whence both $Ay^t \geq -b$ and $-Ay^t \geq b$ hold, which is equivalent to $Ay^t + b = 0$, the matrix form of (8.3.4), and $y^t = m$ is the same as (8.3.3) with $m = v$. The relation (8.3.9) becomes

$$x^1 A - x^2 A \leq c \tag{8.3.11}$$

or $-zA \leq c$, which we may write

$$zA + c \geq 0, \tag{8.3.12}$$

the matrix form of (8.3.2). And finally, (8.3.8) becomes

$$-x^1b^t + x^2b^t = M \tag{8.3.13}$$

or

$$zb^t = M, \tag{8.3.14}$$

which is (8.3.1) with $M = u$. Thus, in the new form, Problem II is Problem III and Problem I is Problem IV. Hence, the duality Theorem 8.2.3 is immediately translated into Theorem 8.3.1, which is now proved.

Having proved Theorem 8.3.1, we no longer need the variables x_i used in its proof and we now introduce different variables, which we designate as $x_1, \ldots, x_n$ for the left-hand sides of the inequalities (8.3.2), writing

$$\begin{aligned} x_1 &= a_{11}z_1 + a_{21}z_2 + \cdots + a_{m1}z_m + c_1 \geq 0, \\ x_2 &= a_{12}z_1 + a_{22}z_2 + \cdots + a_{m2}z_m + c_2 \geq 0, \\ &\cdots\cdots\cdots\cdots\cdots\cdots\cdots\cdots \\ x_n &= a_{1n}z_1 + a_{2n}z_2 + \cdots + a_{mn}z_m + c_n \geq 0. \end{aligned} \tag{8.3.15}$$

Here, of course, $x_i \geq 0$, $i = 1, \ldots, n$. There is a simple identity that does not involve the z's explicitly. We state this as

THEOREM 8.3.2. *For the x's of* (8.3.15) *and y's of* (8.3.4) *the following identity holds:*

$$x_1y_1 + x_2y_2 + \cdots + x_ny_n = -u + v. \tag{8.3.16}$$

Proof: ■ Replacing the x's in the left-hand side of (8.3.16) by their values from (8.3.15) and using (8.3.4), this becomes

$$\begin{aligned} \sum_{j=1}^{n} y_j\left(\sum_{i=1}^{m} a_{ij}z_i + c_j\right) &= \sum_{i=1}^{m} z_i\left(\sum_{j=1}^{n} a_{ij}y_j\right) + \sum_{j=1}^{n} y_jc_j \\ &= -\sum_{i=1}^{m} b_iz_i + \sum_{j=1}^{n} c_jy_j = -u + v. \end{aligned} \tag{8.3.17}$$

COROLLARY. *For a solution of Problems III and IV we must have* $x_iy_i = 0$, $i = 1, \ldots, n$, *in* (8.3.16).

This is immediate, since for a solution we know that $u = v$, and as $x_i \geq 0$, $y_i \geq 0$, $i = 1, \ldots, n$, this requires that every term x_iy_i on the left of (8.3.16) must be zero. Hence, for a solution for every $i = 1, \ldots, n$, either $x_i = 0$ or $y_i = 0$, or both: $x_i = y_i = 0$. Furthermore, (8.3.16) shows that for any feasible values for Problems III and IV, we have

$$b_1z_1 + b_2z_2 + \cdots + b_mz_m = u < v = c_1y_1 + c_2y_2 + \cdots + c_nv_n. \qquad (8.3.18)$$

Thus, any feasible set of y's gives an upper bound (say v_0) for all possible u's and any feasible set of z's gives a lower bound (say u_0) for all possible v's. Hence, whenever both problems are feasible and $u = v$, we have a solution to both problems. Thus, our problems have been reduced to the combinatorial issue of finding a subset $I = \{i_1, \ldots, i_r\}$ of $1, 2, \ldots, n$ such that putting $x_i = 0$ for $i \in I$ and $y_j = 0$ for $j \notin I$ gives feasible values in both problems.

We follow A. Tucker [1] and Princeton lectures by M. L. Balinski in presenting this simpler method of finding the solution. The schema

$$\begin{array}{c|ccccc|c}
 & \cdots & \eta^* & \cdots & \eta & \cdots & \\
\hline
\vdots & & \vdots & & \vdots & & \vdots \\
x^* & \cdots & a & \cdots & b & \cdots & = -y^* \\
\vdots & & \vdots & & \vdots & & \vdots \\
x & \cdots & c & \cdots & d & \cdots & = -y \\
\vdots & & \vdots & & \vdots & & \vdots \\
\hline
 & \cdots = & \xi^* & \cdots = & \xi & \cdots &
\end{array} \qquad (8.3.19)$$

is a convenient form of presenting simultaneously the following two systems of linear equations:

1. A row system,

$$\begin{array}{ccc}
\vdots & \vdots & \vdots \\
\cdots + a\eta^* + & \cdots + b\eta \cdots = & -y^* \\
\vdots & \vdots & \vdots \\
\cdots + c\eta^* + & \cdots + d\eta \cdots = & -y \\
\vdots & \vdots & \vdots
\end{array}, \qquad (8.3.20a)$$

in which the dependent $-y$ *labels* (variables or numbers) are expressed as a linear combination of the independent η labels.

2. A column system,

$$\begin{array}{ccc} \vdots & \vdots & \vdots \\ \cdots + x^*a + \cdots & + \; xc \cdots & = \xi^* \\ \vdots & \vdots & \vdots \\ \cdots + x^*b + \cdots & + \; xd \cdots & = \xi \\ \vdots & \vdots & \vdots \end{array}, \qquad (8.3.20b)$$

in which the dependent ξ labels are expressed as linear combinations of the independent x labels.

A *pivot transformation* with *pivot entry* $a \neq 0$ re-expresses our two systems of linear equations, solving the row system for η^* and the column system for x^*, making η^* and x^* independent labels and y^* and ξ^* dependent labels. In the row system, solving for $-\eta^*$ in terms of y^* and the other row labels, we have

$$\cdots a^{-1}y^* + \cdots + ba^{-1}\eta \cdots = -\eta^*, \qquad (8.3.21)$$

and substituting for η^* in the row equation, we have

$$\begin{array}{cccc} \vdots & & \vdots & \vdots \\ \cdots \; a^{-1}y^* + \cdots + & & ba^{-1}\eta & \cdots = -\eta^* \\ \vdots & & \vdots & \vdots \\ \cdots \; ca^{-1}y^* + \cdots + & & (d - ca^{-1}b)\eta & \cdots = -y \\ \vdots & & \vdots & \vdots \end{array} \qquad (8.3.22)$$

Similarly solving the column system for x^* in terms of ξ^* and the other row labels, we have

$$\cdots \xi^*a^{-1} \cdots - xca^{-1} \cdots = x^*, \qquad (8.3.23)$$

and substituting for x^* in the column equations, we have

$$\begin{array}{ccccccc} & \cdot & & \cdot & & & \cdot \\ & \cdot & & \cdot & & & \cdot \\ & \cdot & & \cdot & & & \cdot \\ \cdots & \xi^* a^{-1} & \cdots - & x c a^{-1} & \cdots & = & x^* \\ & \cdot & & \cdot & & & \cdot \\ & \cdot & & \cdot & & & \cdot \\ & \cdot & & \cdot & & & \cdot \\ \cdots & \xi^* b a^{-1} & \cdots + & x(d - c a^{-1} b) & \cdots & = & \xi \\ & \cdot & & \cdot & & & \cdot \\ & \cdot & & \cdot & & & \cdot \\ & \cdot & & \cdot & & & \cdot \end{array} \tag{8.3.24}$$

The pivot transformation with pivot entry $a \neq 0$ is the transformation that replaces the schema (8.3.19) by the schema

$$\begin{array}{c|ccccc|c} & \cdots & y^* & \cdots & \eta & \cdots & \\ \hline \cdot & & \cdot & & \cdot & & \cdot \\ \cdot & & \cdot & & \cdot & & \cdot \\ \cdot & & \cdot & & \cdot & & \cdot \\ \xi^* & \cdots & a^{-1} & \cdots & a^{-1} b & \cdots & = -\eta^* \\ \cdot & & \cdot & & \cdot & & \cdot \\ \cdot & & \cdot & & \cdot & & \cdot \\ \cdot & & \cdot & & \cdot & & \cdot \\ x & \cdots & -c a^{-1} & \cdots & d - c a^{-1} b & \cdots & = -y \\ \cdot & & \cdot & & \cdot & & \cdot \\ \cdot & & \cdot & & \cdot & & \cdot \\ \cdot & & \cdot & & \cdot & & \cdot \\ \hline & \cdots & = x^* & \cdots & = \xi & \cdots & \end{array} \quad . \tag{8.3.25}$$

The transformed schema (8.3.25) corresponds to the row equations (8.3.22) and column equations (8.3.24). Here the row equations (8.3.20a) have been changed into (8.3.22) by solving for $-\eta^*$ in terms of y^* and the remaining row labels, and the column equations (8.3.20b) have been changed into (8.3.24) by solving for x^* in terms of ξ^* and the remaining column labels.

In other words, the schema (8.3.25) is obtained from the schema (8.3.19) by

1. Interchanging labels corresponding to the pivot entry a; namely, replacing η^* with y^* and $-y^*$ with $-\eta^*$ in the row system, and replacing x^* with ξ^*, and ξ^* with x^* in the column system.
2. Keeping all remaining labels unchanged.
3. Replacing the pivot a by its reciprocal a^{-1}; replacing each remaining entry in the pivot row, b by $a^{-1}b$; replacing each remaining entry in the pivot

column, c by $-ca^{-1}$; replacing an entry d not in the pivot row or column by $d - ca^{-1}b$, where b is the entry in the same row as a and in the same column as d, and c is the entry in the same row as d and in the same column as a, so that a, b, c, d form a rectangle as in (8.3.19).

It is to be emphasized that the *only* effect of a pivot transformation is to express a pair of linear systems in terms of different sets of independent and dependent labels. We shall use the notation of the schemata and pivot transformations to solve a pair of dual problems of the form of Problems III and IV, and then discuss in general the simplex method, which is essentially an algorithm for solving such problems. This can also be used to provide a direct proof of Theorem 8.3.1.

Problem (of Type III): ■ Given the inequalities

$$\begin{aligned} x_1 &= -z_1 - z_2 + 2 \geq 0, \\ x_2 &= z_1 + z_2 - 1 \geq 0, \\ x_3 &= z_1 + 2z_2 - 2 \geq 0, \\ x_4 &= -z_1 + 1 \geq 0, \\ x_5 &= -6z_1 - 5z_2 + 10 \geq 0, \end{aligned} \tag{8.3.26}$$

maximize

$$u = 3z_1 + 2z_2. \tag{8.3.27}$$

We can now write down the dual problem.

Dual Problem (of Type IV): ■ Given the equations

$$\begin{aligned} -y_1 + y_2 + y_3 - y_4 - 6y_5 + 3 &= 0, \\ -y_1 + y_2 + 2y_3 - 5y_5 + 2 &= 0, \end{aligned} \tag{8.3.28}$$

subject to

$$y_1 \geq 0, \quad y_2 \geq 0, \quad y_3 \geq 0, \quad y_4 \geq 0, \quad y_5 \geq 0, \tag{8.3.29}$$

minimize

$$v = 2y_1 - y_2 - 2y_3 + y_4 + 10y_5. \tag{8.3.30}$$

A single schema describes both problems simultaneously:

	y_1	y_2	y_3	y_4	y_5	1	
z_1	-1^*	1	1	-1	-6	3	$= 0$
z_2	-1	1	2	0	-5	2	$= 0$
1	2	-1	-2	1	10	0	$= v = \min$
	$= x_1$	$= x_2$	$= x_3$	$= x_4$	$= x_5$	$= u$	$= \max$

$x_i \geq 0, \quad y_i \geq 0.$ (8.3.31)

Here we can verify directly the relation of Theorem 8.3.2, namely,

$$x_1y_1 + x_2y_2 + x_3y_3 + x_4y_4 + x_5y_5 + u = v. \tag{8.3.32}$$

First we eliminate the unrestricted variables z_1 and z_2. Eliminating z_1 by pivoting on the starred -1 in (8.3.31), we have

$$\begin{array}{c|cccccc|l} & 0 & y_2 & y_3 & y_4 & y_5 & 1 & \\ \hline x_1 & -1 & -1 & -1 & 1 & 6 & -3 & = -y_1 \\ z_2 & -1 & 0 & 1^* & 1 & 1 & -1 & = 0 \\ 1 & 2 & 1 & 0 & -1 & -2 & 6 & = v = \min \\ \hline & = z_1 & = x_2 & = x_3 & = x_4 & = x_5 & = u & = \max \end{array} \qquad x_i \geq 0, \quad y_i \geq 0. \tag{8.3.33}$$

Now eliminate z_2, pivoting on the starred 1 in (8.3.33) to obtain

$$\begin{array}{c|cccccc|l} & 0 & y_2 & 0 & y_4 & y_5 & 1 & \\ \hline x_1 & -2 & -1 & 1 & 2 & 7 & -4 & = -y_1 \\ x_3 & -1 & 0 & 1 & 1 & 1 & -1 & = -y_3 \\ 1 & 2 & 1 & 0 & -1 & -2 & 6 & = v = \min \\ \hline & = z_1 & = x_2 & = z_2 & = x_4 & = x_5 & = u & = \max \end{array} \qquad x_i \geq 0, \quad y_i \geq 0. \tag{8.3.34}$$

The zero labels do not contribute to the row equations, but their columns serve to express the unrestricted variables z_1 and z_2 in terms of the restricted variables x_1 and x_3.

We can now put the problem in a *canonical form*, with the unrestricted variables expressed in terms of restricted variables and a further schema representing a pair of linear systems on restricted variables alone. Thus, we rewrite (8.3.34) in the canonical form of linear equations

$$\begin{aligned} z_1 &= -2x_1 - x_3 + 2 \\ z_2 &= x_1 + x_3 \end{aligned} \tag{8.3.35}$$

and the schema

$$\begin{array}{c|cccc|l} & y_2 & y_4 & y_5 & 1 & \\ \hline x & -1 & 2 & 7 & -4 & = -y_1 \\ x_3 & 0 & 1 & 1 & -1 & = -y_3 \\ 1 & 1 & -1 & -2 & 6 & = v = \min \\ \hline & = x_2 & = x_4 & = x_5 & = u & = \max \end{array} \qquad x_i \geq 0, \quad y_i \geq 0. \tag{8.3.36}$$

If we put all the independent variables equal to zero in the systems that this schema represents, putting $x_1 = x_3 = 0$, $y_2 = y_4 = y_5 = 0$, we have what is called a *basic solution* of (8.3.32) with $u = v$. We can read off the values of the other variables from the bordering row and column of our schema, namely $x_2 = 1$, $x_4 = -1$, $x_5 = -2$, $y_1 = 4$, $y_3 = 1$, and $u = v = 6$. Here, since all $y_i \geq 0$ for the solution, this is a feasible solution for the row program. It is not, however, feasible in the column program, since we have negative x's. The values $y_1 = 4$, $y_3 = 1$, $y_2 = y_4 = y_5 = 0$ are feasible for the row program and give $v = 6$ as a value. But as the corresponding set of values for the x's is not feasible, we have not attained the minimum v for the row program.

The simplex method uses a pivot transformation that will retain the feasibility of the row program for its basic solution and will reduce the value of v. We consider the effect of a pivot transformation on a pivot entry a_{rs} on its own column and on the right-hand column:

$$\begin{array}{ccc} a_{1s} & \cdots & b_1 \\ \cdot & & \cdot \\ \cdot & & \cdot \\ \cdot & & \cdot \\ a_{is} & \cdots & b_i \\ \cdot & & \cdot \\ \cdot & & \cdot \\ \cdot & & \cdot \\ a_{rs}^* & \cdots & b_r \\ \cdot & & \cdot \\ \cdot & & \cdot \\ \cdot & & \cdot \\ a_{ms} & \cdots & b_m \\ c_s & \cdots & d \end{array} \quad \rightarrow \quad \begin{array}{ccc} -a_{1s}a_{rs}^{-1} & \cdots & b_1 - a_{1s}a_{rs}^{-1}b_r \\ \cdot & & \\ \cdot & & \\ \cdot & & \\ -a_{is}a_{rs}^{-1} & \cdots & b_i - a_{is}a_{rs}^{-1}b_r \\ \cdot & & \\ \cdot & & \\ \cdot & & \\ a_{rs}^{-1} & \cdots & b_r a_{rs}^{-1} \\ \cdot & & \\ \cdot & & \\ \cdot & & \\ -a_{ms}a_{rs}^{-1} & \cdots & b_m - a_{ms}a_{rs}^{-1}b_r \\ -c_s a_{rs}^{-1} & \cdots & d - c_s a_{rs}^{-1} b_r \end{array} \tag{8.3.37}$$

Here, $b_1, \ldots, b_m$ are nonpositive, and we wish the new b's to be nonpositive. Also we wish to have the new value of $u = v$ less than the previous one, or $d - c_s a_{rs}^{-1} b_r \leq d$. This will be achieved if $a_{rs} > 0$, $c_s \leq 0$, and for $i \neq r$, $b_i - a_{is}a_{rs}^{-1}b_r \leq 0$, which certainly holds if $a_{is} \leq 0$, but for $a_{is} > 0$, only if $a_{is}^{-1}b_i \leq a_{rs}^{-1}b_r$. Note that if $c_s \geq 0$ in every column, then our basic solution is also column-feasible, and we have a solution to our dual problems. Furthermore, if $c_s < 0$ and $a_{1s}, \ldots, a_{ms} \leq 0$, we have

$$x_s = a_{1s}xi_1 + \cdots + a_{ms}xi_m + c_s,$$

and so $x_s < 0$ for *any* choice of the other x's ≥ 0; in this case, our column program has no solution. This gives us a rule for a pivot choice when we have a row-feasible basic solution.

Rule: Pivot Choice for Row-Feasible Schemata. Choose as pivot entry a positive a_{rs} in a column, with $c_s < 0$, such that $a_{rs}^{-1}b_r \geq a_{is}^{-1}b_i$ for $a_{is} > 0$, $i \neq s$.

In (8.3.36) such a choice is the 1 in the x_3 row and y_4 column. Pivoting on this, we have

$$
\begin{array}{c|cccc|l}
 & y_2 & y_3 & y_5 & 1 & \\
\hline
x_1 & -1 & -2 & 5^* & -2 & = -y_1 \\
x_4 & 0 & 1 & 1 & -1 & = -y_4 \\
1 & 1 & 1 & -1 & 5 & = v = \min \\
\hline
 & = x_2 & = x_3 & = x_5 & = u & = \max
\end{array}
\qquad (8.3.38)
$$

The starred 5 here is the appropriate pivot, and we find

$$
\begin{array}{c|cccc|l}
 & y_2 & y_3 & y_1 & 1 & \\
\hline
x_5 & \frac{-1}{5} & \frac{-2}{5} & \frac{1}{5} & \frac{-2}{5} & = -y_5 \\
x_4 & \frac{1}{5} & \frac{7}{5} & \frac{-1}{5} & \frac{-3}{5} & = -y_4 \\
1 & \frac{4}{5} & \frac{3}{5} & \frac{1}{5} & \frac{23}{5} & = v = \min \\
\hline
 & = x_2 & = x_3 & = x_1 & = u & = \max
\end{array}
\qquad (8.3.39)
$$

In (8.3.39) both row and column values are feasible in this basic solution, and we have solved our dual problems. The solutions are

$$
\begin{aligned}
&x_1 = \tfrac{1}{5}, \quad x_2 = \tfrac{4}{5}, \quad x_3 = \tfrac{3}{5}, \quad x_4 = 0, \quad x_5 = 0. \\
&y_1 = 0, \quad y_2 = 0, \quad y_3 = 0, \quad y_4 = \tfrac{3}{5}, \quad y_5 = \tfrac{2}{5}. \\
&u = v = \tfrac{23}{5}. \\
&z_1 = 1, \quad z_2 = \tfrac{4}{5}.
\end{aligned}
\qquad (8.3.40)
$$

The values of z_1 and z_2 are, of course, obtained by substituting in (8.3.35).

In general, the schema representing the dual Problems III and IV is of the form

$$
\begin{array}{c|cccc|c|l}
 & y_1 & y_2 & \cdots & y_N & 1 & \\
\hline
z_1 & a_{11} & a_{12} & \cdots & a_{1N} & b_1 & = 0 \\
z_2 & a_{21} & a_{22} & \cdots & a_{2N} & b_2 & = 0 \\
\vdots & \vdots & \vdots & & \vdots & \vdots & \\
z_M & a_{M2} & a_{M2} & \cdots & a_{MN} & b_M & = 0 \\
1 & c_1 & c_2 & \cdots & c_N & d & = v = \min \\
\hline
 & = x_1 & = x_2 & \cdots & = x_N & & = u = \max
\end{array}
\qquad x_i \geq 0, \quad y_i \geq 0. \qquad (8.3.41)
$$

Our first procedure is to eliminate as many as possible of the unrestricted variables z_i. This process may be carried out by pivot transformations as long as there is a nonzero entry in a column with a y at the top and a zero at the right. This terminates with a schema of the form

$$
\begin{array}{c|ccc:ccc|c|l}
 & y'_{m+1} & \cdots & y'_{m+n} & 0 & \cdots & 0 & 1 & \\
\hline
x'_1 & a_{11} & \cdots & a_{1n} & a'_{1n+1} & \cdots & a'_{1N} & b_1 & = -y_1 \\
\vdots & \vdots & & \vdots & \vdots & & \vdots & & \\
x'_m & a'_{m1} & \cdots & a'_{mn} & a'_{mn+1} & \cdots & a_{mN} & b'_m & = -y'_m \\
\hdashline
z'_{m+1} & 0 & \cdots & 0 & a'_{m+1,n+1} & \cdots & a'_{m+1N} & b_{m+1} & = 0 \\
\vdots & \vdots & & \vdots & \vdots & & \vdots & & \\
z'_M & 0 & \cdots & 0 & a'_{Mn+1} & \cdots & a'_{MN} & b'_M & = 0 \\
\hline
1 & c'_1 & \cdots & c'_n & c'_{n+1} & \cdots & c'_N & d' & = v = \min \\
\hline
 & = x'_{m+1} & & = x_{m+n} & = z'_1 & & = z'_m & = u = \max &
\end{array}
\qquad x'_i \geq 0,\ y'_i \geq 0. \tag{8.3.42}
$$

In this schema the primed variables are a rearrangement of the original variables in (8.3.41) and the rows and columns have been permuted according to the pattern of the zero entries and zero labels. The equations for the rows $m + 1, \ldots, M$ read $0 = b'_j$, $j = m + 1, \ldots, M$. Hence, if any one of b'_j, $j = m + 1, \ldots, M$ is different from zero, we have the conflict $0 = b'_j \neq 0$, and so the row system is incompatible.

On the other hand, if $b'_j = 0$ for $j = m + 1, \ldots, M$, these row equations reduce to the trivial equation $0 = 0$ and may be discarded from the row system. Columns $n + 1$ to N now are linear equations expressing $z'_1, \ldots, z'_m$ in terms of the restricted variables $x'_1, \ldots, x'_m$ and the variables $z'_{m+1}, \ldots, z'_M$, which are now arbitrary parameters. There remain to be considered only the relations on $x'_1, \ldots, x'_{m+n}$ and $y'_1, \ldots, y'_{m+n}$, all of which are restricted, and we have the *canonical form* for our dual problems (or programs):

$$
\begin{array}{c|ccc|c|l}
 & y_{m+1} & \cdots & y_{m+n} & 1 & \\
\hline
x_1 & a_{11} & \cdots & a_{1n} & b_1 & = -y_1 \\
\vdots & \vdots & & \vdots & \vdots & \\
x_M & a_{1m} & \cdots & a_{mn} & b_m & = -y_m \\
\hline
1 & c_1 & \cdots & c_n & d & = v = \min \\
\hline
 & = x_{m+1} & & = x_{m+n} & = u & = \max
\end{array}
\qquad x_i \geq 0, \quad y_i \geq 0. \tag{8.3.43}
$$

For simplicity we have dropped the primes on the variables, and the key identity

$$x_1 y_1 + \cdots + x_{m+n} y_{m+n} + u = v \tag{8.3.44}$$

will hold. Our problem now is to find, if it exists, a schema obtained from (8.3.44) by a succession of pivot steps, a feasible solution to both problems so that $x_i y_i = 0$, $i = 1, \ldots, m + n$, and $x_i \geq 0, y_i \geq 0, i = 1, \ldots, m + n$. To simplify our notation for part of our argument, let us write $+$ for a positive number and $\oplus$ for a number that is positive or zero; and similarly, $-$ and $\ominus$ for a negative number and a number negative or zero. If we can derive a schema from (8.3.43) of the form

$$\begin{array}{c|ccc|c|l} & & & & 1 & \\ \hline & & & & \ominus & \\ & & & & \vdots & \\ & & & & \ominus & \\ \hline 1 & \oplus & \cdots & \oplus & & = v = \min \\ \hline & & & & = u & = \max \end{array}$$

putting the dependent variables all equal to zero, we have a feasible basic solution to both programs and thus a solution to both dual problems. If at any stage we have a row of the form

$$\begin{array}{|ccc|c|} \multicolumn{3}{c}{} & \multicolumn{1}{c}{1} \\ \hline \oplus & \cdots & \oplus & + \\ \hline \end{array} \tag{8.3.45}$$

or a column of the form

$$\begin{array}{c|c|} \hline & \ominus \\ & \vdots \\ & \ominus \\ \hline 1 & - \\ \hline \end{array}\,, \tag{8.3.46}$$

the row system or the column system is incompatible with the requirement that $-y_i \leq 0$ in all cases and $+x_i \geq 0$ in all cases for row or column feasibility, respectively.

If, in (8.3.43), $b_1 \leq 0, \ldots, b_m \leq 0$, the system is row-feasible. If it is not also column-feasible, some $c_s < 0$, and in the column of c_s, there must be some positive entries, for if not, the column has the pattern of (8.3.46) and the column program is not feasible. Hence, suppose that there are positive entries in the column of c_s. Choose a_{rs} as pivot entry in this column, where

$$a_{rs} > 0 \quad \text{and} \quad a_{rs}^{-1} b_r \geq a_{is}^{-1} b_i \quad \text{for any} \quad a_{is} > 0. \qquad (8.3.47)$$

As we observed in our example, such a choice preserves row feasibility and, if $b_r < 0$, gives a new value to v strictly less than the previously exhibited value of v.

Similarly, if $c_1 \geq 0, \ldots, c_n \geq 0$, our system is row-feasible, and if every b is nonpositive, we have feasible solutions to both problems and we are finished. Suppose that $b_r > 0$ and that its row is not of the form (8.3.45). Then there are negative entries in this row and our pivot choice is a_{rs}, where

$$a_{rs} < 0, \quad a_{rs}^{-1} c_s \geq a_{rj}^{-1} c_j \quad \text{for any} \quad a_{rj} < 0. \qquad (8.3.48)$$

for which, if $c_s > 0$, the new value of u is strictly greater than the previous value. For a schema that is either row- or column-feasible, a succession of such pivot steps either exhibits a nonfeasible column or row, and no solution exists, or leads to the pattern (8.3.44) and solutions to our dual problems.

Of course our canonical form (8.3.43) may be neither row- nor column-feasible. Permute the rows to put the k nonpositive b's at the top and we have the schema

$$\begin{array}{c|c|c|l} & y_{m+1} \quad \cdots \quad y_{m+n} & 1 & \\ \hline x_1 & & \ominus & = -y_1 \\ \vdots & & \vdots & \\ x_k & & \ominus & = -y_k \\ \hline x_{k+1} & & + & = -y_{k+1} \\ \hline \vdots & & \vdots & \\ x_m & & + & = y_m \\ \hline 1 & c_1 \quad \cdots \quad c_n & & = v \\ \hline & & = u & \end{array} \qquad (8.3.49)$$

Now treat the first $(k + 1)$ rows of this schema as a row-feasible program, seeking to minimize $-y_{k+1}$. If we ever attain a nonpositive value for $-y_{k+1}$, we have $(k + 1)$ rows rather than k in row-feasible form and we continue to put more rows into feasible form. If not, we reach one of two patterns,

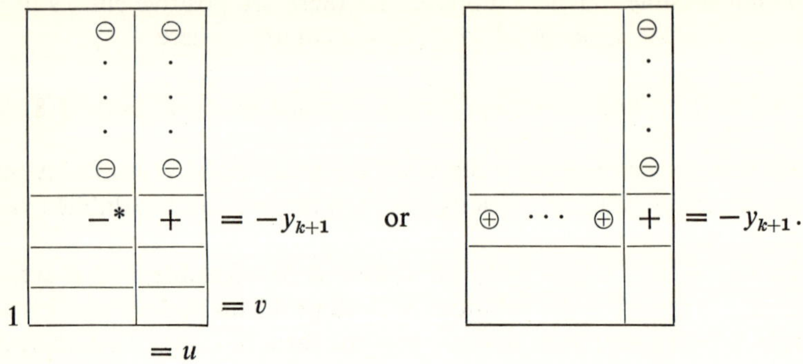

(8.3.50)

The second of these patterns is the conflicting row pattern of (8.3.45). If we pivot on the starred negative number in the first one, we easily verify that this gives $(k + 1)$ or more feasible rows. Thus, we have procedures to cover all cases.

9

Graphical Methods. DeBruijn Sequences

9.1 Complete Cycles

A cycle is a sequence of letters $a_1a_2 \cdots a_r$ taken in a circular order—that is, a_1 follows a_r, and $a_2 \cdots a_ra_1, \ldots, a_ra_1 \cdots a_{r-1}$ are all the same cycle as $a_1a_2 \cdots a_r$. There are many combinatorial problems on cycles, but we shall consider only one here. If n is a positive integer and $N = 2^n$, a cycle of zeros and 1's, $a_1a_2 \cdots a_N$ will be called a *complete cycle* if the subsequences $a_ia_{i+1} \cdots a_{i+n-1}$, $i = 1, \ldots, N$ consist of all possible $N = 2^n$ ordered sequences $b_1 \cdots b_n$ of zeros and 1's. For $n = 1, 2, 3$, we find the following complete cycles:

$$\begin{aligned} n &= 1, & &01, \\ n &= 2, & &0011, \\ n &= 3, & &00010111, \\ & & &00011101. \end{aligned} \tag{9.1.1}$$

For $n = 4$, there are exactly 16 complete cycles. Our problem is to prove the existence of complete cycles of each length $N = 2^n$, and if possible, to find their number.

Both parts of this problem are answered in a theorem due to N. DeBruijn [1]. Because of this theorem, complete cycles are sometimes called DeBruijn sequences.

THEOREM OF DEBRUIJN. *For each positive integer n there are exactly* $2^{2^{n-1}-n}$ *complete cycles of length* $N = 2^n$.

The proof of this theorem will be given in Section 9.3. In this section we give a graphical interpretation of the complete cycles.

DEFINITION. *A graph G is a set of points* $\{p_i\}$ *and arcs* $\{A_j\}$. *Each arc* A_j *joins two points* p_i *and* p_k, *its end points.*

We do not require that two end points of an arc be distinct. If the end points of an arc are the same point, the arc is called a *loop*. There may be no arc joining two points, one arc or many arcs. We shall consider only finite graphs—that is, those having a finite number of points and arcs. In a graph G, let $p_1, \ldots, p_n$ be a succession of points such that there are arcs A_i joining p_i and p_{i+1} for $i = 1, \ldots, n-1$. The succession of arcs $A_1, \ldots, A_{n-1}$ is called a *path*, and we say that p_1 and p_n are *connected*. G is called "connected" if every pair of points in G is connected.

In addition we may associate a direction with an arc A_i joining p_i and p_k if one of the points, p_i, is called the beginning of A_i and the other, p_k, its end. A directed graph is a graph, all of whose arcs are directed. For a path A_1, $A_2, \ldots, A_n$ in a directed graph, we require that for each $i = 1, \ldots, n-1$, the end point of A_i shall be the beginning point of A_{i+1}. Such a path is called a cycle if the end of A_n is the beginning of A_1. A cycle is taken to be in circular order, and $A_2, \ldots, A_n, A_1$ is the same as $A_1, A_2, \ldots, A_n$.

Let each of the 2^{n-1} sequences of zeros and 1's of length $n-1$, $c_1, \ldots, c_{n-1}$, be associated with a point $p_i = c_1, \ldots, c_{n-1}$, and let each sequence of length

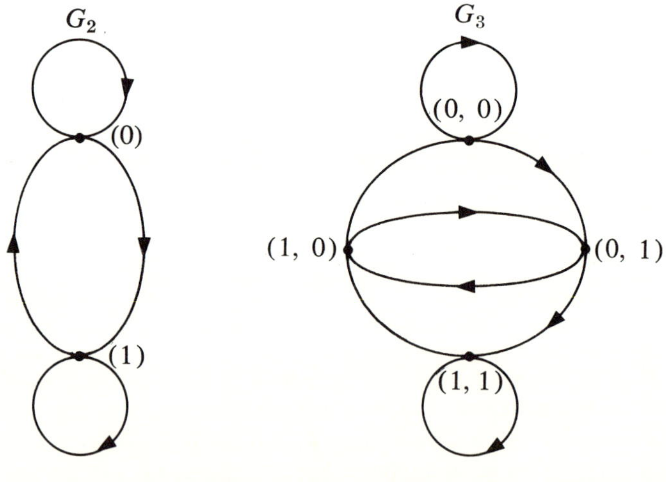

FIGURE 9.1. Graphs of G_2 and G_3.

n, $b_1, b_2, \ldots, b_{n-1}, b_n$ be associated with a directed arc whose beginning is $(b_1, b_2, \ldots, b_{n-1})$ and whose end is $(b_2, \ldots, b_{n-1}, b_n)$. Let us call this graph G_n. G_2 and G_3 are represented in Figure 9.1, where the arrows indicate the direction of the arcs.

We note that G_n is connected, since the sequences of length n in $c_1 \cdots c_{n-1}d_1 \cdots d_{n-1}$ give a path leading from $(c_1, \ldots, c_{n-1})$ to $(d_1, \ldots, d_{n-1})$. In the set of all sequences $b_1b_2 \cdots b_{n-1}b_n$, every sequence of length $n - 1$ occurs twice as $b_1b_2 \cdots b_{n-1}$ and twice as $b_2 \cdots b_{n-1}b_n$. Hence, every point of G_n is the beginning of exactly two arcs and the end of exactly two arcs. If $N = 2^n$, a complete cycle $a_1a_2 \cdots a_N$ may be associated with the path in G_n whose arcs are A_i, $a_ia_{i+1} \cdots a_{i+n-1}$, $i = 1, \ldots, N$. This is a path, since for each i the end point of A_i is the beginning point of A_{i+1}. Furthermore, it is a cycle, since the end point of A_N is the beginning of A_1. Thus, this is a cycle that uses each arc of G_n exactly once. Conversely, a directed cycle in G_n including each arc exactly once leads immediately to the construction of a complete cycle $a_1a_2 \cdots a_N$ if we take the arcs in order as $a_ia_{i+1} \cdots a_{i+N-1}$, $i = 1, 2, \ldots, N$. Thus, the problem of constructing complete cycles $a_1a_2 \cdots a_N$ of zeros and 1's is precisely the equivalent of finding cycles in G_n, which use each arc exactly once, and we shall also call these complete cycles.

9.2 Theorems on Graphs

First we shall give some properties of undirected graphs, some of which are clearly applicable to directed graphs as well. A point p of a graph is called an m node if there are m ends of arcs that are the point p. We call p an even node if m is even.

LEMMA 9.2.1. *Any graph contains an even number of odd nodes.*

Proof: ■ If there are x_i points that are i nodes, then $x_1 + 2x_2 + \cdots + kx_k = 2A$, where A is the number of arcs (since the left-hand side counts the number of ends of arcs and each arc has exactly two ends). From this equation, $x_1 + x_3 + x_5 + \cdots + x_{2y+1}$ is an even number, which is the assertion of our lemma.

THEOREM 9.2.1. (EULER). *There is a cycle going through each arc of a graph exactly once if and only if the graph is connected and all its nodes are even.*

Proof: ■ If there is a cycle going through each arc of a graph G exactly once, then, trivially, G is connected. Furthermore, each time the cycle goes through a point, it accounts for one arc end going in and another going out, and hence every node is even. Now let us suppose that G is connected and that every node is even. Let P be the path in G that is the longest (in the sense of using the largest possible number of arcs). We shall show that P must be the desired cycle using all arcs of the graph. First, P is a cycle. For if P is $A_1, A_2, \ldots, A_m$ and if A_m ends in a point q, which is not the beginning of A_1, then at every previous time that q was traversed, two arc ends at q were used. But this accounts for an odd number of arc ends at q, and as q is an even node, there must be an additional arc at q, and this may be added as A_{m+1} to P to form a longer path. Thus, P is a cycle.

Second, suppose there is some point q in P and an arc B with an end at q that has not been used in P. Renumbering if necessary, we may take q to be the end of A_m and the beginning of A_1. Further, write $B = B_1$ joining $q = q_1$ to a point q_2. Then $A_1, A_2, \ldots, A_m, B_1$ is a path longer than P, contrary to our choice of P. We have shown that P is a cycle and that every arc through every point of P is in P. Now we assert that P includes every point of G, for if p is a point of G not in P, let q be a point of P. Since G is connected, there is a path $U_1, U_2, \ldots, U_w$, where U_i joins points x_i and x_{i+1} and $x_1 = q$, $x_{w+1} = p$. But as q is a point of P, the arc U_1 joining $x_1 = q$ and x_2 is an arc of P. Thus, x_2 is a point of P, and so U_2 is an arc of P. Continuing $x_3, \ldots,$ and finally $x_{w+1} = p$ is a point of P, contrary to our choice of p as point of G not in P. Hence, P contains all points of G and so also all arcs of G, and our theorem is proved.

THEOREM 9.2.2. *If G is connected and has exactly* $(2s > 0)$ *odd nodes, there are s paths $P_1, \ldots, P_s$ and no fewer that together use all arcs of G exactly once. Each of $P_1, \ldots, P_s$ begins at an odd node and ends at another odd node.*

Proof: ■ According to Lemma 9.2.1, if there are any odd nodes their number must be even. If this number is $2s$, we divide them into s pairs and form a new graph G_1 by adjoining to G s new arcs, each joining a pair of these nodes. Then G_1 is a connected graph, all of whose vertices are even. By Theorem 9.2.1 there

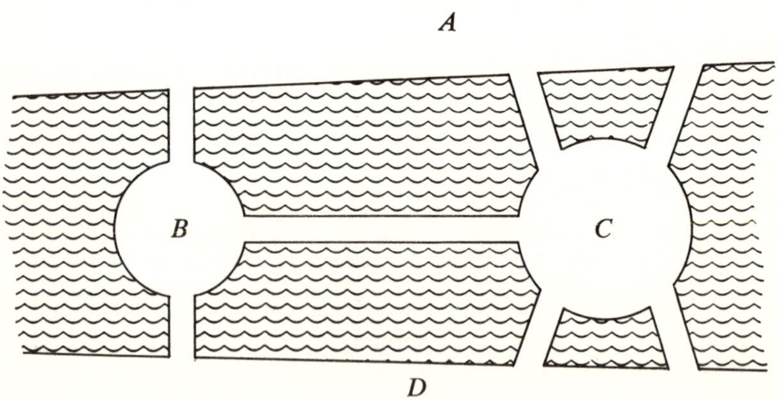

FIGURE 9.2. The Königsberg bridge problem.

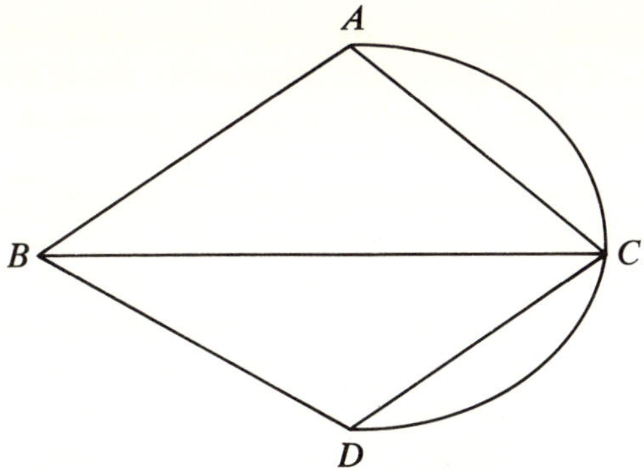

FIGURE 9.3. Graph of the bridge problem.

is a cycle C in G_1 traversing all arcs of G_1 exactly once. If we remove from C the s arcs not in G, the remaining ones consist of s paths using all arcs of G, each path beginning in an odd node and ending in another. We observe that in any collection of paths P_i using all arcs of G, every odd node must be the end of a path, and so with $2s$ odd nodes we cannot cover all arcs of G using fewer than s paths.

An easy application of this theorem is the celebrated Königsberg bridge problem, originally solved by Euler: A river ran through the city, and there were two islands in the river. There was a bridge between the two islands and there were also two bridges to each shore from the larger island and one to each shore from the smaller island. The problem was to walk over all seven bridges without using any bridge twice. Figure 9.2 shows the arrangement of the bridges.

If we take base points on the shores and on the two islands we see that the problem is that of traversing all arcs of Figure 9.3 in a single path. Since all four nodes of this graph are odd, Theorem 9.2.2 shows that it takes at least two separate paths to traverse all arcs, and so the original problem of walking over all seven bridges without crossing any bridge twice is impossible.

9.3 Proof of the DeBruijn Theorem

A generalization of Theorem 9.2.1, due to I. J. Good [1], shows that there is always at least one complete cycle $a_1a_2 \cdots a_N$, with $N = 2^n$.

THEOREM 9.3.1 (I. J. GOOD). *If G is a connected directed graph, and if at each point of G there are the same number of arcs going out as coming in, then there is a directed cycle in G that goes through every arc of G in its given direction, and uses no arc twice.*

Proof: ■ The proof is essentially the same as that of Theorem 9.2.1. Let $P = A_1, A_2, \ldots, A_m$ be a maximal directed path in G, where A_i is an arc beginning at p_i and ending at p_{i+1} for $i = 1, \ldots, m$ and all arcs $A_1, \ldots, A_m$ are distinct. If $p_{m+1} \neq p_1$, then every time p_{m+1} has been traversed in P, the path has come in on one arc and gone out on another. But then in P there are more arcs coming in to p_{m+1} than going out, and so, by our hypothesis, there must be another arc A_{m+1} going out from p_{m+1}, and thus P is not maximal. Hence, $p_{m+1} = p_1$ and P is a cycle. If for a point q of P there is an arc B_1 that is not in P, this may be supposed to be an arc going out from q; renumbering if necessary, we have $q = p_1 = p_{m+1}$. But then $A_1, A_2, \ldots, A_m, B_1$ is a longer path, conflicting with our choice of P. Hence, for the maximal P, P is a directed cycle containing all arcs through any point of P. But since G is connected, this means that P contains all arcs of G. Hence, P is the desired directed cycle going through every arc of G in its given direction and using every arc exactly once.

In Section 9.1 we observed that the graph, G_n, associated with complete cycles is connected and that at every point there are two arcs coming in and two arcs going out. Hence, Theorem 9.3.1 tells us that for every positive integer n, there is at least one complete cycle $a_1a_2 \cdots a_N$ with $N = 2^n$. But this does not tell us the number of complete cycles. The crux of DeBruijn's proof lies in the recognition of a relationship between the graphs G_n and G_{n+1} and the fact that this implies a relation between the number of complete cycles in G_n and the number in G_{n+1}.

Let us call a directed graph G a 2-graph if at every point there are two arcs coming in and two going out. Our graphs G_n are 2-graphs. Given a 2-graph G, we shall define a *doubled graph* G^* according to the following rules:

1. To each arc B_0 of G we associate a point b_0 of G^*.
2. If B_1 is an arc of G whose beginning point is the end point of B_0, then in G^* there is an arc U_{01} whose beginning is b_0 and whose end is b_1.
3. The points and arcs of G^* are those determined from G by the rules 1 and 2 and no others.

It is not difficult to see that G^* is also a 2-graph with twice as many points as G (this being the number of arcs in G). Furthermore, and this is important in DeBruijn's proof, *G_{n+1} is the doubled graph of G_n*, for we have in an arbitrary sequence of $(n + 1)$ zeros and 1's,

$$b_1b_2, b_3, \ldots, b_{n-1}, b_n, b_{n+1}.$$

Points in G_n:

$$(b_1, b_2, b_3, \ldots, b_{n-1}) = x_1,$$
$$(b_2, b_3, \ldots, b_{n-1}, b_n) = x_2,$$
$$(b_3, \ldots, b_{n-1}, b_n, b_{n+1}) = x_3.$$

Arcs in G_n = points in G_{n+1}:

$$(b_1, b_2, b_3, \ldots, b_{n-1}, b_n) = B_0 = b_0,$$
$$(b_2, b_3, \ldots, b_n, b_{n+1}) = B_1 = b_1.$$

Arc in G_{n+1}:

$$(b_1, b_2, b_3, \ldots, b_{n-1}, b_n, b_{n+1}) = U_{01}.$$

In this context, DeBruijn's theorem is a corollary of:

THEOREM 9.3.2. *If G is a 2-graph with m points and has exactly M complete cycles, then G^* has exactly $2^{m-1}M$ complete cycles.*

COROLLARY (THEOREM OF DEBRUIJN). *G_n has exactly $2^{2^{n-1}-n}$ complete cycles.*

Proof of Corollary from Theorem 9.3.2: ■ G_n has 2^{n-1} points, the points $c_1, \ldots, c_{n-1}$. G_2 has exactly one complete cycle. From this and the fact that $G_{n+1} = G_n^*$, the corollary follows by induction on n from the theorem.

Proof of Theorem: ■ We follow DeBruijn's notation, writing $|G|$ for the number of complete cycles in G. The cycles are counted in circular fashion, a cycle $A_1, A_2, \ldots, A_m$ being considered the same as $A_2, \ldots, A_m, A_1$. Note that $|G| = 0$ if G is not connected. Note also that Theorem 9.3.2 is a generalization of the corollary. Here, as so often is the case, the proof of an important result depends upon the proof of an appropriate generalization.

The proof will be by induction on m. If $m = 1$, G has a single point p and two loops A_1 and A_2 from p to p. Here, G^* is the graph G_2 and has $2^{1-1} = 1$ complete cycle. We suppose as a first case that there is a loop at every point. Then (assuming G connected) G is of a simple type, having points $p_1, p_2, \ldots, p_m$, and at p_i a loop $A_i: p_i \to p_i$ and also arcs $B_i: p_i \to p_{i+1}$, taking subscripts modulo m. Then G^* has $2m$ points $a_i, b_i, i = 1, \ldots, m$, and arcs $U_i: a_i \to a_i$, $V_i: a_i \to b_i$, $W_i: b_i \to a_{i+1}$, and $X_i: b_i \to b_{i+1}$. In a complete cycle of G^* we see that U_i must be preceded by W_{i-1} and followed by V_i. But W_{i-1}, U_i, V_i lead from b_{i-1} to b_i, as also does X_{i-1}. Hence, a complete cycle of G^* involves a path going through $b_1, b_2, \ldots, b_m$ twice, in this cyclical order, with a choice of either of two routes from b_{i-1} to b_i, one the first time around and the other the second time around. This gives exactly 2^{m-1} complete cycles.

Now suppose that G has a point x not on a loop. Let P, Q be the arcs coming into x and R, S be the arcs going out. Then

$$G$$

$$\begin{array}{ll} P: a \to x & R: x \to c \\ Q: b \to x & S: x \to d \end{array}.$$

Here, P, Q, R, S are different, but some of the points a, b, c, d may coincide with each other, though not with x.

We may suppress x to form a 2-graph from G in two ways, either making $P = R, Q = S$ or $P = S, Q = R$. We call these two graphs respectively G_1 and G_2. Every complete cycle in G corresponds to a complete cycle in either G_1 or G_2 (but not both), depending on the way the point x is traversed. Thus, $|G| = |G_1| + |G_2|$:

$$\begin{array}{ll} G_1 & G_2 \\ P = R\colon a \to c & P = S\colon a \to d \\ Q = S\colon b \to d & Q = R\colon b \to c \end{array}.$$

Each of G_1 and G_2 has $m - 1$ points, and so by our induction,

$$|G_1^*| = 2^{m-2}|G_1|, \quad |G_2^*| = 2^{m-2}|G_2|.$$

We shall prove $|G^*| = 2\,|G_1^*| + 2\,|G_2^*|$. Let p, q, r, s be the points of G^* corresponding to P, Q, R, S in G. Then we have in G^*, G_1^*, G_2^*:

$$\begin{array}{llll} & G^* & G_1^* & G_2^* \\ X_1\colon & p \to r & p = r & p = s \\ X_2\colon & p \to s & q = s & q = r \\ X_3\colon & q \to r & & \\ X_4\colon & q \to s & & \end{array}$$

Here, G_1^* and G_2^* may be obtained from G^* by deleting the four arcs X_1, X_2, X_3, X_4 and identifying p and q with r and s as indicated. A complete cycle in G^* uses the four X arcs and also four paths leading from r and s to p and q. This may happen in three ways:

Case 1	Case 2	Case 3
$C_1\colon r \to p$	$D_1\colon r \to p$	$E_1\colon r \to q$
$C_2\colon r \to q$	$D_2\colon r \to p$	$E_2\colon r \to q$
$C_3\colon s \to p$	$D_3\colon s \to q$	$E_3\colon s \to p$
$C_4\colon s \to q$	$D_4\colon s \to q$	$E_4\colon s \to p$

Using the particular paths of case 1, we have four complete cycles of G^*, namely:

$$\begin{array}{l} X_1,\ C_1,\ X_2,\ C_4,\ X_3,\ C_2,\ X_4,\ C_3; \\ X_1,\ C_2,\ X_3,\ C_1,\ X_2,\ C_4,\ X_4,\ C_3; \\ X_1,\ C_2,\ X_4,\ C_3,\ X_2,\ C_4,\ X_3,\ C_1; \\ X_1,\ C_2,\ X_4,\ C_4,\ X_3,\ C_1,\ X_2,\ C_3. \end{array}$$

We may use the C paths to form complete cycles in G_1^* and G_2^* thus:

$$\begin{array}{l} G_1^*\colon\ C_1,\ C_2,\ C_4,\ C_3; \\ G_2^*\colon\ C_1,\ C_3,\ C_4,\ C_2. \end{array}$$

Using the paths of case 2, we have four complete cycles of G^*, namely:

$$X_1,\ D_1,\ X_2,\ D_3,\ X_4,\ D_4,\ X_3,\ D_2;$$
$$X_1,\ D_1,\ X_2,\ D_4,\ X_4,\ D_3,\ X_3,\ D_2;$$
$$X_1,\ D_2,\ X_2,\ D_3,\ X_4,\ D_4,\ X_3,\ D_1;$$
$$X_1,\ D_2,\ X_2,\ D_4,\ X_4,\ D_3,\ X_3,\ D_1.$$

These paths do not form a connected cycle in G_1^*, but in G_2^* we have two complete cycles:

$$G_2^*: \begin{matrix} D_1D_3D_2D_4 \\ D_1D_4D_2D_3 \end{matrix}.$$

Similarly, the paths of case 3 give four complete cycles for G^*, two for G_1^* and none for G_2^*. We note that all complete cycles for G_1^* and G_2^* are made up of paths of C's, D's, or E's. Hence, in every case we have

$$|G^*| = 2\,|G_1^*| + 2\,|G_2^*|.$$

By induction,

$$|G_1^*| = 2^{m-2}\,|G_1| \quad \text{and} \quad |G_2^*| = 2^{m-2}\,|G_2|.$$

Since $|G| = |G_1| + |G_2|$, we find

$$|G^*| = 2^{m-1}\,|G_1| + 2^{m-1}\,|G_2| = 2^{m-1}\,|G|,$$

proving our theorem.

10

Block Designs

10.1 General Discussion

A problem of great generality in combinatorial analysis is that of arranging objects into a specified number of sets so that the ith object appears r_i times in all the sets taken together, the jth set contains k_j objects, and so that pairs, triples, and similar groups of objects occur a specified number of times. This may be called an "incidence system" or "tactical configuration."

We shall investigate here a special kind of incidence system, which is called an incomplete balanced block design, though for the most part we shall refer to this more briefly as a block design, or merely a design.

DEFINITION. *A balanced incomplete block design is an arrangement of v distinct objects into b blocks such that each block contains exactly k distinct objects, each object occurs in exactly r different blocks, and every pair of distinct objects a_i, a_j occurs together in exactly λ blocks.*

Each of the b blocks $B_1, \ldots, B_b$ contains k objects, but different blocks B_i and B_j may contain the same objects. Thus, a block design is not merely a collection of subsets of a set, but is an array of objects and blocks with a relation of incidence telling which objects belong to which blocks. The term *balanced incomplete block design* comes from the theory of design of experiments as applied in statistics. The set of all ${}_vC_k$ combinations of v objects taken k at a time as blocks would be a complete block design. But a portion of these, in which each pair of objects a_i, a_j occurs the same number of times, is incomplete, but is balanced so far as comparisons between pairs are concerned.

There are two elementary relations on the five parameters for a block design:

$$bk = vr, \tag{10.1.1a}$$

$$r(k - 1) = \lambda(v - 1). \tag{10.1.1b}$$

These are easily proved. The first counts the total number of incidences in two ways, each of b blocks containing k objects and each of v objects being contained in r blocks. The second counts the occurrences of pairs containing a particular object a_1. The object a_1 occurs in r blocks, and in each of these is a pair with the $(k - 1)$ remaining objects, while on the other hand, a_1 is paired λ times with each of the remaining $(v - 1)$ objects.

A block design is described by its incidence matrix. This is a matrix $A = (a_{ij})$, $i = 1, \ldots, v$, $j = 1, \ldots, b$, where if $a_1, \ldots, a_v$ are the objects and $B_1, \ldots, B_b$ are the blocks, we have

$$\begin{aligned} a_{ij} &= 1, \qquad \text{if } a_i \in B_j, \\ a_{ij} &= 0, \qquad \text{if } a_i \notin B_j. \end{aligned} \tag{10.1.2}$$

The basic requirements for the block design are expressed by the matrix equations

$$AA^T = B = \begin{pmatrix} r & \lambda & \cdots & \lambda \\ \lambda & r & & \\ \cdot & & \cdot & \cdot \\ \cdot & & \cdot & \cdot \\ \cdot & & \cdot & \cdot \\ \lambda & & \cdots & r \end{pmatrix} = (r - \lambda)I_v + \lambda J_v, \qquad w_v A = k w_b. \tag{10.1.3}$$

Here, J_v is the $v \times v$ matrix of all 1's and w_v and w_b are respectively the vectors of v and b 1's. The element b_{ij} of the matrix B is the inner product of the ith row of A with the jth row of A. Thus, $b_{ii} = r$ counts the number of 1's in the ith row of A. But if $j \neq i$, then both the ith and jth rows have a 1 in column t if and only if both a_i and a_j belong to B_t. Thus, an off-diagonal element b_{ij} of B counts the occurrences of the pair a_i, a_j, and so is λ in every instance. The relation $w_v A = k w_b$ expresses the fact that there are k 1's in each column. Conversely, a $v \times b$ matrix A of zeros and 1's satisfying (10.1.3) is the incidence matrix of a block design with parameters b, v, r, k, λ.

The relation $AA^T = B$ may, of course, also be reformulated as a relation on quadratic forms. Let us associate indeterminates $x_1, \ldots, x_v$ with the objects, and the linear form L_j with the block B_j, where the a_{ij} are the incidence numbers of (10.1.2):

$$L_j = \sum_{i=1}^{v} a_{ij}x_i, \qquad j = 1, \ldots, b. \tag{10.1.4}$$

Here, (10.1.3) takes the form

$$L_1^2 + \cdots + L_b^2 = (r - \lambda)(x_1^2 + \cdots + x_v^2) + \lambda(x_1 + \cdots + x_v)^2. \tag{10.1.5}$$

Three examples of block designs are given below.

1. For $v = b = 7, r = k = 3, \lambda = 1$:

B_0 : 0, 1, 3
B_1 : 1, 2, 4
B_2 : 2, 3, 5
B_3 : 3, 4, 6
B_4 : 4, 5, 0
B_5 : 5, 6, 1
B_6 : 6, 0, 2

2. For $b = 12, v = 9, r = 4, k = 3, \lambda = 1$:

B_1 : 1, 2, 3	B_4 : 1, 4, 7	B_7 : 1, 5, 9	B_{10} : 1, 6, 8
B_2 : 4, 5, 6	B_5 : 2, 5, 8	B_8 : 2, 6, 7	B_{11} : 2, 4, 9
B_3 : 7, 8, 9	B_6 : 3, 6, 9	B_9 : 3, 4, 8	B_{12} : 3, 5, 7

3. For $v = b = 15, r = k = 7, \lambda = 3$:

B_0:	0, 1, 2,	3, 4, 5, 6
B_1:	0, 1, 2,	7, 8, 9, 10
B_2:	0, 1, 2,	11, 12, 13, 14
B_3:	0, 3, 4,	7, 8, 11, 12
B_4:	0, 3, 4,	9, 10, 13, 14
B_5:	0, 5, 6,	7, 8, 13, 14
B_6:	0, 5, 6,	9, 10, 11, 12
B_7:	1, 3, 5,	7, 9, 11, 13
B_8:	1, 3, 6,	7, 10, 12, 14
B_9:	1, 4, 5,	8, 10, 11, 14
B_{10}:	1, 4, 6,	8, 9, 12, 14
B_{11}:	2, 3, 5,	8, 10, 12, 13
B_{12}:	2, 3, 6,	8, 9, 11, 14
B_{13}:	2, 4, 5,	7, 9, 12, 14
B_{14}:	2, 4, 6,	7, 10, 11, 13

A block design is called *symmetric* if $v = b$ (and so, of course, also $k = r$), and the first and third designs listed here are of this kind. The second design is not symmetric, but has an additional property of another kind. As written, the blocks in example 2 are divided into four groups of three blocks each, each group containing all nine objects. This is the property of being a *group-divisible* design. We shall show in the next section that a symmetric design always has one more property—any two different blocks have exactly λ objects in common. This leads to a method of constructing two further designs from a symmetric block design, as may be illustrated from the third example.

We take one of the blocks (say, B_0) and construct blocks $B_1', \ldots, B_{b-1}$, where B_i' contains the λ objects common to B_0 and B_i, $i = 1, \ldots, b-1$. This is the *derived design*. If the original design had parameters $v = b$, $r = k$, and λ, the derived design has parameters $v' = k$, $b' = b - 1 = v - 1$, $r' = r - 1 = k - 1$, $k' = \lambda$, $\lambda' = \lambda - 1$. Here, the derived design has parameters $v' = 7$, $b' = 14$, $r' = 6$, $k' = 3$, $\lambda' = 2$, the objects being 0, 1, 2, 3, 4, 5, 6 as they occur in blocks $B_1, \ldots, B_{14}$. If we delete B_0 and its objects from blocks $B_1, \ldots, B_{14}$, we have blocks $B_1^*, \ldots, B_{14}^*$, where the objects are 7, 8, 9, 10, 11, 12, 13, 14, and obtain a *residual design*. In the residual design the parameters are $v^* = v - k$, $b^* = b - 1 = v - 1$, $k^* = k - \lambda$, $r^* = r = k$, $\lambda^* = \lambda$. In this instance we have $v^* = 8$, $b^* = 14$, $k^* = 4$, $r^* = 7$, $\lambda^* = 3$.

10.2 Elementary Theorems on Block Designs

Our basic relation on the incidence matrix of a block design is

$$AA^T = B = (r - \lambda)I + \lambda J = \begin{pmatrix} r & \lambda & \cdots & \lambda \\ \lambda & r & & \\ \cdot & & \cdot & \cdot \\ \cdot & & \cdot & \cdot \\ \cdot & & \cdot & \cdot \\ \lambda & & \cdots & r \end{pmatrix} \tag{10.2.1}$$

We can easily evaluate the determinant of B and find

$$\det B = (r - \lambda)^{v-1}(v\lambda - \lambda + r). \tag{10.2.2}$$

To see this, subtract the first column from all others, and then add rows 2, 3, ..., v to the first. Then all elements above the main diagonal are zero; on the main diagonal our first entry is $r + (v - 1)\lambda$, and the rest are $r - \lambda$. This gives us det B as in (10.2.2). If $r = \lambda$, the design is trivial, since then an object a_i in each of its r occurrences is paired with every other a_j and each block contains all v objects. Otherwise, $r > \lambda$ and B is nonsingular. From this we may derive Fisher's inequality

$$b \geq v \qquad \text{and hence} \qquad r \geq k, \tag{10.2.3}$$

for A is of rank at most b while B is of rank v. But the rank of a product of matrices cannot exceed the rank of either factor, and so we have $b \geq v$.

The following theorem, though easy to prove, is very strong in its consequences.

THEOREM 10.2.1. *In a symmetric design, if v is even, then $k - \lambda$ is a square.*

Proof: ■ Since $b = v$, A is a square matrix, and so from (10.2.1)

$$(\det A)^2 = \det B = (k - \lambda)^{v-1}(v\lambda - \lambda + k). \tag{10.2.4}$$

Since $k(k - 1) = \lambda(v - 1)$, we have $v\lambda - \lambda + k = k^2$.

But then the other factor of det B, $(k - \lambda)^{v-1}$, must also be a square, and as v is even, this means that $k - \lambda$ is a square, proving our theorem. This theorem tells us that the conditions (10.1.1) on the parameters of a block design are not sufficient for its existence. For example, parameters $b = v = 22$, $r = k = 7$, $\lambda = 2$ satisfy conditions (10.1.1), but here the design (if it existed) would be symmetric. But $v = 22$ is even and $k - \lambda = 5$ is not a square, and so no such design exists.

The next theorem gives a fundamental property of symmetric designs.

THEOREM 10.2.2. *If A is the incidence matrix of a symmetric block design, then A satisfies the four following relations:*

$$AA^T = B = (k - \lambda)I + \lambda J, \tag{10.2.5}$$

$$A^TA = B = (k - \lambda)I + \lambda J, \tag{10.2.6}$$

$$AJ = kJ, \tag{10.2.7}$$

$$JA = kJ. \tag{10.2.8}$$

Of these, (10.2.5) is the form taken by (10.2.1), recalling that $b = v$, $k = r$, and (10.2.7) says that every row of A contains k 1's, this being the number $r = k$ of blocks containing each object. The equation (10.2.8) says that every column of A contains k 1's, this being the number k of objects in each block. Thus, it remains to show only that (10.2.6) also holds. It will be sufficient to prove that (10.2.6) is a consequence of (10.2.5), (10.2.7), and (10.2.8). This is purely a theorem on matrices, as we now prove in the following theorem, which assumes considerably less.

THEOREM 10.2.3 (RYSER). *Let A be a nonsingular $v \times v$ matrix that satisfies either* (10.2.5) *or* (10.2.6) *and also either* (10.2.7) *or* (10.2.8). *Then A satisfies all four equations,* (10.2.5), (10.2.6), (10.2.7), *and* (10.2.8). *Furthermore v, k, λ satisfy the relation $k^2 - k = \lambda(v - 1)$.*

Proof: ■ Since $\det B = (k - \lambda)^{v-1}(v\lambda - \lambda + k)$, the nonsingularity of A means merely that $k - \lambda \neq 0$, $v\lambda - \lambda + k \neq 0$. Let us assume that (10.2.5) and (10.2.7) hold. Thus, we are given A nonsingular and

$$AA^T = (k - \lambda)I + \lambda J, \qquad AJ = kJ. \tag{10.2.9}$$

We now find $A^{-1}(AJ) = A^{-1}(kJ)$, whence $J = kA^{-1}J$, and so $k \neq 0$ and $A^{-1}J = k^{-1}J$. Also, $(AJ)^T = (kJ)^T$ or $JA^T = kJ$, since $J^T = J$. Noting that $J^2 = vJ$, we have

$$\begin{aligned} A^T &= A^{-1}(AA^T) = (k - \lambda)A^{-1} + \lambda A^{-1}J \\ &= (k - \lambda)A^{-1} + \lambda k^{-1}J. \\ kJ &= JA^T = (k - \lambda)JA^{-1} + \lambda k^{-1}J^2 \\ &= (k - \lambda)JA^{-1} + \lambda k^{-1}vJ. \end{aligned}$$

Thus,

$$JA^{-1} = \frac{k - \lambda k^{-1}v}{k - \lambda} J = mJ,$$

writing m for the constant. Thus,

$$\begin{aligned} J &= mJA, \\ vJ &= J^2 = (mJA)J = mJ(AJ) = mJ(kJ) \\ &= mkJ^2 = mkvJ. \end{aligned}$$

This gives $v = mkv$, whence $mk = 1$, $m = k^{-1}$. But we had $m(k - \lambda) = k - \lambda k^{-1}v$. Putting $m = k^{-1}$ gives $k^{-1}(k - \lambda) = k - \lambda k^{-1}v$. Multiplying by k gives $k - \lambda = k^2 - \lambda v$, which is the relation

$$k^2 - k = \lambda(v - 1), \tag{10.2.10}$$

which we wished to prove. Also, $JA^{-1} = mJ = k^{-1}J$ gives $J = k^{-1}JA$ or

$$JA = kJ, \tag{10.2.11}$$

the relation (10.2.8) we wished to prove. We may now return to

$$A^T = (k - \lambda)A^{-1} + \lambda k^{-1}J$$

and multiply to get

$$\begin{aligned} A^TA &= (k - \lambda)I + \lambda k^{-1}JA \\ &= (k - \lambda)I + \lambda k^{-1}(kJ) \\ &= (k - \lambda)I + \lambda J, \end{aligned} \tag{10.2.12}$$

which is the relation (10.2.6) we wished to prove.

We have now shown that for A nonsingular, (10.2.5) and (10.2.7) imply (10.2.6), (10.2.8), and the relation $k^2 - k = \lambda(v - 1)$.

Now let us suppose that (10.2.5) and (10.2.8) hold and that A is nonsingular. Then

$$AA^T = (k - \lambda)I + \lambda J,$$
$$JA = kJ.$$

This is somewhat simpler than our first case, as

$$\begin{aligned} J(AA^T) &= (k - \lambda)J + \lambda J^2, \\ kJA^T &= (k - \lambda)J + \lambda vJ = mJ, \qquad m = k - \lambda + \lambda v, \\ kJ(A^TJ) &= mJ^2, \\ kJ(JA)^T &= mJ^2, \\ kJ(kJ)^T &= mJ^2, \\ k^2J^2 &= mJ^2, \end{aligned}$$

whence $k^2 = m = k - \lambda + \lambda v$, which is the relation $k^2 - k = \lambda(v - 1)$ we wish to prove. Also, $kJA^T = mJ = k^2J$, and $k \neq 0$, since A is nonsingular. Then $JA^T = kJ$, which gives

$$AJ = (JA^T)^T = (kJ)^T = kJ,$$

which is the relation (10.2.7) we wish to prove. Finally,

$$A^TA = A^{-1}(AA^T)A = (k - \lambda)I + \lambda A^{-1}JA.$$

But with $AJ = kJ = JA$, we have $A^{-1}JA = J$, whence $A^TA = (k - \lambda)I + \lambda J$, the relation (10.2.6). Thus, (10.2.5) and (10.2.8) imply the other relations.

Now, with A^T replacing A, it follows from the preceding discussion that (10.2.6) and either (10.2.7) or (10.2.8) imply the other relations, and all parts of our theorem are proved.

There are a number of important consequences of Theorems 10.2.2 and 10.2.3. If A is the incidence matrix of a symmetric design, the relation $A^TA = (k - \lambda)I + \lambda J$ asserts that any two distinct blocks of the design have exactly λ objects in common, the fact mentioned in Section 10.1 whose proof we promised in this section. In a more precise way, the relations say that A^T is also the incidence matrix of a block design. If D is our original design with objects $a_1, \ldots, a_v$ and blocks $B_1, \ldots, B^v$, A^T is the incidence matrix of a *dual design* D' with objects $b_1, \ldots, b_v$ and blocks $A_1, \ldots, A_v$. We may set up a correspondence between D and D', writing

$$\begin{aligned} a_i &\rightleftarrows A_i, \\ B_j &\rightleftarrows b_j, \end{aligned} \qquad (10.2.13)$$

and defining D' by the rule $b_j \in A_i$ if and only if $a_i \in B_j$ in D.

In the case of the first of the examples at the end of Section 10.1, the dual D' is easily seen to be equivalent to the original design, if we renumber appropriately. But in the third example, although the dual design D' has parameters $v = b = 15, r = k = 7, \lambda = 3$, it would not be equivalent to the original design D by any renumbering of objects and blocks, for in D the triples of objects 0, 1, 2; 0, 3, 4; 0, 5, 6; 0, 7, 8; 0, 9, 10; 0, 11, 12; 0, 13, 14 have the property of being the objects in common to *three* blocks, and no other triples have this property. Note that all of these contain the object 0. In D', triples of objects b_0, b_1, b_2; b_0, b_3, b_4; b_0, b_5, b_6; b_1, b_3, b_5; b_1, b_4, b_6; b_2, b_3, b_6; b_2, b_4, b_5 are the objects common to three blocks, and no other triples have this property. But these sets of triples do not have any object in common. But a renumbering of objects and blocks must take such triples of intersections into triples of intersections. Hence, no renumbering of blocks and objects makes D' equivalent to D.

10.3 The Bruck-Ryser-Chowla Theorem

The most fundamental theorem on the existence of symmetric block designs is due to Bruck, Ryser, and Chowla.

THEOREM 10.3.1. *If a symmetric block design exists with parameters v, k, λ, then, writing $n = k - \lambda$:*

1. *If v is even, n is a square.*
2. *If v is odd, $z^2 = nx^2 + (-1)^{(v-1)/2}\lambda y^2$ has a solution in integers x, y, z not all zero.*

We have already proved condition 1 as Theorem 10.2.1. This theorem was first proved in the case $\lambda = 1$, when the design is a finite projective plane, by Bruck and Ryser [1] in 1949. In 1950 the theorem in its full form was proved by Chowla and Ryser [1], and Shrikhande [1] also proved condition 1 independently in the same year. The arithmetical considerations exclude many parameters that satisfy the elementary condition $k(k - 1) = \lambda(v - 1)$. For example, if $v = 43$, $k = 7$, $\lambda = 1$, condition 2 applies and gives $z^2 = 6x^2 - y^2$ or $z^2 + y^2 = 6x^2$, which has no solution in integers x, y, z not all zero. Hence, no symmetric design with parameters $v = 43$, $k = 7$, $\lambda = 1$ exists. It is worth remarking that this is the strongest known condition on the existence of symmetric designs. Indeed it is even conceivable that the necessary conditions of the Bruck-Ryser-Chowla theorem are even sufficient for the existence of the design. In no instance has it been shown by any other theorem, or by exhaustive search, that a symmetric design fails to exist whose parameters v, k, λ satisfy $k(k - 1) = \lambda(v - 1)$ and the Bruck-Ryser-Chowla conditions. There are, of course, many parameters for which the existence of the design is undecided. The projective plane of order 10 with parameters $v = 111$, $k = 11$, $\lambda = 1$, $n = 10$ is a particularly interesting case.

The basic observation of Bruck and Ryser was that for symmetric designs, the relation (10.1.5) takes the form

$$L_1^2 + \cdots + L_v^2 = (k - \lambda)(x_1^2 + \cdots + x_v^2) + \lambda(x_1 + \cdots + x_v)^2 = Q, \quad (10.3.1)$$

where

$$L_j = \sum_{i=1}^{v} a_{ij} x_j, \quad (10.3.2)$$

the a_{ij} being the incidence numbers 0 or 1. Using only the fact that the L's have rational coefficients, the existence of a symmetric design then implies that the quadratic form Q given by the right-hand side of (10.3.1) is rationally equivalent to $L_1^2 + \cdots + L_v^2$, a sum of squares. For the rational equivalence of quadratic forms, there are available the powerful and deep results of Hasse and Minkowski. Application of these results to (10.3.1) yields Theorem 10.3.1. The Chowla-Ryser paper, which we shall follow in the main, gives a more elementary derivation of Theorem 10.3.1, though for other purposes we need the full strength of the Hasse-Minkowski results.

For the proof of the theorem we need an arithmetical theorem and an identity. The identity is

$$(b_1^2 + b_2^2 + b_3^2 + b_4^2)(x_1^2 + x_2^2 + x_3^2 + x_4^2) = y_1^2 + y_2^2 + y_3^2 + y_4^2, \quad (10.3.3)$$

where

$$\begin{aligned} y_1 &= b_1x_1 - b_2x_2 - b_3x_3 - b_4x_4, \\ y_2 &= b_2x_1 + b_1x_2 - b_4x_3 + b_3x_4, \\ y_3 &= b_3x_1 + b_4x_2 + b_1x_3 - b_2x_4, \\ y_4 &= b_4x_1 - b_3x_2 + b_2x_3 + b_1x_4. \end{aligned} \quad (10.3.4)$$

This identity is most easily remembered in terms of quaternions, though historically the identity antedates Hamilton's discovery of quaternions. If we have a system of units 1, i, j, k, where $i^2 = j^2 = k^2 = -1$, $ij = k$, $jk = i$, $ki = j$, $ji = -k$, $kj = -i$, $ik = -j$, the multiplication rule for quaternions becomes

$$(b_1 + b_2i + b_3j + b_4k)(x_1 + x_2i + x_3j + x_4k) = y_1 + y_2i + y_3j + y_4k, \quad (10.3.5)$$

where y_1, y_2, y_3, y_4 are given as in (10.3.4), and if we define the norm $N(\beta)$ of a quaternion $\beta = b_1 + b_2i + b_3j + b_4k$ by the rule

$$N(\beta) = b_1^2 + b_2^2 + b_3^2 + b_4^2, \quad (10.3.6)$$

the identity (10.3.3) becomes

$$N(\beta)N(\xi) = N(\beta\xi). \qquad (10.3.7)$$

The arithmetical theorem is due to Lagrange. For a proof the reader is referred to Hardy and Wright [1, p. 302].

THEOREM OF LAGRANGE. *Every positive integer n can be represented as the sum of the square of four integers* $n = b_1^2 + b_2^2 + b_3^2 + b_4^2$.

We note that if b_1, b_2, b_3, b_4 are integers and $n = b_1^2 + b_2^2 + b_3^2 + b_4^2$, then in (10.3.4) the determinant of the linear forms in the x's is n^2, whence we can solve for the x's as linear forms in the y's with rational coefficients, the denominators being n^2.

We can now proceed with our proof of the Bruck-Ryser-Chowla theorem (10.3.1). Write $n = k - \lambda$, and from the theorem of Lagrange we have $n = b_1^2 + b_2^2 + b_3^2 + b_4^2$ for integers b_1, b_2, b_3, b_4. Then (10.3.1) becomes

$$L_1^2 + \cdots + L_v^2 = n(x_1^2 + \cdots + x_v^2) + \lambda(x_1 + \cdots + x_v)^2. \qquad (10.3.8)$$

Theorem 10.2.1 proved the theorem for v even. Hence, take v odd, and suppose first that $v \equiv 1 \pmod 4$. On the right-hand side we can write

$$n(x_i^2 + x_{i+1}^2 + x_{i+2}^2 + x_{i+3}^2) = y_i^2 + y_{i+1}^2 + y_{i+2}^2 + y_{i+3}^2, \qquad (10.3.9)$$

using the representation $n = b_1^2 + b_2^2 + b_3^2 + b_4^2$ and the identity (10.3.3), where the y's and x's are related as in (10.3.4). We apply this to the x's four at a time, and as $v \equiv 1 \pmod 4$, one x will be left over.

$$L_1^2 + \cdots + L_v^2 = y_1^2 + y_2^2 + \cdots + y_{v-1}^2 + nx_v^2 + \lambda(x_1 + \cdots + x_v)^2. \qquad (10.3.10)$$

Now write $y_v = x_v$ and, expressing the x's in terms of the y's, we have a rational identity in independent indeterminates $y_1, \ldots, y_v$.

$$L_1^2 + \cdots + L_v^2 = y_1^2 + \cdots + y_{v-1}^2 + ny_v^2 + \lambda w^2, \qquad (10.3.11)$$

where $L_1, \ldots, L_v$ and $w = x_1 + x_2 + \cdots + x_v$ are rational linear forms in $y_1, \ldots, y_v$. Suppose $L_1 = c_{11}y_1 + \cdots + c_{1v}y_v$. If $c_{11} \neq 1$, we may put $L_1 = y_1$ as a relation specializing y_1 as a rational linear form in $y_2, \ldots, y_v$. If $c_{11} = 1$, we put $L_1 = -y_1$. In either case we have specialized y_1 as a rational linear form in $y_2, \ldots, y_v$ and also have $L_1^2 = y_1^2$. Then (10.3.11) becomes

$$L_2^2 + \cdots + L_v^2 = y_2^2 + \cdots + y_{v-1}^2 + ny_v^2 + \lambda w^2, \qquad (10.3.12)$$

an identity in $y_2, \ldots, y_v$. Similarly, we may specialize y_2 as a linear form in $y_3, \ldots, y_v$, putting $L_2 = \pm y_2$. Continuing the process for $L_3, \ldots, L_{v-1}$, we finally have

$$L_v^2 = ny_v^2 + \lambda w^2, \qquad (10.3.13)$$

where L_v and w are rational multiples of y_v, which has preserved its independence throughout. This is important, since the relation (10.3.12) would tell us nothing if L_v, y_v, and w were identically zero. Taking y_v as an integer x, which is a multiple of the denominators appearing in L_v and w, we have in integers x, y, z with $x \neq 0$:

$$z^2 = nx^2 + \lambda y^2. \tag{10.3.14}$$

This equation is a consequence of the existence of the design with $v \equiv 1 (\text{mod } 4)$. When $v \equiv 3 (\text{mod } 4)$, let x_{v+1} be a new indeterminate and add nx_{v+1}^2 to both sides of (10.3.8) and use (10.3.9) to get the form

$$L_1^2 + \cdots + L_v^2 + nx_{v+1}^2 = y_1^2 + \cdots + y_{v+1}^2 + \lambda w^2. \tag{10.3.15}$$

Proceeding as before, we find

$$nx^2 = y_{v+1}^2 + \lambda w^2, \tag{10.3.16}$$

with x, w rationally dependent on y_{v+1}, which is independent. We can take y_{v+1} as a multiple of the denominators in x and w and this gives a solution in integers not all zero of

$$z^2 = nx^2 - \lambda y^2. \tag{10.3.17}$$

We may combine (10.3.14) and (10.3.17) into the single relation

$$z^2 = nx^2 + (-1)^{(v-1)/2} \lambda y^2. \tag{10.3.18}$$

This is part 2 of Theorem 10.3.1, whose proof is now complete.

The basic incidence equation for a symmetric design is

$$AA^T = (k - \lambda)I + \lambda J. \tag{10.3.19}$$

In terms of quadratic forms, this has been written as

$$L_1^2 + \cdots + L_v^2 = (k - \lambda)(x_1^2 + \cdots + x_v^2) + \lambda(x_1 + \cdots + x_v)^2, \tag{10.3.20}$$

where

$$L_j = \sum_{i=1}^{v} a_{ij} x_i. \tag{10.3.21}$$

The Bruck-Ryser-Chowla theorem gives necessary conditions for the existence of a rational matrix A satisfying (10.3.19), or what is the same thing, for the existence of rational linear forms L_j satisfying (10.3.20). These conditions are in fact sufficient for the rational solution of (10.3.19) or (10.3.20), though the proof of this requires the deep Hasse-Minkowski results, which will be summarized in the next section. But in two instances the sufficiency can be shown directly. First, if $k - \lambda = n$ is a square, then we easily verify that

$$
\begin{aligned}
A &= \sqrt{n}\, I + \frac{k - \sqrt{n}}{vJ}, \\
L_j &= \sqrt{n}\, x_j + \frac{k - \sqrt{n}}{v}(x_1 + \cdots + x_v), \qquad j = 1, \ldots, v
\end{aligned}
\tag{10.3.22}
$$

are rational solutions respectively of (10.3.19) and (10.3.20). This proves the sufficiency of condition 1 of Theorem 10.3.1 when v is even, and of course it includes cases when v is odd and n happens to be a square. When $\lambda = 1$, we may write $k = n + 1$, and then $v = n^2 + n + 1$. The identity (10.3.20) may be rewritten as

$$
L_1^2 + \cdots + L_v^2 = n\left(x_2 + \frac{x_1}{n}\right)^2 + n\left(x_3 + \frac{x_1}{n}\right)^2 + \cdots + n\left(x_v + \frac{x_1}{n}\right)^2 + (x_2 + \cdots + x_v)^2. \tag{10.3.23}
$$

It is necessary to note here that on the right-hand side of (10.3.23), the coefficient of x_1^2 is $(v - 1)/n = (n^2 + n)/n = n + 1$, as it should be to agree with (10.3.20). When $v \equiv 1 \pmod 4$, the identity

$$
n(u_i^2 + u_{i+1}^2 + u_{i+2}^2 + u_{i+3}^2) = y_i^2 + y_{i+1}^2 + y_{i+2}^2 + y_{i+3}^2,
$$

applied to the right-hand side of (10.3.23), puts it in the form

$$
L_1^2 + \cdots + L_v^2 = y_1^2 + \cdots + y_v^2, \tag{10.3.24}
$$

where $y_v = x_2 + \cdots + x_v$, and we have immediately $L_i = y_i$, $i = 1, \ldots, v$ as a rational solution of (10.3.20). We have $v \equiv 1 \pmod 4$ when $n \equiv 0, 3 \pmod 4$. When $n \equiv 1, 2 \pmod 4$, we have $v \equiv 3 \pmod 4$ and condition 2 of the theorem tells us that

$$
z^2 = nx^2 - y^2 \tag{10.3.25}
$$

has solutions in integers not all zero. Then $x \neq 0$ and $n = r^2 + s^2$, $r = y/x$, $s = z/x$, and we may use the identity

$$
(r^2 + s^2)(u_i^2 + u_{i+1}^2) = (ru_i + su_{i+1})^2 + (ru_{i+1} - su_i)^2 \tag{10.3.26}
$$

to convert (10.3.23) to the form

$$
L_1^2 + \cdots + L_v^2 = y_1^2 + \cdots + y_v^2, \tag{10.3.27}
$$

and again $L_i = y_i$, $i = 1, \ldots, v$ gives a rational solution to (10.3.8). When

v is odd and $\lambda > 1$, there does not appear to be an analog of (10.3.23) to give an elementary proof of the sufficiency of the Bruck-Ryser-Chowla conditions for the rational solution of (10.3.8)."

10.4 Statement of the Hasse-Minkowski Theorem. Applications

In this section we shall state some theorems from number theory that are needed for our work in combinatorial analysis. No proofs will be given, but references will be given to books where the proofs may be found.

Let p be an odd prime. The integers $b \not\equiv 0 (\text{mod}\, p)$ are divided into two classes called *quadratic residues* and *quadratic nonresidues* according as $x^2 \equiv b (\text{mod}\, p)$ does or does not have a solution $x (\text{mod}\, p)$. Thus, modulo 7, 2 is a quadratic residue, and 3 is a quadratic nonresidue, since $4^2 \equiv 2 (\text{mod } 7)$, but $x^2 \equiv 3 (\text{mod } 7)$ does not have a solution. This property is expressed in terms of the *Legendre symbol* $(\frac{b}{p})$ by the rules

$$\left(\frac{b}{p}\right) = +1, \quad \text{if } b \text{ is a quadratic residue modulo } p, \tag{10.4.1a}$$

$$\left(\frac{b}{p}\right) = -1, \quad \text{if } b \text{ is a quadratic nonresidue modulo } p. \tag{10.4.1b}$$

Thus $(\frac{2}{7}) = +1$ and $(\frac{3}{7}) = -1$. If $b \equiv 0 (\text{mod}\, p)$, we may write $(\frac{b}{p}) = 0$. The following theorems express some of the main properties of quadratic residues and nonresidues. For proofs the reader is referred to Hardy and Wright [1, p. 68].

THEOREM 10.4.1. *If p is an odd prime, then*

$$b \equiv c (\text{mod } p) \quad \text{implies} \left(\frac{b}{p}\right) = \left(\frac{c}{p}\right), \tag{10.4.2}$$

$$b^{(p-1)/2} \equiv \left(\frac{b}{p}\right) (\text{mod } p), \tag{10.4.3}$$

$$\left(\frac{bc}{p}\right) = \left(\frac{b}{p}\right)\left(\frac{c}{p}\right). \tag{10.4.4}$$

This theorem includes the statement that the product of two nonresidues is a residue. Thus, $(\frac{3}{7}) = -1$, $(\frac{6}{7}) = -1$, and $3 \cdot 6 = 18 \equiv 4 (\text{mod } 7)$ and $2^2 \equiv 4 (\text{mod } 7)$.

There are also relationships of a deeper kind between symbols $(\frac{b}{p})$ for different primes p. These are due to Gauss, and the second and third of the following are known as the "law of quadratic reciprocity."

THEOREM 10.4.2. *If p and q are two odd primes,*

$$\left(\frac{-1}{p}\right) = (-1)^{(p-1)/2}, \tag{10.4.5}$$

$$\left(\frac{2}{p}\right) = (-1)^{(p^2-1)/8}, \tag{10.4.6}$$

$$\left(\frac{p}{q}\right)\left(\frac{q}{p}\right) = (-1)^{((p-1)/2)((q-1)/2)}. \tag{10.4.7}$$

Theorems 10.4.1 and 10.4.2 not only express deep properties of quadratic residues, but also make the evaluation of the symbol $(\frac{p}{b})$ relatively easy. Thus, $(\frac{26}{83}) = (\frac{2}{83})(\frac{13}{83})$. Here, $(\frac{2}{83}) = -1$ by (10.4.6). By (10.4.7), $(\frac{13}{83})(\frac{83}{13}) = +1$. Hence, $(\frac{13}{83}) = (\frac{83}{13})$, and by (10.4.2), $(\frac{83}{13}) = (\frac{5}{13})$. By (10.4.7), $(\frac{5}{13})(\frac{13}{5}) = +1$ whence $(\frac{5}{13}) = (\frac{13}{5}) = (\frac{3}{5})$. We may now find by trial that $(\frac{3}{5}) = -1$, or may use (10.4.6) and (10.4.7) to find $(\frac{3}{5}) = (\frac{5}{3}) = (\frac{2}{3}) = -1$. This gives

$$(\tfrac{13}{83}) = (\tfrac{83}{13}) = (\tfrac{5}{13}) = (\tfrac{3}{5}) = -1.$$

Then $(\frac{26}{83}) = (-1)(-1) = +1$, and so $x^2 \equiv 26 (\text{mod } 83)$ has a solution. This method does not tell us how to find the solution, which by trial we find to be $x \equiv 21 (\text{mod } 83)$.

For congruences modulo powers of a prime p, we have:

THEOREM 10.4.3. *Let p be a prime and let $b = p^a b_1$, where p does not divide b_1. Then, for an arbitrarily high power of p, p^n, the congruence*

$$x^2 \equiv b (\text{mod } p^n) \tag{10.4.8}$$

is solvable if and only if (1) *a is even, and* (2) *for p odd, $(b_1/p) = +1$, for $p = 2$, $b_1 \equiv 1 (\text{mod } 8)$.*

In Section 10.3 the Bruck-Ryser-Chowla theorem showed that a necessary condition for the existence of a rational $v \times v$ matrix A satisfying

$$AA^T = B = (k - \lambda)I + \lambda J \tag{10.4.9}$$

is that (1), for v even, $n = k - \lambda$ is a square; and (2), for v odd,

$$nx^2 + (-1)^{(v-1)/2}\lambda y^2 = z^2 \tag{10.4.10}$$

has solutions in integers x, y, z not all zero. These conditions are also sufficient. If n is a square, we find that

$$A = \sqrt{n}\, I + \frac{k - \sqrt{n}}{v} J \tag{10.4.11}$$

is a rational solution of (10.4.9). This covers the cases when v is even. But the sufficiency of (10.4.10) when v is odd requires the deep Hasse-Minkowski theory of quadratic forms. If b and c are integers, the Hilbert norm-residue symbol $(b, c)_p$ is defined to be $+1$ or -1, according as the congruence

$$bx^2 + cy^2 \equiv z^2 (\text{mod } p^m) \tag{10.4.12}$$

does or does not have solutions in integers x, y, z, not all multiples of p, for arbitrarily high powers of p. We also include $p = \infty$ to mean that

$$bx^2 + cy^2 = z^2 \tag{10.4.13}$$

does or does not have solutions in real numbers x, y, z, not all zero. Thus, $(b, c)\infty = +1$ unless both b and c are negative.

For a detailed account of this theory, see Burton W. Jones [1, pp. 17–81].

THEOREM 10.4.4. *The equation*

$$bx^2 + cy^2 = z^2 \tag{10.4.14}$$

has solutions in integers x, y, z not all zero if and only if $(b, c)_p = +1$ for all primes p, including $p = \infty$.

As we observed, the matrix equation (10.4.9) is equivalent to the equation

$$L_1^2 + \cdots + L_v^2 = (k - \lambda)(x_1^2 + \cdots + x_v^2) + \lambda(x_1 + \cdots + x_v)^2. \tag{10.4.15}$$

Thus, the existence of a rational matrix A satisfying (10.4.9) is precisely equivalent to the existence of a rational transformation, taking the form $x_1^2 + \cdots + x_v^2$ into the form $(k - \lambda)(x_1^2 + \cdots + x_v^2) + \lambda(x_1 + \cdots + x_v)^2$. The Hasse-Minkowski theory gives conditions for the rational equivalence of any two rational quadratic forms. This is expressible in terms of the Hasse symbol $c_p(f)$, defined for a quadratic form

$$f = \sum_{i,j=1}^{m} b_{ij} x_i x_j, \qquad b_{ij} = b_{ji}$$

and primes p, including $p = \infty$.

If

$$f = \sum_{i,j=1}^{m} b_{ij} x_i x_j, \qquad b_{ij} = b_{ji},$$

define for $r = 1, \ldots, m$,

$$D_r = \det(b_{ij}), \qquad i, j = 1, \ldots, r.$$

D_r is called the rth leading principal minor of $B = (b_{ij})$. We shall suppose that the b's are all integers and that $D_1, \ldots, D_m$ are all different from zero. We shall not need the more general theory.

We define for f and every prime p, including $p = \infty$:

$$c_p(f) = (-1, -D_m)_p \prod_{i=1}^{m-1} (D_i, -D_{i+1})_p. \qquad (10.4.16)$$

The major theorem to which we shall appeal is:

THEOREM 10.4.5 (HASSE-MINKOWSKI). *If f_1 and f_2 are integral quadratic forms in m variables, none of whose leading principal minors vanish, then a necessary and sufficient condition that f_1 and f_2 be rationally equivalent is that $c_p(f_1) = c_p(f_2)$ for all odd primes p and also $p = \infty$.*

The following theorem gives the main computational properties of the Hilbert norm-residue symbol $(b, c)_p$.

THEOREM 10.4.6. *The Hilbert norm-residue symbol has the following properties, where b, c, d, e are integers:*

$$\prod (b, c)_p = 1, \quad \text{the product being over all primes including } p = \infty. \qquad (10.4.17)$$

$$(b, c)_\infty = 1, \quad \text{unless both } b \text{ and } c \text{ are negative}. \qquad (10.4.18)$$

$$(b, c)_p = (c, b)_p. \qquad (10.4.19)$$

$$(b_1 b_2, c)_p = (b_1, c)_p (b_2, c)_p. \qquad (10.4.20)$$

$$(bd^2, ce^2)_p = (b, c)_p. \qquad (10.4.21)$$

$$(b, -b)_p = 1, \quad (b^2, c)_p = 1. \qquad (10.4.22)$$

If p is an odd prime,

$$(b, c)_p = 1, \quad \text{if } b \text{ and } c \text{ are prime to } p. \qquad (10.4.23)$$

$$(b, p)_p = \left(\frac{b}{p}\right), \quad \text{if } b \not\equiv 0 (\text{mod } p). \qquad (10.4.24)$$

$$(p, p)_p = (-1, p)_p. \qquad (10.4.25)$$

$$\text{If } b_1 \equiv b_2 \not\equiv 0 (\text{mod } p), \quad \text{then } (b_1, c)_p = (b_2, c)_p. \qquad (10.4.26)$$

We are enabled by (10.4.17) to exclude $p = 2$ from consideration in Theorem 10.4.4, since $c_2(f_1) = c_2(f_2)$ is a consequence of $c_p(f_1) = c_p(f_2)$ for all odd primes p and $p = \infty$.

The problem with which we are concerned is to show that the rational solution of (10.4.15), when v is odd, is equivalent to the existence of integers

x, y, z not all zero satisfying (10.4.10). Here the rational forms are

$$f_1 = n(x_1^2 + \cdots + x_v^2) + \lambda(x_1 + \cdots + x_v)^2,$$
$$f_2 = x_1^2 + \cdots + x_v^2.$$

For f_1,

$$D_r = n^{r-1}[k + (r-1)\lambda], \qquad r = 1, \ldots, v,$$

and for f_2,

$$D_r = 1, \qquad r = 1, \ldots, v,$$

Here the calculations are easy for f_2, since $c_\infty(f_2) = -1$ and $c_p(f_2) = +1$ for p finite and odd. Since for f_1 all the D's are positive, we have $c_\infty(f_1) = -1$. Hence, by Theorem 10.4.4, f_1 and f_2 are rationally equivalent if and only if $c_p(f_1) = +1$ for all finite odd primes. This is the requirement that for finite odd p,

$$c_p(f_1) = (-1, -D_v)_p \prod_{i=1}^{v-1} (D_i, -D_{i+1})_p = 1, \qquad (10.4.27)$$
$$D_i = n^{i-1}[k + (i-1)\lambda], \qquad i = 1, \ldots, v.$$

The evaluation of D_i is exactly similar to the evaluation in equation (10.2.2).

Detailed and troublesome calculation shows that for v odd, (10.4.27) leads to

$$c_p(f_1) = (n, (-1)^{(v-1)/2}\lambda)_p = +1, \qquad (10.4.28)$$

for all finite odd primes p. Since n is positive, it is trivial that

$$(n, (-1)^{(v-1)/2}\lambda)_\infty = +1. \qquad (10.4.29)$$

Thus, using (10.4.17) to obtain also $(n, (-1)^{(v-1)/2}\lambda)_2 = +1$, we may appeal to Theorem 10.4.4 to conclude that the rational equivalence of f_1 and f_2 for v odd is equivalent to the existence of rational integers x, y, z not all zero satisfying

$$nx^2 + (-1)^{(v-1)/2}\lambda y^2 = z^2. \qquad (10.4.30)$$

This proves that (10.4.30) is not only necessary but also sufficient for the rational solution of (10.4.15) or, equivalently, (10.4.9). This proves the Bruck-Ryser-Chowla theorem and its rational converse.

There remains only the troublesome task of showing that (10.4.27) reduces to (10.4.28) for odd primes p. Let us write

$$E_i = k + (i-1)\lambda, \qquad i = 1, \ldots, v$$

and note that $E_v = k^2$, since $k^2 - k = \lambda(v-1)$. Since v is odd, n^{v-1} is a square, and so for odd primes p,

$$(-1, -D_v)_p = (-1, -n^{v-1}k^2)_p = (-1, -1)_p = +1. \qquad (10.4.31)$$

In the following calculations we omit for convenience the subscript p. Using (10.4.31), $c(f_1)$ becomes

$$
\begin{aligned}
c(f_1) &= \prod_{i=1}^{v-1} (D_i, -D_{i+1}) \\
&= \prod_{i=1}^{v-1} (n^{i-1}E_i, -n^iE_{i+1}) \\
&= \prod_{i=1}^{v-1} (n^{i-1}, -n^i)(n^{i-1}, E_{i+1})(E_i, n^i)(E_i, -E_{i+1}). \qquad (10.4.32)
\end{aligned}
$$

If i is even, $(n^{i-1}, -n^i) = (n, -1)$. If i is odd, $(n^{i-1}, -n^i) = (1, -n) = 1$. Hence

$$\prod_{i=1}^{v-1} (n^{i-1}, -n^i) = (n, -1)^{(v-1)/2}. \qquad (10.4.33)$$

Also,

$$
\begin{aligned}
\prod_{i=1}^{v-1} (n^{i-1}, E_{i+1}) &= \prod_{i=1}^{v-1} (n^2 \cdot n^{i-1}, E_{i+1}) \\
&= \prod_{i=1}^{v-1} (n^{i+1}, E_{i+1}) \\
&= (n^v, E_v) \prod_{j=2}^{v-1} (n^j, E_j) \\
&= \prod_{i=2}^{v-1} (n^i, E_i), \qquad (10.4.34)
\end{aligned}
$$

since $E_v = k^2$. Substituting (10.4.33) and (10.4.34) into (10.4.32), we have

$$
\begin{aligned}
c(f_1) &= (n, -1)^{(v-1)/2}(E_1, n) \prod_{i=2}^{v-1} (n^i, E_i)^2 \prod_{i=1}^{v-1} (E_i, -E_{i+1}) \\
&= (n, -1)^{(v-1)/2}(k, n) \prod_{i=1}^{v-1} (E_i, -E_{i+1}). \qquad (10.4.35)
\end{aligned}
$$

We now wish to prove

$$P = \prod_{i=1}^{v-1} (E_i, -E_{i+1}) = (k, n)(\lambda, n) \qquad (10.4.36)$$

for every odd prime p. If (10.4.36) is true, then substitution into (10.4.35) gives (10.4.28), which as we have already observed, yields the Bruck-Ryser-Chowla theorem and its rational converse.

Let p^a (possibly $p^0 = 1$) be the highest power of p dividing both k and λ

so that $k = p^a k_1$, $\lambda = p^a \lambda_1$, $p \nmid (k_1, \lambda_1)$, $k_1 + (v-1)\lambda_1 = p^a k_1^2$. Then

$$\begin{aligned}
P &= \prod_{i=1}^{v-1} (p^a[k_1 + (i-1)\lambda_1], -p^a[k_1 + i\lambda_1]) \\
&= \prod (p^a, -p^a)(p^a, k_1 + i\lambda_1)(k_1 + (i-1)\lambda_1, p^a) \\
&\qquad \times (k_1 + (i-1)\lambda_1, -(k_1 + i\lambda_1)) \\
&= (k_1, p^a)(k_1 + (v-1)\lambda_1, p^a) \cdot \prod_{i=1}^{v-2} (p^a, k_1 + i\lambda_1)^2 \\
&\qquad \times \prod_{i=1}^{v-1} ([k_1 + (i-1)\lambda_1], -(k_1 + i\lambda_1)). \qquad (10.4.37)
\end{aligned}$$

Now $(k_1 + (v-1)\lambda_1, p^a) = (p^a k_1^2, p^a) = (p^a, p^a)$. Thus, from (10.4.37) we have

$$P = (k_1, p^a)(p^a, p^a) \prod_{i=1}^{v-1} (k_1 + (i-1)\lambda_1, -(k_1 + i\lambda_1)). \qquad (10.4.38)$$

First suppose that $p \mid \lambda_1$. From $p(k-1) = \lambda(v-1)$ we obtain $k_1(p^a k_1 - 1) = \lambda_1(v-1)$, whence as $p \nmid k_1$, we must have $a = 0$ and $k_1 = k$. Also $p \nmid n = k - \lambda$ and $p \nmid k + i\lambda$ for all i. In this case, $P = 1$ and

$$(k, n)(\lambda, n) = 1 \cdot (\lambda, k - \lambda) = (\lambda, k) = (\lambda, (k^2 - k) + k) = (\lambda, k^2) = 1,$$

since $k^2 - k \equiv 0 \pmod p$. In this case, $P = (k, n)(\lambda, n)$.

In the remaining cases, $p \nmid \lambda_1$. Then at most one of $k_1 + (i-1)\lambda_1$, $k_1 + i\lambda_1$ is divisible by p, and if neither is, then $(k_1 + (i-1)\lambda_1, -(k_1 + i\lambda_1)) = 1$. Let r be an integer in the range $0 \le r \le v-1$ such that $k_1 + r\lambda_1 \equiv 0 \pmod p$. If $1 \le r \le v-2$, the product for P contains exactly two terms involving $k_1 + r\lambda_1$, and their product is

$$\begin{aligned}
&(k_1 + (r-1)\lambda_1, -(k_1 + r\lambda_1))(k_1 + r\lambda_1, -(k_1 + (r+1)\lambda_1)) \\
&\quad = (-\lambda_1, -(k_1 + r\lambda_1))(k_1 + r\lambda_1, -\lambda_1) \\
&\quad = (-\lambda_1, -(k_1 + r\lambda_1)^2) \\
&\quad = (-\lambda_1, -1) = 1, \qquad (10.4.39)
\end{aligned}$$

since $p \nmid \lambda_1$. Hence $P = (k_1, p^a)(p^a, p^a)$ unless one or both of k_1, $k_1 + (v-1)\lambda$ is divisible by p. When $a = 0$, then $k_1 = k$ and $k_1 + (v-1)\lambda_1 = k^2$ and p divides neither of these. In this case $P = 1$, and as $p \nmid k$, $p \nmid \lambda$, $(k, n)(\lambda, n) = 1 \cdot 1$ if $p \nmid n = k - \lambda$, while if $p \mid n = k - \lambda$, then $(k, n)(\lambda, n) = (k, n)^2 = 1$. Thus, in case $a = 0$, we have $P = (k, n)(\lambda, n)$. If $a > 0$, then $k_1 + (v-1)\lambda_1 = p^a k_1^2$, and

$$\begin{aligned}
&(k_1 + (v-2)\lambda_1, -(k_1 + (v-1)\lambda_1)) \\
&\quad = (-\lambda_1, -p^a k_1^2) \\
&\quad = (-\lambda_1, -p^a) = (-\lambda_1, -1)(-\lambda_1, p^a) \\
&\quad = (-1, p^a)(\lambda_1, p^a) = (p^a, p^a)(\lambda_1, p^a). \qquad (10.4.40)
\end{aligned}$$

But we may or may not have $p \mid k_1$. Using (10.4.40),

$$P = (k_1, p^a)(\lambda_1, p^a), \qquad \text{if } p \nmid k_1. \tag{10.4.41}$$

$$P = (k_1, p^a)(\lambda_1, p^a)(k_1, -k_1 - \lambda_1), \qquad \text{if } p \mid k_1. \tag{10.4.42}$$

We have, on expanding and simplifying,

$$\begin{aligned}(k, n)(\lambda, n) &= (p^a k_1, p^a(k_1 - \lambda_1))(p^a\lambda_1, p^a(k_1 - \lambda_1)) \\ &= (k_1, p^a)(k_1, k_1 - \lambda_1)(\lambda, p^a)(\lambda_1, k_1 - \lambda_1).\end{aligned} \tag{10.4.43}$$

If $p \nmid k_1$, then $(k_1, k_1 - \lambda_1) = (\lambda_1, k_1 - \lambda_1)$, since if $p \nmid k_1 - \lambda_1$, they are both 1, but are equal if $p \mid k_1 - \lambda_1$. Thus, from (10.4.41) and (10.4.43), $P = (k, n)(\lambda, n)$ if $p \nmid k_1$. If $p \mid k_1$,

$$(k_1, -k_1 - \lambda_1) = (k_1, k_1 - \lambda_1) \qquad \text{and} \qquad (\lambda_1, k_1 - \lambda_1) = 1,$$

since both terms are prime to p. Here, again, comparing (10.4.42) and (10.4.43), we find $P = (k, n)(\lambda, n)$. Thus, we have proved that $P = (k, n)(\lambda, n)$ in all cases. As was remarked earlier, on substitution into (10.4.35) this yields (10.4.28), and so yields the Bruck-Ryser-Chowla theorem and its rational converse.

As an application of this theorem, consider the parameters $v = 29$, $k = 8$, $\lambda = 2$. Here $k^2 - k = \lambda(v - 1)$ is satisfied. The Bruck-Ryser-Chowla theorem says that a necessary condition for the existence of a solution is that there exist integers x, y, z not all zero satisfying

$$6x^2 + 2y^2 = z^2. \tag{10.4.44}$$

For this to be solvable, the Hilbert symbol $(6, 2)_p$ must be $+1$ for all odd primes p and for $p = \infty$. This is certainly true for $p = \infty$ and for odd primes p not dividing 6 or 2—that is, for $p \neq 3$. Thus, the critical case is $p = 3$. Here $(6, 2)_3 = (2, 2)_3(3, 2)_3 = (3, 2)_3 = \left(\frac{2}{3}\right) = -1$, using the rules of Theorem 10.4.6 and of Theorem 10.4.2. Hence, (10.4.44) does not have solutions in integers not all zero, and there is no design with parameters $v = 29$, $k = 8$, $\lambda = 2$.

11

Difference Sets

11.1 Examples and Definitions

DEFINITION. *Two block designs B and B' are said to be isomorphic if there is a one-to-one mapping α of objects and blocks of B onto those of B' such that if x_i is an object, B_j a block of B,*

$$\alpha\colon x_i \to x_i' = (x_i)\alpha, \qquad \textit{an object of } B',$$

$$\alpha\colon B_j \to B_j' = (B_j)\alpha, \qquad \textit{a block of } B',$$

then $x_i \in B_j$ if and only if $(x_i)\alpha \in (B_j)\alpha$.

Thus, two block designs are isomorphic if they have the same incidence relationships and so are essentially the same. If $B' = B$, then the mapping α is called an *automorphism* of B. The automorphisms of any block design B always form a group, since if α_1 and α_2 are two automorphisms of B, the product mapping $\alpha_1\alpha_2$ is also an automorphism and also the inverse mappings α_1^{-1} and α_2^{-1} are automorphisms.

We shall be interested here in block designs with special kinds of automorphisms. In the following example, B is a block design with objects the residues $0, \ldots, 12 \pmod{13}$, and blocks B_i, $i \equiv 0, \ldots, 12 \pmod{13}$. This is

$$
\begin{array}{llll}
B_0\colon & 0, & 1, & 3, & 9 \\
B_1\colon & 1, & 2, & 4, & 10 \\
B_2\colon & 2, & 3, & 5, & 11 \\
B_3\colon & 3, & 4, & 6, & 12 \\
B_4\colon & 4, & 5, & 7, & 0 \\
B_5\colon & 5, & 6, & 8, & 1 \\
B_6\colon & 6, & 7, & 9, & 2 \\
B_7\colon & 7, & 8, & 10, & 3 \\
B_8\colon & 8, & 9, & 11, & 4 \\
B_9\colon & 9, & 10, & 12, & 5 \\
B_{10}\colon & 10, & 11, & 0, & 6 \\
B_{11}\colon & 11, & 12, & 1, & 7 \\
B_{12}\colon & 12, & 0, & 2, & 8
\end{array}
\tag{11.1.1}
$$

Here, $\alpha\colon i \to i + 1$, $B_i \to B_{i+1}$ is an automorphism of B, which permutes the objects and blocks each in a cycle of 13. In general, a symmetric (v, k, λ) block design is called B-cyclic if B has an automorphism α that permutes the objects and also the blocks in a cycle of length v. Note that the cyclic automorphism α and the set of objects in a single block completely determine the entire design, as is the case in the design of (11.1.1).

The block B_5: 5, 6, 8, 1 in (11.1.1) has the property that the $(4 \cdot 3 = 12)$ differences of distinct elements in the block yield every difference except $0 \pmod{13}$ exactly once. Thus,

$$
\left.
\begin{array}{ll}
1 \equiv 6 - 5, & 7 \equiv 8 - 1, \\
2 \equiv 8 - 6, & 8 \equiv 1 - 6, \\
3 \equiv 8 - 5, & 9 \equiv 1 - 5, \\
4 \equiv 5 - 1, & 10 \equiv 5 - 8, \\
5 \equiv 6 - 1, & 11 \equiv 6 - 8, \\
6 \equiv 1 - 8, & 12 \equiv 5 - 6,
\end{array}
\right\} \pmod{13}. \tag{11.1.2}
$$

DEFINITION. *A set of k residues D: $\{a_1, \ldots, a_k\}$ modulo v is called a (v, k, λ)-difference set if for every $d \not\equiv 0 \pmod{v}$ there are exactly λ ordered pairs (a_i, a_j), $a_i, a_j \in D$ such that $a_i - a_j \equiv d \pmod{v}$.*

Thus, 5, 6, 8, 1 is a 13, 4, 1 difference set from (11.1.2).

THEOREM 11.1.1. *A set of k residues D: $\{a_1, \ldots, a_k\}$ modulo v is a (v, k, λ)-difference set if and only if the sets B_i: $\{a_1 + i, a_2 + i, \ldots, a_k + i\}$ modulo v, $i = 0, \ldots, v - 1$ are a cyclic (v, k, λ) block design B.*

Proof: ∎ Suppose first that the sets B_i: $\{a_1 + i, \ldots, a_k + i\}$ of residues modulo v, for $i = 0, \ldots, v - 1$ form a cyclic block design B, where obviously the cyclic automorphism is $\alpha: i \to i + 1;\ B_i \to B_{i+1},\ i = 0, \ldots, v - 1$ modulo v. Here, if $d \not\equiv 0 \pmod{v}$, the objects $0, d$ occur together in exactly λ blocks—that is, for λ choices $a_i, a_j \in D$ and t, we have

$$d \equiv a_i + t, \qquad 0 \equiv a_j + t \pmod{v}, \tag{11.1.3}$$

whence there are exactly λ ordered pairs (a_i, a_j), $a_i, a_j \in D$ with

$$a_i - a_j \equiv d \pmod{v}, \tag{11.1.4}$$

since $t \equiv -a_j$ is uniquely determined by a_j. Hence, D is a difference set. Conversely, suppose that D: $\{a_1, \ldots, a_k\}$ is a (v, k, λ) difference set. Then in the sets B_i: $\{a_1 + i, \ldots, a_k + i\}$ of residues modulo v, $i = 0, \ldots, v - 1$, every pair of distinct residues r, s occurs together λ times. For let $r - s \equiv d \not\equiv 0 \pmod{v}$. Then for exactly λ ordered pairs (a_i, a_j), $a_i, a_j \in D$, we have

$$r - s \equiv a_i - a_j \pmod{v},$$

and so both $r \equiv a_i + t$ and $s \equiv a_j + t$ belong to B_t, where t is determined by $t \equiv r - a_i \equiv s - a_j \pmod{v}$. Thus, the sets B_i are the blocks of a cyclic (v, k, λ) block design B, and $\alpha: i \to i + 1;\ B_i \to B_{i+1}$, objects and block subscripts being residues modulo v, is the corresponding automorphism.

R. H. Bruck [1] has generalized the idea of a cyclic difference set to that of a group difference set D, which may be based on any finite group G. Let G be a finite group of order v and suppose that a (v, k, λ) block design B admits G as a regular group of automorphisms. By this we mean that if x is a particular object and B_0 is a particular block, then as g runs over the elements of G, $(x)g$ and $(B_0)g$ run over all the objects and blocks of B. Here we call x the base point. If $y = (x)g_1$ is another object, then $(x)g = (y)g_1^{-1}g$. Thus, we may identify the objects with the group elements, but a change of base point replaces g by $g_1^{-1}g$ for an appropriate g_1.

For example, if G is the Abelian group of order 16 generated by a, b, c, d, where $a^2 = b^2 = c^2 = d^2 = 1$, then $D = \{a, b, c, d, ab, cd\}$ is a (16, 6, 2)-group difference set, and if we take $(B_0)g$ as the set $\{ag, bg, cg, dg, abg, cdg\}$ g running over the 16 elements of G, we have a (16, 6, 2) block design that has G as a regular group of automorphisms. For difference sets in groups that may be non-Abelian, we use the following definition.

DEFINITION. *A set of k different elements D: $\{a_1, \ldots, a_k\}$ from a group G of order v is called a (v, k, λ)-group difference set if either of the following conditions holds:*

1. *For every $d \neq 1$, $d \in G$, there are exactly λ ordered pairs (a_i, a_j), $a_i, a_j \in D$, such that $a_i a_j^{-1} = d$.*
2. *For every $d \neq 1$, $d \in G$, there are exactly λ ordered pairs (a_i, a_j), $a_i, a_j \in D$, such that $a_i^{-1} a_j = d$.*

The two conditions are obviously equivalent if G is Abelian, but we shall see that they are in fact equivalent in every finite group G. We note from the definition that it is immediate that $k(k-1) = \lambda(v-1)$, since there are $k(k-1)$ ordered pairs and we are to have λ representations $a_i a_j^{-1}$ or $a_i^{-1} a_j$ of each of $v - 1$ elements.

THEOREM 11.1.2. *The properties* 1 *and* 2 *of group difference sets D are equivalent. If B is a (v, k, λ) block design admitting the group G of order v as a regular group of automorphisms and if $(x)a_1, \ldots, (x)a_k$ are the objects of a block B_0, then D: $\{a_1, \ldots, a_k\}$ is a (v, k, λ)-group difference set. Conversely, if D: $\{a_1, \ldots, a_k\}$ is a (v, k, λ)-group difference set of elements from the group G of order v, then the sets $B(g)$: $\{a_1 g, \ldots, a_k g\}$, as g runs over G, form a (v, k, λ) block design admitting G as a regular group of automorphisms.*

Proof: ■ Suppose that B is a (v, k, λ) block design admitting the group G of order v as a regular group of automorphisms. If x is any object of B, taking x as a base point, all objects will be $(x)g$, $g \in G$ and we may identify $(x)g$ with g. Then, if $a_1, \ldots, a_k$ are the objects of a block B_0, $(B_0)g$ contains $a_1 g, \ldots, a_k g$. With $d \neq 1$, an element of G, there are exactly λ blocks containing d and 1, and so there are exactly λ ordered pairs (a_i, a_j) such that for some g, $a_i g = d$, $a_j g = 1$, whence $a_i a_j^{-1} = d$. But here $g = a_j^{-1}$ is determined by a_j. Hence, $D = \{a_1, \ldots, a_k\}$ is a (v, k, λ)-difference set satisfying property 1. Conversely, if $D = \{a_1, \ldots, a_k\}$ is a (v, k, λ)-difference set of elements from G of order v, satisfying property 1, let us construct the sets $B(g)$: $\{a_1 g, \ldots, a_k g\}$ as g runs over G. If r, s are two different elements of G, let $rs^{-1} = d \neq 1$ determine d, and from λ ordered pairs (a_i, a_j) such that $a_i a_j^{-1} = d$, we determine g from such a pair by $a_i^{-1} r = a_j^{-1} s = g$, whence both $r = a_i g$ and $s = a_j g$ belong to $B(g)$.

Thus, the sets $B(g)$ are the blocks of a design B admitting G as a regular group of automorphisms. We have shown the equivalence of difference sets with property 1 to block designs B, admitting G regularly. What about property 2? From Section 10.2, if $d \neq 1$, the blocks $a_1, a_2, \ldots, a_k$ and $a_1 d, a_2 d, \ldots, a_k d$ have exactly λ elements in common. But this says that for exactly λ ordered pairs (a_i, a_j), we have $a_i d = a_j$, whence $d = a_i^{-1} a_j$. This is property 2 and is thus a consequence of property 1. Conversely, if we assume property 2, the sets $B(r)$: $\{a_1 r, \ldots, a_k r\}$ and $B(s)$: $\{a_1 s, \ldots, a_k s\}$ have exactly λ elements in common, whence, from Section 10.2, they are the blocks of a (v, k, λ) design.

11.2 Finite Fields

A field F is a set of elements with an addition $a + b$ and a multiplication ab satisfying the following familiar axioms:

A0. *For* $a, b \in F$, $a + b = c$ *exists and is a unique element of* F.

A1. $(a + b) + c = a + (b + c)$.

A2. $b + a = a + b$.

A3. *An element* 0 *belongs to* F *such that* $a + 0 = 0 + a = a$ *for every* $a \in F$.

A4. *For every* $a \in F$ *there is an element* $-a \in F$ *such that* $a + (-a) = (-a) + a = 0$.

M0. *For* $a, b \in F$, $ab = c$ *exists and is a unique element of* F.

M1. $(ab)c = a(bc)$.

M2. $ba = ab$.

M3. *An element* 1 *belongs to* F *such that* $a1 = 1a = a$ *for every* $a \in F$.

M4. *For every* $a \neq 0$ *of* F *there is an element* $a^{-1} \in F$ *such that* $aa^{-1} = a^{-1}a = 1$.

D. $a(b + c) = ab + ac$ *and* $(a + b)c = ac + bc$.

Familiar fields are the field of rational numbers, the field of real numbers, and the field of complex numbers. But there are also fields with a finite number of elements. For example, the four elements 0, 1, a, b form a field if we use the following tables of addition and multiplication. Here, $x + y$ and xy are in their tables in the row labeled x and column labeled y.

Addition

	0	1	a	b
0	0	1	a	b
1	1	0	b	a
a	a	b	0	1
b	b	a	1	0

Multiplication

	0	1	a	b
0	0	0	0	0
1	0	1	a	b
a	0	a	b	1
b	0	b	1	a

(11.2.1)

We shall give some of the basic facts about finite fields, but for more detail and for proofs, the reader is referred to Albert [2] or Dean [1].

A congruence of the rational integers modulo m, written as

$$x \equiv y(\text{mod } m), \tag{11.2.2}$$

is defined to be equivalent to the equation

$$x - y = tm \tag{11.2.3}$$

for an appropriate integer t. Similarly, a congruence can be defined for polynomials $A(x)$ over a field F (that is, polynomials with coefficients from F), writing

$$A(x) \equiv B(x)(\text{mod } f(x)) \tag{11.2.4}$$

to be equivalent to the equation

$$A(x) - B(x) = t(x)f(x) \tag{11.2.5}$$

for an appropriate polynomial $t(x)$. All integers x such that $x \equiv b(\text{mod } m)$, for a fixed b, form a "residue class" modulo m, which we may designate as $\{b\}$ or

$b(\text{mod } m)$. Since $x \equiv b(\text{mod } m)$, $y \equiv c(\text{mod } m)$ imply $x + y \equiv b + c(\text{mod } m)$ and $xy \equiv bc(\text{mod } m)$, the congruences thus define addition and multiplication of residue classes modulo m:

$$\{b\} + \{c\} = \{b + c\} \quad \text{and} \quad \{b\}\{c\} = \{bc\}. \tag{11.2.6}$$

The same considerations apply to residue classes of polynomials over a field F modulo $f(x)$. In both cases we easily verify that all field axioms except M4 are satisfied by the addition and multiplication of residue classes as given by (11.2.6). But, in general, M4 is not satisfied. There are usually zero divisors. Thus,

$$4 \cdot 3 \equiv 0(\text{mod } 6), \tag{11.2.7}$$

and for polynomials over the rational field,

$$(3x + 3)(5x - 5) \equiv 0(\text{mod } x^2 - 1). \tag{11.2.8}$$

It is easy to see from M4 that if $xy = 0$ and $x \neq 0$, then $y = x^{-1}(xy) = x^{-1}0 = 0$.

Now, if p is a rational prime, a fundamental property of primes (not proved here) is that if $p \mid xy$ (read p divides xy), then either $p \mid x$ or $p \mid y$. Hence, in terms of congruences, if

$$xy \equiv 0(\text{mod } p), \tag{11.2.9}$$

then

$$x \equiv 0(\text{mod } p) \quad \text{or} \quad y \equiv 0(\text{mod } p). \tag{11.2.10}$$

From the definition (11.2.3) this implies that if

$$ax \equiv ay(\text{mod } p), \quad a \not\equiv 0(\text{mod } p), \tag{11.2.11}$$

then

$$x \equiv y(\text{mod } p). \tag{11.2.12}$$

There are exactly p residue classes modulo p, and we may take $0, 1, \ldots, p - 1$ as representatives of these classes. Then, if $a \not\equiv 0(\text{mod } p)$, the numbers $a \cdot 1, a \cdot 2, \ldots, a(p - 1)$ are, by (11.2.11) and (11.2.12), all in different residue classes and all $\not\equiv 0(\text{mod } p)$. We may appeal to the "box principle," which says that if a finite number n of objects is placed in n boxes and no box contains two objects, then each box contains exactly one object. For comparison of $a \cdot 1$, $a \cdot 2, \ldots, a(p - 1)(\text{mod } p)$ and $1, 2, \ldots, p - 1(\text{mod } p)$, the box principle shows that they are the same residue classes in some order. This gives two valuable consequences: First, for some $x = 1, \ldots, p - 1$, we have

$$ax \equiv 1(\text{mod } p), \tag{11.2.13}$$

and so in the system of residue classes modulo p, x is the inverse of a, and M4 is satisfied. Second, on multiplying together, we have

$$(a \cdot 1)(a \cdot 2) \cdots (a(p - 1)) \equiv 1 \cdot 2 \cdots (p - 1)(\text{mod } p), \tag{11.2.14}$$

and dividing by $(p - 1)!$, since $(p - 1)! \not\equiv 0(\text{mod } p)$,

$$a^{p-1} \equiv 1(\text{mod } p), \tag{11.2.15}$$

the celebrated theorem of Fermat. Thus, (11.2.15) is satisfied for $a \not\equiv 0(\text{mod } p)$.

If we multiply by a, we have

$$a^p \equiv a (\text{mod } p), \tag{11.2.16}$$

satisfied by every integer $a(\text{mod } p)$. Thus, the residues mod p form a finite field with p elements, which we denote by J_p or GF(p). (GF stands for Galois field.)

If we consider polynomials over a field F, the analog of a prime is an *irreducible* polynomial. We say that $f(x)$ of degree $n \geq 1$ is irreducible over F if there do not exist two polynomials $g(x)$ and $h(x)$ over F, both of degree less than n such that $f(x) = g(x)h(x)$. A fundamental property of irreducible polynomials (not proved here) is that if an irreducible $f(x)$ divides a product $A(x)B(x)$, then $f(x)$ divides $A(x)$ or $f(x)$ divides $B(x)$.

Suppose that $f(x)$ is irreducible over J_p. Then, for polynomials $A(x)$ modulo $f(x)$, if

$$A(x) = f(x)q(x) + r(x), \tag{11.2.17}$$

we have

$$A(x) \equiv r(x), \qquad (\text{mod } f(x)), \tag{11.2.18}$$

and if $f(x)$ is of degree n, we may take $r(x)$ to be of degree at most $n - 1$. Hence, modulo $f(x)$, we have as representative of residue classes polynomials

$$A(x) \equiv a_0 + a_1 x + \cdots + a_{n-1}x^{n-1} (\text{mod } f(x)), \tag{11.2.19}$$

where each of $a_0, \ldots, a_{n-1}$ may be any one of the p elements of J_p. Hence, there are p^n residue classes modulo $f(x)$ over J_p.

Writing $q = p^n$, let $u_0 = 0$, $u_1 = 1$, $u_2, \ldots, u_{q-1}$ be representatives of the distinct residue classes modulo $f(x)$. If $y \equiv u_j \not\equiv 0(\text{mod } f(x))$, consider $yu_1, \ldots, yu_{q-1}(\text{mod } f(x))$.

If $yu_i \equiv yu_k(\text{mod } f(x))$, then $y(u_i - u_k)$ is divisible by $f(x)$; and by the irreducibility of $f(x)$ this requires $y \equiv 0(\text{mod } f(x))$ or $u_i \equiv u_k(\text{mod } f(x))$, neither of which can hold. Hence, $yu_i \not\equiv yu_k(\text{mod } f(x))$ for $i \neq k$, and similarly $yu_i \not\equiv 0(\text{mod } f(x))$ for $i \neq 0$. Thus, as with the residues mod p, we may apply the box principle and conclude that $yu_1, \ldots, yu_{q-1}(\text{mod } f(x))$ are the residue classes $u_1, \ldots, u_{q-1}(\text{mod } f(x))$ in some order. In particular, for some u_i, we have

$$yu_i \equiv u_1 \equiv 1(\text{mod } f(x)), \tag{11.2.20}$$

and here u_i is an inverse of y in the residue classes; so axiom M4 holds and the $q = p^n$ residues form a finite field, which we denote by GF(p^n). As a further consequence we have

$$(yu_1)(yu_2) \cdots (yu_{q-1}) \equiv u_1 u_2 \cdots u_{q-1}(\text{mod } f(x)), \tag{11.2.21}$$

and dividing out $u_1 u_2 \cdots u_{q-1}$ by the fundamental property of the irreducible $f(x)$, we have

$$y^{q-1} \equiv 1(\text{mod } f(x)), \tag{11.2.22}$$

holding for any polynomial $y \not\equiv 0(\text{mod } f(x))$.

Multiplying by y, we have

$$y^q \equiv y(\text{mod } f(x)), \tag{11.2.23}$$

holding for any polynomial $y = A(x)$ over J_p.

As an example, the polynomial $f(x) = x^2 + x + 1$ is irreducible over J_2, since the only polynomials of degree 1 over J_2 are x and $x + 1$, and neither of these divides $x^2 + x + 1$. Thus, the residues $0, 1, x, x + 1 (\text{mod } x^2 + x + 1)$ form a finite field $\text{GF}(2^2)$. Indeed, these residues correspond to the elements $0, 1, a, b$ in the example at the beginning of this section, whose addition and multiplication are given in (11.2.1).

The fields $\text{GF}(p^n)$ are indeed all the finite fields, and there is a field $\text{GF}(p^n)$ for every prime power p^n. The following theorems give the main facts about finite fields. For proof the reader again is referred to the books by Albert [2] or Dean [1].

THEOREM 11.2.1. *The number of elements in a finite field is a prime power p^n. For every prime power p^n there is a finite field* $\text{GF}(p^n)$, *unique to within isomorphism.* $\text{GF}(p^n)$ *may be represented as the residue classes modulo any polynomial $f(x)$ irreducible of degree n over J_p.*

THEOREM 11.2.2. *The $p^n - 1$ elements of* $\text{GF}(p^n)$ *different from* 0 *form a cyclic group under multiplication. A generator of this multiplicative group is called a primitive root of* $\text{GF}(p^n)$.

THEOREM 11.2.3. *The automorphisms of* $\text{GF}(p^n)$ *form a cyclic group of order n generated by the automorphism $\alpha: x \to x^p$ for every $x \in$* $\text{GF}(p^n)$.

THEOREM 11.2.4. *The subfields of* $\text{GF}(p^n)$ *are precisely the fields* $\text{GF}(p^m)$ *where m divides n. For each m dividing n,* $\text{GF}(p^n)$ *has a unique subfield* $\text{GF}(p^m)$ *consisting of those elements satisfying $z^{p^m} = z$. A primitive root x of* $\text{GF}(p^n)$ *satisfies $g(x) = 0$, $g(x)$ of degree n/m irreducible over* $\text{GF}(p^m)$.

Some examples will illustrate the content of these theorems. Since $25 = 5^2$ is a prime power, there is a field $\text{GF}(5^2)$ with 25 elements, and it can be represented by the residue classes modulo $f(x)$, where $f(x)$ is irreducible of degree 2 over J_5. Two such irreducible polynomials are $f_1(x) = x^2 - 2$ and $f_2(x) = x^2 + x + 1$. Thus, we have two fields, F_1 and F_2, with 25 elements. Theorem 11.2.1 asserts that F_1 and F_2 are isomorphic. The element $y \equiv 2x + 1 (\text{mod } x^2 + x + 1)$ satisfies

$$\begin{aligned} y^2 &\equiv 4x^2 + 4x + 1 \\ &\equiv -4 + 1 \equiv 2 (\text{mod } x^2 + x + 1). \end{aligned}$$

The mapping

$$ax + b (\text{mod } x^2 - 2) \to a(2x + 1) + b (\text{mod } x^2 + x + 1)$$

can be verified to be an isomorphism between F_1 and F_2.

The polynomial $f(x) = x^6 + x^5 + x^3 + x^2 + 1$ is irreducible over J_2, and so the residues $A(x) \text{ mod } f(x)$ form the field $\text{GF}(2^6)$ with 64 elements. In this field the element x is a primitive root and the powers of x, $1, x, x^2, \ldots, x^{62}$ yield the 63 nonzero elements of $\text{GF}(2^6)$. The subfields of $\text{GF}(2^6)$ are $\text{GF}(2)$, $\text{GF}(2^2)$ and

GF(2^3). The elements of GF(2) are 0, 1; the elements of GF(2^2) are 0, 1, x^{21}, x^{42} satisfying $z^4 = z$; the elements of GF(2^3) are 0, 1, x^9, x^{18}, x^{27}, x^{36}, x^{45}, x^{54} satisfying $z^8 = z$. The automorphisms are a cyclic group of order 6, generated by α: $z \to z^2 = (z)\alpha$ for every $z \in$ GF(2^6). For example,

$$\begin{aligned}(x + x^2 + x^3)\alpha &= x^2 + x^4 + x^6 \\ &= x^2 + x^4 + (x^5 + x^3 + x^2 + 1) \\ &= x^5 + x^4 + x^3 + 1.\end{aligned}$$

If we take $w = x^{21} = x^4 + x^2 + x + 1$, we find $x^3 + wx^2 + wx + w = 0$, and the polynomial $F(x) = x^3 + wx^2 + wx + w$ is irreducible over GF(2^2), whose elements are 0, 1, w, $w + 1$, where $w^2 + w + 1 = 0$. We can check the irreducibility of $F(x)$ over GF(2^2) by noting that if $F(x)$ had a linear factor, it would have to have 0, 1, w or $w + 1$ as a root, but this is seen not to be the case.

11.3 The Theorem of Singer

Suppose F is any field. The space of all vectors $(a_0, \ldots, a_n)$, $a_i \in F$ is called the *projective geometry* of dimension n over F, denoted by PG(n, F). The zero vector $0 = (0, \ldots, 0)$ is the void space, and we say that the void space has dimension -1. A point P, of dimension 0, is the set of vectors

$$bx = (bx_0, \ldots, bx_n),$$

where $x = (x_0, \ldots, x_n) \neq 0$ and b ranges over all elements of F. More generally, if $y_0, \ldots, y_k$ are $k + 1$ independent vectors, the set of all vectors $b_0y_0 + \cdots + b_ky_k$, $b_i \in F$ is a subspace S_k of dimension k. A subspace of dimension $n - 1$ is called a *hyperplane*. It is not difficult to show that the points in common to two distinct hyperplanes form a subspace of dimension $n - 2$. If $(c_0, \ldots, c_n) \neq (0, \ldots, 0)$, it can be shown that the set of all vectors $(x_0, \ldots, x_n)$ satisfying

$$c_0x_0 + \cdots + c_nx_n = 0 \tag{11.3.1}$$

is a hyperplane, and conversely, that every hyperplane can be defined in this way, $(c_0, \ldots, c_n)$ and $(sc_0, \ldots, sc_n)$, $s \neq 0$, defining the same plane.

If the field F is the finite field GF(p^r), then writing $q = p^r$, there are q^{n+1} vectors $(x_0, \ldots, x_n)$, $x_i \in$ GF(p^r). Each of the $(q^{n+1} - 1)$ vectors different from 0 determines a point, and since $(x_0, \ldots, x_n)$ and $(bx_0, \ldots, bx_n)$, $b \neq 0$, determine the same point, there are $v = (q^{n+1} - 1)/(q - 1)$ points. In the same way, since $(c_0, \ldots, c_n) \neq (0, \ldots, 0)$ and $(sc_0, \ldots, sc_n)$, $s \neq 0$, determine the same hyperplane, there are v hyperplanes. If $y_0, \ldots, y_t$ are $(t + 1)$ independent vectors, then $b_0y_0 + \cdots + b_ty_t$, $b_i \in$ GF(p^r) gives q^{t+1} different vectors. Excluding the zero vector, we have $(q^{s+1} - 1)/(q - 1)$ different points in this space S_t of dimension t. Thus, a hyperplane has $k = (q^n - 1)/(q - 1)$ different points and a space S_{n-2} has $\lambda = (q^{n-1} - 1)/(q - 1)$ points. We denote the geometry by $PG(n, p^r)$.

THEOREM 11.3.1 (THEOREM OF SINGER [1]). *The hyperplanes of* $PG(n, p^r), q = p^r$ *as blocks, points as objects, form a symmetric block design with*

$$v = \frac{q^{n+1} - 1}{q - 1}, \qquad k = \frac{q^n - 1}{q - 1}, \qquad \textit{and} \qquad \lambda = \frac{q^{n-1} - 1}{q - 1}.$$

This design is cyclic, and the points in any hyperplane determine a (v, k, λ)*-difference set.*

Proof: ∎ We note that with v, k, λ as in the theorem, $PG(n, p')$ has v points and v hyperplanes, each hyperplane containing k points, and two distinct hyperplanes intersect in an S_{n-2}, which has λ points. Thus, the hyperplanes as blocks of points form a (v, k, λ) block design. This much is easy. The main part of the theorem is to prove that the design is cyclic.

Let x be a primitive root of the field $GF(q^{n+1})$. Then (Theorem 11.2.4) x is a root of a polynomial $F(y)$ irreducible, of degree $n + 1$ over $GF(q)$. Let this be

$$F(y) = y^{n+1} + c_n y^n + \cdots + c_1 y + c_0, \qquad c_i \in GF(q). \tag{11.3.2}$$

Since $F(x) = 0$, $x^{n+1} = -c_0 - c_1 x - \cdots - c_n x^n$. From this it follows that for any power of x we may write

$$x^i = a_0 + a_1 x + \cdots + a_n x^n, \qquad a_j \in GF(q). \tag{11.3.3}$$

This establishes a correspondence between x^i and the vector $(a_0, \ldots, a_n)$ over $GF(q)$. The $(q^{n+1} - 1)$ different powers of x correspond to the $(q^{n+1} - 1)$ different vectors $(a_0, \ldots, a_n)$ over $GF(q)$, not the zero vector. If we write $v = (q^{n+1} - 1)/(q - 1)$, then $0, 1, x^v, \ldots, x^{(q-2)v}$ are the q solutions of $z^q = z$ and so are the subfield $GF(q)$ of $GF(q^{n+1})$. If $x^{sv} = t$, then $t \in GF(q)$, and so if (11.3.3) holds, then

$$x^{i+sv} = ta_0 + ta_1 x + \cdots + ta_n x^n. \tag{11.3.4}$$

Hence, x^i and x^j correspond to the same point of $PG(n, q)$ if and only if $i \equiv j \pmod{v}$. Thus, if a mapping α of $GF(q^{n+1})$ is given by

$$\begin{aligned} \alpha &: 0 \to 0, \\ \alpha &: x^i \to x^{i+1}, \end{aligned} \tag{11.3.5}$$

then from (11.3.3) and (11.3.2), the mapping of vectors over $GF(q)$ is

$$\alpha: (a_0, a_1, \ldots, a_n) \to (-a_n c_0, a_0 - a_n c_1, \ldots, a_{n-1} - a_n c_n). \tag{11.3.6}$$

This mapping α from the form (11.3.6) is in fact a mapping of points onto points and indeed of spaces S_t onto spaces S_t, since if $u_0, \ldots, u_t$ are independent vectors, we find, with $b_0, \ldots, b_t \in GF(q)$,

$$(b_0 u_0 + \cdots + b_t u_t)\alpha = b_0(u_0)\alpha + \cdots + b_t(u_t)\alpha. \tag{11.3.7}$$

Since $x^{v(q-1)} = 1$, the mapping (11.3.5) is one-to-one and so also is the corresponding mapping (11.3.6). Furthermore, since x^i and x^{i+v} correspond to the same point and $1, x, \ldots, x^{v-1}$ correspond to different points, the mapping α in (11.3.6) permutes the v points in a single cycle. Moreover, since $v - qk = 1$, k and v are relatively prime. Thus, if α^j takes the points of a hyperplane into themselves, then α^j permutes these in cycles of length t, where t divides k. But then, since $\alpha^v = 1$, t must also be a divisor of v, and, since $v - qk = 1$, we must have $t = 1$ and $\alpha^j = 1$. Then, because no power of α except $\alpha^v = 1$ fixes a hyperplane, α must permute the hyperplanes in a cycle of length v. This proves that the block design B of v points in v hyperplanes, each containing k points, is a cyclic block design, and our theorem is proved.

As an example of the application of the theorem of Singer, let $q = 2^2$, $n = 2$. Here $GF(q^{n+1}) = GF(2^6)$. $GF(q) = GF(2^2)$ can be taken as the field over J_2 with elements $0, 1, w, w + 1$, where $w^2 + w + 1 = 0$. As we noted in Section 11.2, a primitive root x of $GF(2^6)$ satisfies the irreducible equation $x^3 + wx^2 + wx + w = 0$ over $GF(2^2)$. This gives, for $1, x, \ldots, x^{21}$:

$$
\begin{aligned}
1 &= 1 \\
x &= x, \\
x^2 &= x^2 \\
x^3 &= w + wx + wx^2 \\
x^4 &= w + 1 + x + x^2 \\
x^5 &= w + x + (w+1)x^2 \\
x^6 &= 1 + (w+1)x \\
x^7 &= x + (w+1)x^2 \\
x^8 &= 1 + x \\
x^9 &= x + x^2 \\
x^{10} &= w + wx + (w+1)x^2 \\
x^{11} &= 1 + (w+1)x + (w+1)x^2 \\
x^{12} &= 1 + wx^2 \\
x^{13} &= w + 1 + wx + (w+1)x^2 \\
x^{14} &= 1 + wx + (w+1)x^2 \\
x^{15} &= 1 + (w+1)x^2 \\
x^{16} &= 1 + x^2 \\
x^{17} &= w + (w+1)x + wx^2 \\
x^{18} &= w + 1 + x \\
x^{19} &= (w+1)x + x^2 \\
x^{20} &= w + wx + x^2 \\
x^{21} &= w
\end{aligned}
\tag{11.3.8}
$$

Since $x^v = x^{21} = w \in GF(2^2)$, the mapping $x^i \to x^{i+1}$ permutes the points of $PG(2, 2^2)$ in a single cycle of length 21. If $x^i = a_0 + a_1 x + a_2 x^2$, the points (a_0, a_1, a_2) with $a_2 = 0$ lie on a hyperplane ("hyperplane" = "line" in a plane). If the points corresponding to x^i, $i = 0, \ldots, 20$ are replaced by the residue $i(\text{mod } 21)$, the points with $a_2 = 0$ are

$$0, 1, 6, 8, 18(\text{mod } 21). \tag{11.3.9}$$

These residues form a (21, 5, 1) difference set, as may easily be shown directly, and give a (21, 5, 1) block design, with blocks

$$B_i: \{i, i+1, i+6, i+8, i+18\}(\text{mod } 21).$$

Similarly, x, the primitive root found for $GF(2^6)$ satisfies over $GF(2)$,

$$f(x) = x^6 + x^5 + x^3 + x^2 + 1 = 0, \tag{11.3.10}$$

and we have

$$x^i = p_i + q_i x + r_i x^2 + s_i x^3 + t_i x^4 + u_i x^5. \tag{11.3.11}$$

The points in $PG(5, 2)$ with $u_i = 0$ form a hyperplane of dimension 4. From (11.3.10) we can verify that for any n,

$$u_{n+6} = u_{n+5} + u_{n+3} + u_{n+2} + u_n, \tag{11.3.12}$$

and from (11.3.11) that

$$u_0 = u_1 = u_2 = u_3 = u_4 = 0, \qquad u_5 = 1. \tag{11.3.13}$$

From (11.3.12) and (11.3.13), we can calculate those subscripts i modulo $63 = v$ such that $u_i = 0$. There are $k = 31$ of these, and $\lambda = 15$. Thus, a (63, 31, 15) difference set is

$$\begin{gathered} 0, 1, 2, 3, 4, 8, 9, 10, 11, 13, 14, 16, 17, 18, 21, 24, 25, 27, 30, \\ 32, 36, 41, 42, 45, 46, 47, 49, 51, 53, 54, 61(\text{mod } 63). \end{gathered} \tag{11.3.14}$$

In the application of Singer's theorem it is sufficient to find an equation such as (11.3.10) for a primitive root of $GF(q^{n+1})$ and to compute the appropriate u's from the recurrence corresponding to (11.3.12). It is not necessary to have a primitive root of $GF(q^{n+1})$, but we must have an element x such that x^v is the first power of x in $GF(q)$.

11.4 The Multiplier Theorem

The difference set 0, 1, 3, 9(mod 13) has the property that, when multiplied by 3, its elements are merely permuted among themselves. Multiplication of $0 + i$, $1 + i$, $3 + i$, $9 + i(\text{mod } 13)$ by 3 takes it into $0 + 3i$, $3 + 3i$, $9 + 3i$, $1 + 3i(\text{mod } 13)$—takes the block B_i into B_{3i}, as $i = 0, \ldots, 12(\text{mod } 13)$—of the cyclic block design given in (11.1.1). This shows that the mapping $x \to 3x(\text{mod } 13)$ of the residues modulo 13 and $B_i \to B_{3i}$ gives an automorphism of the design. This is, of course, an automorphism of the design in addition to the cyclic automorphism $i \rightarrow i + 1$, $B_i \rightarrow B_{i+1}$, which determined the entire design from the difference set 0, 1, 3, 9(mod 13).

DEFINITION. *An integer t is a multiplier of the cyclic difference set $a_1, \ldots, a_k \pmod{v}$ if $x \rightarrow xt \pmod{v}$ is an automorphism of the cyclic block design.*

We note that t is a multiplier if and only if $x \rightarrow xt$ maps the difference set $a_1, \ldots, a_k$ onto another block—that is, $ta_1, \ldots, ta_k \pmod{v}$ are $a_1 + s, \ldots, a_k + s \pmod{v}$ in some order for an appropriate s. Since 1 must occur as a difference, $1 = a_i - a_j$ and also $1 \equiv ta_r - ta_s \pmod{v}$, we see that necessarily a multiplier t must be prime to the modulus v. It readily follows that the multipliers must form a multiplicative group modulo v. As an example we note that 2 is a multiplier of the difference set $a_1, \ldots, a_{31} \pmod{63}$ given in (11.3.14). We find that $2a_1, \ldots, 2a_{31} \pmod{63}$ are a rearrangement of $a_1 + 18$, $a_2 + 18, \ldots, a_{31} + 18 \pmod{63}$.

R. H. Bruck [1], who extended the idea of difference set from the cyclic case to the case of group difference sets, observed that a multiplier t of a cyclic difference set is in fact an automorphism $x \rightarrow xt \pmod{v}$ of the underlying group (here the cyclic group of order v), which is also an automorphism of the block design. In exact form, his definition of a multiplier for general groups follows.

DEFINITION. *If $x \rightarrow (x)\theta$ is an automorphism of the group G, then θ is a multiplier of the group difference set $D = \{a_1, \ldots, a_k\}$ if $D\theta = aDb$ for appropriate $a, b \in G$.*

For example, with $D = a, b, c, d, ab, cd$ in the Abelian group G generated by a, b, c, d, where $a^2 = b^2 = c^2 = d^2 = 1$, we have $D\theta_i = D$ for

$$\theta_1\colon a \rightarrow b,\ b \rightarrow a,\ c \rightarrow d,\ d \rightarrow c,$$

$$\theta_2\colon a \rightarrow c,\ b \rightarrow d,\ c \rightarrow a,\ d \rightarrow b,$$

$$\theta_3\colon a \rightarrow d,\ b \rightarrow c,\ c \rightarrow b,\ d \rightarrow a,$$

$$\theta_4\colon a \rightarrow b,\ b \rightarrow a,\ c \rightarrow c,\ d \rightarrow d.$$

The knowledge of a multiplier is of great assistance in constructing difference sets or in proving their nonexistence. It is a remarkable fact that no cyclic difference known does not possess some nontrivial multiplier. The following theorem, due to Hall and Ryser [1], proves the existence of multipliers in many cases.

THEOREM 11.4.1 (MULTIPLIER THEOREM). *If $a_1, \ldots, a_k \pmod{v}$ are a cyclic difference set, where $k(k-1) = \lambda(v-1)$, and if p is a prime dividing $n = k - \lambda$ such that $(p, v) = 1$ and $p > \lambda$, then p is a multiplier of the difference set.*

Proof: ■ We consider the ring $R(x, x^{-1})$ of polynomials in x and x^{-1} with integral coefficients. We note that in R, $x^i \equiv x^j (\text{mod } x^v - 1)$ if and only if $i \equiv j (\text{mod } v)$. We can associate with the difference set D: $\{a_1, \ldots, a_k\} (\text{mod } v)$ the polynomial $\theta(x) = x^{a_1} + x^{a_2} + \cdots + x^{a_k}$. Then, since D is a difference set, we have

$$\theta(x)\theta(x^{-1}) \equiv k + \lambda(x + \cdots + x^{v-1})(\text{mod } x^v - 1). \qquad (11.4.1)$$

The left-hand side is

$$\sum_{i,j=1}^{k} x^{a_i - a_j}$$

and we have 0 as a difference $a_i - a_j$ for k times, and every other difference as $d \not\equiv 0 (\text{mod } v)$ for λ times. If we write $n = k - \lambda$ and $T(x) = 1 + x + \cdots + x^{v-1}$, then (11.4.1) takes the form

$$\theta(x)\theta(x^{-1}) \equiv n + \lambda T(x)(\text{mod } x^v - 1). \qquad (11.4.2)$$

If t is a multiplier, the fact that $ta_1, \ldots, ta_k (\text{mod } v)$ are $a_1 + s, \ldots, a_k + s$ for an appropriate s, can be expressed in the form

$$\theta(x^t) \equiv x^s \theta(x)(\text{mod } x^v - 1). \qquad (11.4.3)$$

Since $x^v - 1 = (x - 1)T(x)$ and $p \mid n$ by our hypothesis, a consequence of (11.4.2) is

$$\theta(x)\theta(x^{-1}) \equiv 0(\text{modd } p, T(x)), \qquad (11.4.4)$$

where $A(x) \equiv B(x)(\text{modd } p, T(x))$ means $A(x) - B(x) = pU(x) + T(x)V(x)$ for appropriate polynomials $U(x)$ and $V(x)$. Multiplying (11.4.4) by $\theta(x)^{p-1}$, we have

$$\theta(x)^p \theta(x^{-1}) \equiv 0(\text{modd } p, T(x)). \qquad (11.4.5)$$

Since the binomial coefficients $\binom{p}{i}$, $i = 1, \ldots, p - 1$ are all multiples of p, we have for any polynomials $A(x)$, $B(x)$,

$$(A(x) + B(x))^p \equiv A(x)^p + B(x)^p (\text{mod } p), \qquad (11.4.6)$$

and by repeated application of this we find

$$\theta(x)^p \equiv \theta(x^p)(\text{mod } p), \qquad (11.4.7)$$

whence, substituting in (11.4.5), we have

$$\theta(x^p)\theta(x^{-1}) \equiv 0(\text{modd } p, T(x)). \qquad (11.4.8)$$

This becomes modulo $x^v - 1$:

$$\theta(x^p)\theta(x^{-1}) \equiv pR(x) + A(x)T(x)(\text{mod } x^v - 1). \qquad (11.4.9)$$

Since $x^i T(x) \equiv T(x)(\text{mod } x^v - 1)$, we have $A(x)T(x) = A(1)T(x)(\text{mod } x^v - 1)$, and so

$$\theta(x^p)\theta(x^{-1}) \equiv pR(x) + A(1)T(x)(\text{mod } x^v - 1). \qquad (11.4.10)$$

Putting $x = 1$ in (11.4.10) gives

$$k^2 = pR(1) + A(1)v. \tag{11.4.11}$$

But we have

$$k^2 = k - \lambda + \lambda v = n + \lambda v, \tag{11.4.12}$$

and since $p \mid n$, we have

$$A(1)v \equiv \lambda v (\text{mod } p), \tag{11.4.13}$$

and as $p \nmid v$, this gives

$$A(1) \equiv \lambda (\text{mod } p), \tag{11.4.14}$$

whence $A(1) = \lambda + pm$ for some integer m.

Thus, (11.4.10) becomes

$$\theta(x^p)\theta(x^{-1}) \equiv pS(x) + \lambda T(x)(\text{mod } x^v - 1), \tag{11.4.15}$$

where $S(x) = R(x) + mT(x)$.

Now substituting $x = 1$ in (11.4.15), we have

$$k^2 = pS(1) + \lambda v, \tag{11.4.16}$$

and from (11.4.12) this yields

$$pS(1) = n. \tag{11.4.17}$$

In (11.4.15) replace x by x^{-1} and note that the modulus $x^{-v} - 1 = -x^{-v}(x^v - 1)$, where x^{-v} is a unit, and that $T(x^{-1}) \equiv T(x)(\text{mod } x^v - 1)$. Then we have

$$\theta(x^{-p})\theta(x) \equiv pS(x^{-1}) + \lambda T(x)(\text{mod } x^v - 1). \tag{11.4.18}$$

We now have four relations:

$$\begin{aligned} \theta(x)\theta(x^{-1}) &\equiv n + \lambda T(x)(\text{mod } x^v - 1), \\ \theta(x^p)\theta(x^{-p}) &\equiv n + \lambda T(x)(\text{mod } x^v - 1), \\ \theta(x^p)\theta(x^{-1}) &\equiv pS(x) + \lambda T(x)(\text{mod } x^v - 1), \\ \theta(x^{-p})\theta(x) &\equiv pS(x^{-1}) + \lambda T(x)(\text{mod } x^v - 1). \end{aligned} \tag{11.4.19}$$

The first of these is (11.4.2). The second says that if $a_1, \ldots, a_k (\text{mod } v)$ is a difference set, so is $pa_1, \ldots, pa_k (\text{mod } v)$, since if $a_i - a_j \equiv d$, then $pa_i - pa_j \equiv pd$; also, as d runs over all residues $\not\equiv 0 (\text{mod } v)$ and since p is prime to v, so does pd. From (11.4.19), the product of the left-hand sides of the first two congruences is the same as the product of the left-hand sides of the third and fourth. Hence, the products of the corresponding right-hand sides are congruent and we have

$$\{(pS(x) + \lambda T(x)\}\{pS(x^{-1}) + \lambda T(x)\} \equiv (n + \lambda T(x))^2 (\text{mod } x^v - 1). \tag{11.4.20}$$

Now, since $pS(x)T(x) \equiv pS(1)T(x) \equiv nT(x)(\text{mod } x^v - 1)$, (11.4.20) simplifies to

$$p^2 S(x)S(x^{-1}) \equiv n^2 (\text{mod } x^v - 1). \tag{11.4.21}$$

Now, for the first time, we must appeal to the condition $p > \lambda$ of our hypothesis. In (11.4.15) every coefficient in $\theta(x^p)\theta(x^{-1})$ is a nonnegative integer, and this property is not altered by reduction modulo $x^v - 1$. Hence, if

$$\theta(x^p)\theta(x^{-1}) \equiv a_0 + a_1x + \cdots + a_{v-1}x^{v-1} (\text{mod } x^v - 1), \quad (11.4.22)$$

we have $a_i \geq 0$ for $i = 0, \ldots, v - 1$. From (11.4.15) we have

$$a_i \equiv \lambda (\text{mod } p), \qquad i = 0, \ldots, v - 1.$$

Since we have assumed $p > \lambda$, this congruence requires $a_i \geq \lambda$, whence

$$\theta(x^p)\theta(x^{-1}) - \lambda T(x) \equiv pS(x) (\text{mod } x^v - 1) \quad (11.4.23)$$

has nonnegative coefficients, and so $S(x)$ has nonnegative coefficients. But if

$$S(x) \equiv b_0 + b_1x + \cdots + b_{v-1}x^{v-1} (\text{mod } x^v - 1)$$

has more than one positive term—say, $b_i > 0$ and $b_j > 0$—then on the left-hand side of (11.4.21) there is a term $b_ib_jx^{i-j}$ with $b_ib_j > 0$ and no negative term to cancel it, contrary to the fact that the coefficient of x^{i-j} is zero on the right-hand side. Hence, $S(x)$ is a monomial (say wx^s) and

$$pS(x) \equiv pwx^s \equiv nx^s (\text{mod } x^v - 1), \quad (11.4.24)$$

since $pS(1) = n$ by (11.4.17).

We now substitute (11.4.24) in (11.4.15) and have

$$\theta(x^p)\theta(x^{-1}) \equiv nx^s + \lambda T(x) (\text{mod } x^v - 1). \quad (11.4.25)$$

Let us multiply (11.4.25) by $\theta(x)$:

$$\theta(x^p)\theta(x)\theta(x^{-1}) \equiv nx^s\theta(x) + \lambda\theta(x)T(x) (\text{mod } x^v - 1). \quad (11.4.26)$$

From (11.4.2) and the fact that

$$\theta(x)T(x) \equiv \theta(1)T(x) \equiv kT(x) (\text{mod } x^v - 1), \quad (11.4.27)$$

we find

$$\theta(x^p)(n + \lambda T(x)) \equiv nx^s\theta(x) + \lambda kT(x) (\text{mod } x^v - 1) \quad (11.4.28)$$

and so

$$n\theta(x^p) \equiv nx^s\theta(x) (\text{mod } x^v - 1), \quad (11.4.29)$$

and dividing by n,

$$\theta(x^p) \equiv x^s\theta(x) (\text{mod } x^v - 1). \quad (11.4.30)$$

Comparison with (11.4.3) shows that $t = p$ is a multiplier.

The condition $p > \lambda$, which we used to obtain (11.4.24) from (11.4.15) and (11.4.21), does not appear to be necessary. For every known difference set a prime p is a multiplier if $p \mid n$ and $(p, v) = 1$. But we cannot conclude from (11.4.21) by itself that $S(x)$ is a monomial (mod $x^v - 1$), as we see from the following example:

$$(-1 + 2x^a + 2x^{2a})(-1 + 2x^{-a} + 2x^{-2a}) \equiv 9 (\text{mod } x^{3a} - 1). \quad (11.4.31)$$

11.5 Difference Sets in General Groups

Some of the theory of cyclic difference sets may be generalized to the theory of difference sets in any finite group. Let $D\colon a_1, \ldots, a_k$ be a (v, k, λ) difference set in the group G of order v. Here, a particular object P has been chosen as the base point, and additional objects are Pg, $g \in G$, with the base point P represented by the identity of G. If another point $Q = Pg_1$ is chosen as the base point, then as $Qg = Pg_1g$, an object corresponding to g_1g for the first choice of base point corresponds to g for the second choice. Thus, in our block design, multiplication of all group elements g on the left by a fixed element g_1 merely corresponds to a change of base point. Also $D\colon a_1, \ldots, a_k$ is the base block of our design and $Dg_2\colon \{a_1g_2, \ldots, a_kg_2\}$ is another block. Hence, a change of base point and change of the choice of base block replaces the difference set D by g_1Dg_2, and we may take g_1 and g_2 to be any two elements of the group G.

In the paper already referred to, Bruck [1] observes that a multiplier of a cyclic difference set is an automorphism of the group that is also an automorphism of the block design. Let $\alpha\colon g \to g^\alpha$ be an automorphism of G. We say that α is a multiplier of the difference set if there are elements g_1, g_2 of G such that

$$D^\alpha = g_1Dg_2. \tag{11.5.1}$$

We call α a right multiplier if $D^\alpha = Dg_2$.

THEOREM 11.5.1 (BRUCK). *Let K be the symmetric block design determined by the difference set D of k elements from the group G of order v. Let N be the normalizer of G in the group X of all automorphisms of K. Then a necessary and sufficient condition that the mapping T of K onto itself be contained in N is that*

$$(Px)T = Pa^{-1}x^\alpha, \qquad (By)T = Bby^\alpha, \tag{11.5.2}$$

with P the base point and B the base block for all x, y of G, where α is a multiplier of D such that $D = aDb$. Furthermore, N/G is isomorphic to the group of right multipliers of D.

Proof: ■ First suppose T to be given by (11.5.2). Then T is a one-to-one mapping of the objects and blocks of K onto themselves. If Px is in By, then $xy^{-1} \in D$, and

$$(a^{-1}x^\alpha)(by^\alpha)^{-1} = a^{-1}(xy^{-1})^\alpha b^{-1} \in a^{-1}D^\alpha b^{-1} = D,$$

and so $(Px)T$ is in $(By)T$. Hence, T is an automorphism of K. For any z in G, let z' be defined by $(z')^\alpha = z$. Then

$$(Px)Tz = (Pxz')T, \qquad (By)Tz = (Byz')T$$

for all $x, y, z \in G$, showing that T is in N.

Next suppose that T is in N, and $PT = Pa^{-1}$, $BT = Bb$. Define α by $x^\alpha = T^{-1}xT$. Then α is an automorphism of G and (11.5.2) is immediate. For each $d \in D$, Pd is in B, and so $Pa^{-1}d^\alpha = (Pd)T$ is in $BT = Bb$, whence $a^{-1}d^\alpha b^{-1} \in D$, and $D^\alpha = aDb$.

Finally, since G is regular on the objects of K, the factor group N/G is isomorphic to the subgroup of N fixing P. Here, if $PT = P$, then $a = 1$, and for the corresponding multiplier α, we have $D^\alpha = Db$. Thus, N/G is isomorphic to the group of right multipliers.

For a study of difference sets in general groups it is convenient to use the group ring $G^*(R)$ of the group G over a coefficient ring R, usually the ring Z of rational integers. Here, the elements of G^* are of the form

$$h = a_1x_1 + \cdots + a_gx_g, \qquad x_i \in G, \quad a_i \in R, \tag{11.5.3}$$

Addition in G^* is given by the rule

$$(a_1x_1 + \cdots + a_gx_g) + (b_1x_1 + \cdots + b_gx_g) \\ = (a_1 + b_1)x_1 + \cdots + (a_g + b_g)x_g. \tag{11.5.4}$$

Multiplication in G^* is given by the rule

$$(a_1x_1 + \cdots + a_gx_g)(b_1x_1 + \cdots + b_gx_g) = c_1x_1 + \cdots + c_gx_g, \tag{11.5.5}$$

where

$$c_k = \sum_{i,j} a_ib_j \qquad \text{for all } i, j, \qquad \text{with } x_ix_j = x_k \text{ in } G. \tag{11.5.6}$$

It is easily verified that G^* is an associative ring. If R has a unit, then $G^*(R)$ has a unit, written as 1, which is the product of the unit of R and the unit of G.

Suppose that D is a difference set of k elements in the group G of order v. In the group ring $G^*(Z)$, let us write

$$\theta(d) = d_1 + d_2 + \cdots + d_k, \qquad D = \{d_1, \ldots, d_k\}. \tag{11.5.7}$$

If t is any integer, we also write

$$\theta(d^t) = d_1^t + d_2^t + \cdots + d_k^t. \tag{11.5.8}$$

Let us also write

$$T = \sum_{x \in G} x. \tag{11.5.9}$$

In this notation the relation expressing the fact that D is a difference set over G becomes

$$\theta(d)\theta(d^{-1}) = k - \lambda + \lambda T, \tag{11.5.10}$$

since from the properties of the difference set, the left-hand side must yield the identity k times and every other element of G, λ times.

The multiplier theorem (Theorem 11.4.1) can be generalized to difference sets in an Abelian group. We give here a more general theorem, and in the next section a still more general theorem, whose proof depends on the theory of algebraic numbers and group characters. Both generalizations would be trivial if we could eliminate the condition $p > \lambda$ in the hypothesis of Theorem 11.4.1.

THEOREM 11.5.2. *Let D be a difference set of k elements in the Abelian group G of order v. Let $n_1 = p_1 p_2 \cdots p_s$ be a divisor of $n = k - \lambda$, where $p_1, \ldots, p_s$ are distinct primes. If $(n_1, v) = 1$, $n_1 > \lambda$, and if t is an integer such that $t \equiv p_i^{e_i} (\text{mod } v)$ for an appropriate power $p_i^{e_i}$, $i = 1, \ldots, s$, then the automorphism α of G, defined by $x = x^t$, is a multiplier of the difference set.*

Proof: ■ If $D = \{d_1, \ldots, d_k\}$ are the corresponding elements of the group ring $G(z)$ of G over the integers, then because G is Abelian and the multinomial coefficients are multiples of p, when p is a prime, we have

$$\begin{aligned}\theta(d)^p &= d_1^p + \cdots + d_k^p + pW \\ &= \theta(d^p) + pW,\end{aligned} \tag{11.5.11}$$

where W is an element of $G(z)$.

Moreover, since

$$t \equiv p_i^{e_i} (\text{mod } v), \qquad i = 1, \cdots, s, \tag{11.5.12}$$

we have $x^t = x^{p_i^{e_i}}$ for $i = 1, \ldots, s$ for any $x \in G$. (Incidentally, instead of v in this congruence, we could use v^*, the least common multiple of the orders of elements in G.) Thus, we have

$$\begin{aligned}\theta(d)^{p_i^{e_i}} &= \theta(d^{p_i^{e_i}}) + p_i R_i \\ &= \theta(d^t) + p_i R_i, \qquad i = 1, \ldots, s.\end{aligned} \tag{11.5.13}$$

Also, from

$$\theta(d)\theta(d^{-1}) = n + \lambda T \tag{11.5.14}$$

and multiplying by $\theta(d)^{q-1}$, where $q = p_i^{e_i}$, we have

$$\theta(d)^q \theta(d^{-1}) = n\theta(d)^{q-1} + \lambda\theta(d)^{q-1} T. \tag{11.5.15}$$

Here, since $xT = T$ for any $x \in G$, it follows that $\theta(d)T = kT$, and so (11.5.15) becomes

$$\theta(d)^q \theta(d^{-1}) = n\theta(d)^{q-1} + \lambda(k^{q-1} - 1)T + \lambda T. \tag{11.5.16}$$

Now p_i divides $n = k - \lambda$. If p_i does not divide k, then $k^{q-1} - 1$ is a multiple of p_i, but if p_i divides k, then also p_i divides λ. Hence, in either event, (11.5.16) yields

$$\theta(d)^q \theta(d^{-1}) = p_i V_i + \lambda T. \tag{11.5.17}$$

Now, using (11.5.13) in (11.5.17), we find

$$(\theta(d^t) + p_i R_i)\theta(d^{-1}) = p_i V_i + \lambda T, \tag{11.5.18}$$

and so

$$\theta(d^t)\theta(d^{-1}) = p_i S_i + \lambda T, \qquad i = 1, \ldots, s. \tag{11.5.19}$$

But here, as

$$\theta(d^t)\theta(d^{-1}) - \lambda T = p_i S_i, \qquad i = 1, \ldots, s, \tag{11.5.20}$$

the right-hand side must be a multiple of $n_1 = p_1 p_2 \cdots p_s$. Thus,

$$\theta(d^t)\theta(d^{-1}) = n_1 S + \lambda T. \tag{11.5.21}$$

If $x_1, \ldots, x_v$ are the elements of G, then the left-hand side of (11.5.21) is of the form $a_1 x_1 + \cdots + a_v x_v$, with the a_i nonnegative integers and $\sum a_i = k^2$. Comparison with the right-hand side shows that $a_i \equiv \lambda (\text{mod } n_1)$ for $i = 1, \ldots, v$, and as $n_1 > \lambda$, it follows that $a_i \geq \lambda$ in every instance. Thus, if $S = \sum s_i x_i$, we have $n_1 s_i = a_i - \lambda \geq 0$, $i = 1, \ldots, v$. Then

$$\sum n_1 s_i = \sum a_i - \lambda v = k^2 - \lambda v = k - \lambda = n.$$

Thus, the s_i are nonnegative integers such that $\sum n_1 s_i = n$, whence

$$n_1 S T = \sum_i n_1 s_i T = nT. \tag{11.5.22}$$

By applying the automorphism $x \to x^t$ of G to (11.5.14) and $x \to x^{-1}$ to (11.5.21), we have

$$\theta(d^t)\theta(d^{-t}) = n + \lambda T \quad \text{and} \quad \theta(d)\theta(d^{-t}) = n_1 S^* + \lambda T, \tag{11.5.23}$$

where

$$S^* = \sum s_i x_i^{-1}. \tag{11.5.24}$$

The product of the left-hand sides of (11.5.21) and (11.5.24) is the same as that for (11.5.21) and (11.5.14). Hence, from the right-hand sides,

$$(n_1 S + \lambda T)(n_1 S^* + \lambda T) = (n + \lambda T)^2. \tag{11.5.25}$$

Using (11.5.22) and also

$$n_1 S^* T = \sum_i n_1 s_i T = nT, \tag{11.5.26}$$

the relation (11.5.25) simplifies to

$$n_1^2 S S^* = n^2. \tag{11.5.27}$$

This becomes

$$n_1^2 \sum s_i x_i \sum s_i x_i^{-1} = n^2. \tag{11.5.28}$$

Since the s_i are nonnegative integers, we cannot have a nonzero term $s_i s_j x_i x_j^{-1}$ with $i \neq j$ on the left because only the identity has a nonzero coefficient on the right. Hence, S consists of a single term sg and $n_1 S = n_1 sg = ng$. The equation (11.5.21) now takes the simpler form

$$\theta(d^t)\theta(d^{-1}) = ng + \lambda T. \tag{11.5.29}$$

Multiplying by $\theta(d)$, we have

$$\theta(d^t)(n + \lambda T) = ng\theta(d) + \lambda\theta(d)T. \tag{11.5.30}$$

But $\theta(d^t)T = \theta(d)T = kT$, and so

$$n\theta(d^t) = ng\theta(d), \tag{11.5.31}$$

or dividing by n,

$$\theta(d^t) = g\theta(d). \tag{11.5.32}$$

But this is the statement that the automorphism $x \to x^t$ of G is a multiplier of the difference set, and our theorem is proved.

In the construction of a difference set it is convenient to construct a block fixed by multipliers if we can be sure that such a block exists.

THEOREM 11.5.3. *Let $D = \{d_1, \ldots, d_k\}$ be a difference set in the Abelian group G of order v. Let $x \to x^\alpha$ be a multiplier of the design $D(v, k, \lambda)$ determined by the difference set. Then there is a block fixed by the multiplier. If v and k are relatively prime, there is a block fixed by every multiplier.*

Proof: ■ The first part of this theorem is due to H. Mann, the second to J. Jans, both parts having been communicated verbally to the author.

Let $A = [a_{ij}]$ be the incidence matrix of the block design $D(v, k, \lambda)$ determined by D. Then the multiplier $x \to x^\alpha$ permutes the elements of G. If the elements of G written additively are $g_1 = 0, g_2, \ldots, g_v$, write $P = [p_{rs}]$, where $p_{rs} = 1$ if $g_r^\alpha = g_s$, and $p_{rs} = 1$ otherwise. Similarly, if $B_1, \ldots, B_v$ are the blocks of $D(v, k, \lambda)$ then $x \to x^\alpha$ permutes the B's, and we write $Q = [q_{m,n}]$, where $q_{m,n} = 1$ if $B_m^\alpha = B_n$ and $q_{m,n} = 0$ otherwise. Then the assertion that $x \to x^\alpha$ induces an automorphism of $D(v, k, \lambda)$ is expressed by writing

$$P^{-1}AQ = A. \tag{11.5.33}$$

Here, since A is nonsingular, we have

$$Q = A^{-1}PA. \tag{11.5.34}$$

Hence, for the traces of P and Q, we have

$$\operatorname{trace} Q = \operatorname{trace} P. \tag{11.5.35}$$

But since $g_1^\alpha = 0^\alpha = 0 = g_1$, we have $p_{11} = 1$ and trace $P > 0$. But by (11.5.35), trace $Q > 0$ and trace Q is the number of blocks fixed by the multiplier α. This part is due to Henry Mann and is similar to an argument of E. Parker [1].

In the block $B_i = \{d_1 + g_i, d_2 + g_i, \ldots, d_k + g_i\}$, the sum of the elements is $(d_1 + d_2 + \cdots + d_k) + kg_i$. Hence, if $(v, k) = 1$, there is exactly one choice of g_i so that this sum is the zero element of G. Clearly, this block is fixed by every multiplier of $D(v, k, \lambda)$.

11.6 Some Families of Difference Sets

There are several known families of difference sets. We shall first list these with a brief description and then give further details later. These are considered in Marshall Hall, Jr. [1].

Type S: ■ (Singer difference sets). These are hyperplanes in $PG(n, q)$, $q = p^r$. The parameters are

$$v = \frac{q^{n+1} - 1}{q - 1}, \qquad k = \frac{q^n - 1}{q - 1}, \qquad \lambda = \frac{q^{n-1} - 1}{q - 1}.$$

The existence and construction of these sets were given in Theorem 11.3.1 and its proof.

Type Q: ■ (quadratic residues in $\mathrm{GF}(p^r)$, $p^r \equiv 3 \pmod 4$) :

$$v = p^r = 4t - 1, \qquad k = 2t - 1, \qquad \lambda = t - 1,$$

Type H_6: ■ (p is a prime of the form $p = 4x^2 + 27$). There will exist a primitive root r modulo p such that $\mathrm{Ind}_r(3) \equiv 1 \pmod 6$. The residues $a_1, \ldots, a_{(p-1)/2} \pmod p$ such that $\mathrm{Ind}_r(a_i) \equiv 0$, 1, or $3 \pmod 6$ will form a difference set with

$$v = p = 4t - 1, \qquad k = 2t - 1, \qquad \lambda = t - 1.$$

Note that type H_6 always duplicates the parameters of a difference set of type Q.

Type T: ■ (twin primes). Suppose that p and $q = p + 2$ are both primes. Of the $(p - 1)(q - 1)$ residues modulo pq prime to pq, let $a_1, \ldots, a_m$, $m = (p - 1)(q - 1)/2$ be those for which $(a_i/p) = (a_i/q)$, and also let $a_{m+1}, \ldots, a_{m+p}$ be $0, q, 2q, \ldots, (p - 1)q$. Here, $m + p = (pq - 1)/2 = k$. Then, $a_1, \ldots, a_k$ form a difference set modulo v, $v = pq$. Here the parameters are $v = pq$, $k = (pq - 1)/2$, $\lambda = (pq - 3)/4$. Necessarily, $pq = -1 \pmod 4$, and we have $v = 4t - 1$, $k = 2t - 1$, $\lambda = t - 1$. This also holds for $\mathrm{GF}(p^r)$ and $\mathrm{GF}(q^s)$ if $q^s = p^r + 2$. The three types, Q, H_6, and T, are difference sets of the Hadamard type, since any symmetric design with parameters $v = 4t - 1$, $k = 2t - 1$, $\lambda = t - 1$ corresponds to a Hadamard matrix of order $4t$. Hadamard matrices will be discussed in Chapter 14.

Type B: ■ (biquadratic residues of primes $p = 4x^2 + 1$, x odd). Here,

$$v = p = 4x^2 + 1, \qquad k = x^2, \qquad \lambda = \frac{x^2 - 1}{4}.$$

Type B_0: ■ (biquadratic residues and zero modulo primes $p = 4x^2 + 9$, x odd). Here $v = 4x^2 + 9$, $k = x^2 + 3$, $\lambda = (x^2 + 3)/4$.

Type O: ■ (octic residues of primes $p = 8a^2 + 1 = 64b^2 + 9$ with a, b odd). Here $v = p$, $k = a^2$, $\lambda = b^2$.

Type O_0: ■ (octic residues and zero for primes $p = 8a^2 + 49 = 64b^2 + 441$, a odd, b even). Here $v = p$, $k = a^2 + 6$, $\lambda = b^2 + 7$.

Type W_4: ■ (a generalization of type T developed by Whiteman [1], but which uses biquadratic rather than quadratic residues). Here p and q are two primes such that $(p-1, q-1) = 4$, and we write $d = (p-1)(q-1)/4$. We take g as an appropriate number, which is a primitive root of both p and q. The difference set consists of $1, g, g^2, \ldots, g^{d-1}, 0, q, 2q, \ldots, (p-1)q$ modulo pq, where $v = pq, k = (v-1)/4, \lambda = (v-5)/16$. We must have $q = 3p + 2$ and also $(v-1)/4$, an odd square.

Note that if $a_1, \ldots, a_k$ are a set of residues forming a difference set modulo an odd v, then $-a_1, \ldots, -a_k$ cannot be the same set, for if this were so, then $d \equiv a_i - a_j \equiv (-a_j) - (-a_i)$ modulo v, and a difference d not of the form $2a_i$ occurs an even number of times and a difference of the form $2a_i$ occurs an odd number of times. As a consequence of this observation (if p is an odd prime, $p = ef + 1$), a set of residues modulo p forming a difference set that has the eth power residues as multipliers fixing the set can exist only if f is odd. For $a \equiv b^e (\mathrm{mod}\, p)$ if and only if $a^f \equiv b^{ef} \equiv b^{p-1} \equiv 1 (\mathrm{mod}\, p)$ and if f is even, $(-1)^f \equiv 1$ $(\mathrm{mod}\, p)$ and -1 is an eth power residue, and would be a multiplier fixing the set, which we have shown to be impossible.

If $\alpha: g \rightarrow g^\alpha = g^*$ is an automorphism of a group G, the mapping is one-to-one and so $\alpha^{-1}: g^* \rightarrow g = g^{*\alpha^{-1}}$ is also one-to-one and an automorphism. Also, given two automorphisms $\alpha_1: g \rightarrow g^{\alpha_1}$; $\alpha_2: g \rightarrow g^{\alpha_2}$, the product $\alpha_1\alpha_2$ is $g \rightarrow (g^{\alpha_1})^{\alpha_2} = g^{\alpha_1\alpha_2}$ and is also an automorphism. Thus, automorphisms of a group G themselves form a group $A(G)$. We may combine the group G and its automorphisms $A(G)$ into a larger group, called the holomorph of G and written $H(G)$. To do this, we use the rule

$$(\alpha_1, g_1)(\alpha_2, g_2) = (\alpha_1\alpha_2, g_1^{\alpha_2} g_2), \tag{11.6.1}$$

for $\alpha_1, \alpha_2 \in A(G), g_1, g_2 \in G$.

It is easily verified that the pairs (α, g) form a group $H(G)$ under the product rule (11.6.1). Here, $(\alpha, g)^{-1} = (\alpha^{-1}, g^{-\alpha^{-1}})$. In $H(G)$, if $\alpha_1 = 1$ is the identity automorphism, then $(1, g_1)(1, g_2) = (1, g_1 g_2)$, and so the elements $(1, g)$ of $H(G)$ form a subgroup G isomorphic to G. In fact G is normal in $H(G)$, as is easily checked, and $H(G)/G$ is isomorphic to $A(G)$. Also, in $H(G)$ the automorphism α is induced in G by conjugation by $(\alpha, 1)$; thus,

$$(\alpha^{-1})(1, g)(\alpha, 1) = (\alpha^{-1}, g)(\alpha, 1) = (1, g^\alpha). \tag{11.6.2}$$

Both rules (11.6.1) and (11.6.2) apply if the α's form some group M of automorphisms of G where M is any subgroup of $A(G)$. Let us denote the group by $M(G)$ in this case. Of particular interest to us here is the case in which G is an Abelian group of order v and M is a group of multipliers that, by Theorem 11.5.3 or by some other reason, is known to fix a block of a difference set. For example, if $v = 43, k = 21, \lambda = 10$, then $n = k - \lambda = 11$ is a multiplier by Theorem 11.4.1, and so $(v, k) = 1$. It follows by Theorem 11.5.3 that there is a block fixed by all multipliers and in particular by all powers of the automorphism $\alpha: x \to 11x \pmod{43}$. Here the classes of conjugate elements of G, conjugate by the rule (11.6.2), under powers of the automorphism α are listed.

We write z for the residue $0 \pmod{43}$:

$$
\begin{array}{lrrrrrrr}
z & & & & z & & & \\
C_0: & 1, & 11, & 35, & 41, & 21, & 16, & 4 \\
C_1: & 3, & 33, & 19, & 37, & 20, & 5, & 12 \\
C_2: & 9, & 13, & 14, & 25, & 17, & 15, & 36 \\
C_3: & 27, & 39, & 42, & 32, & 8, & 2, & 22 \\
C_4: & 38, & 31, & 40, & 10, & 24, & 6, & 23 \\
C_5: & 28, & 7, & 34, & 30, & 29, & 18, & 26
\end{array}
\tag{11.6.3}
$$

Here, if we write G in multiplicative form, the elements are $z = 1$, $b = b^1$, $b^2, \ldots, b^{42}, b^{43} = 1$, and the numbers in (11.6.3) are exponents of b. In the notation of (11.5.7) if there is a difference set $D = \{d_1, \ldots, d_{21}\}$, then $\theta(d^{11}) = \theta(d) = d_1 + \cdots + d_{21}$. Hence $\theta(d)$ is a sum of classes. Also, in the notation of (11.5.10), $\theta(d^{-1})$ is a sum of classes. Thus, from the basic equation (11.5.10),

$$\theta(d)\theta(d^{-1}) = n + \lambda T \tag{11.6.4}$$

involves multiplication of classes. Here, since the class z has only a single element and the classes $C_0, \ldots, C_5$ each have seven elements, it is clear that $\theta(d)$ is the sum of three of the six classes $C_0, \ldots, C_5$.

In general, suppose that G is Abelian and that $D = \{d_1, \ldots, d_k\}$, $\theta(d) = d_1 + \cdots + d_k$. If $\alpha_1 = 1, \alpha_2, \ldots, \alpha_s$ form a group M of multipliers fixing the difference set D, then $\theta(d^\alpha) = d_1^\alpha + \cdots + d_k^\alpha = \theta(d) = d_1 + \cdots + d_k$ for each $\alpha = \alpha_1, \ldots, \alpha_s$. In this case, $\theta(d)$ is a sum of classes of conjugate elements of $\bar{G}$ in $M(G)$. Similarly, $\theta(d^{-1})$ is the sum of the classes inverse to those of $\theta(d)$. Thus, if we know the table for multiplication of classes of $\bar{G}$ in $M(G)$, this is sufficient for the evaluation of $\theta(d)\theta(d^{-1})$ and is all that we need know to see whether or not the basic equation (11.6.4) is satisfied.

In any group G let $K_1, K_2, \ldots, K_r$ be the classes of elements, and in the group ring $G(Z)$ let us write

$$C_i = \sum x, \qquad x \in K_i, \qquad i = 1, \ldots, r. \tag{11.6.5}$$

Then it is true that there are nonnegative integers c_{ijk} with

$$C_iC_j = \sum_{1=k}^{r} c_{ijk}C_k, \qquad i, j = 1, \ldots, r, \tag{11.6.6}$$

for in C_iC_j in $G(Z)$, we need only verify that conjugate elements of G appear equally often. For the classes of (11.6.3) we have the multiplication rules given in (11.6.7):

$$zC_i = C_iz = C_i, \qquad i = 0, \ldots, 5$$

$$\begin{aligned}
C_0^2 &= && 2C_1 + 2C_2 + 3C_3 \\
C_0C_1 = C_1C_0 &= 2C_0 + C_1 + C_2 && + 2C_4 + C_5 \\
C_0C_2 = C_2C_0 &= C_0 + C_1 + 2C_2 && + C_4 + 2C_5 \\
C_0C_3 = C_3C_0 &= 7z + 2C_1 + C_2 && + 2C_4 + C_5 \\
C_0C_4 = C_4C_0 &= 2C_0 + C_2 + 2C_3 && + C_4 + C_5 \\
C_0C_5 = C_5C_0 &= C_0 + C_1 + 2C_3 && + C_4 + C_5
\end{aligned} \qquad (11.6.7)$$

In addition, since 3 is a primitive root modulo 43, the map $\beta\colon x \to 3x \pmod{43}$ is an automorphism of G, and indeed $\beta^{30} = \alpha$. Thus, $\{\alpha, G\} = M(G)$ is a subgroup of $\{\beta, G\} = H(G)$, and in fact $[H(G){:}M(G)] = 6$. Here, β induces an automorphism of order 6 on $M(G)$ in which $C_i \to C_{i+1}$, $i = 0, \ldots, 5 \pmod 6$. Thus, from (11.6.7) we may deduce the entire multiplication table for the C's. We may also describe C_i as containing those residues modulo 43 whose indices are congruent to $i \pmod 6$. Here, since $-1 \equiv 42$ belongs to C_3, we have C_i and C_{i+3} as inverse classes in every case. Since $\theta(d)$ is the sum of three C's, if no two C's are consecutive, we take $\theta(d) = C_0 + C_2 + C_4$, and then the difference set consists of the quadratic residues modulo 43. If two C's are consecutive, we take these two to be C_0, C_1. We find, with $T = z + C_0 + C_1 + C_2 + C_3 + C_4 + C_5$,

$$\begin{aligned}
\theta(d) &= C_0 + C_2 + C_4, \\
\theta(d)\theta(d^{-1}) &= 11 + 10T.
\end{aligned} \qquad (11.6.8)$$

We also find, and here the calculations are given in full in (11.6.9), that

$$\begin{aligned}
\theta(d) &= C_0 + C_1 + C_3 \\
\theta(d)\theta(d^{-1}) &= (C_0 + C_1 + C_3)(C_3 + C_4 + C_0) \\
C_0C_3 &= 7z + 2C_1 + C_2 + 2C_4 + C_5 \\
C_0C_4 &= 2C_0 + C_2 + 2C_3 + C_4 + C_5 \\
C_0^2 &= + 2C_1 + 2C_2 + 3C_3 \\
C_1C_3 &= 2C_0 + C_1 + C_2 + 2C_3 + C_5 \\
C_1C_4 &= 7z + C_0 + 2C_2 + C_3 + 2C_5 \\
C_1C_0 &= 2C_0 + C_1 + C_2 + 2C_4 + C_5 \\
C_3^2 &= 3C_0 + 2C_4 + 2C_5 \\
C_3C_4 &= + C_1 + C_2 + 2C_3 + C_4 + C_5 \\
C_3C_0 &= 7z + 2C_1 + C_2 + C_4 + C_5
\end{aligned} \qquad (11.6.9)$$

and

$$\begin{aligned}
\theta(d)\theta(d^{-1}) &= 21z + 10C_0 + 10C_1 + 10C_2 + 10C_3 + 10C_4 + 10C_5 \\
&= 11 + 10T.
\end{aligned}$$

Thus, for $v = 43$, $k = 21$, $\lambda = 10$, we have found that there are, up to equivalence, the two difference sets determined by (11.6.8) and (11.6.9), the first being of the type we have labeled Q, the second of type H_6. There are no others.

Most of the types listed above can be classified as residue difference sets. Here $v = p$ is a prime, G is the cyclic group of order p, and the group M of multipliers is the cyclic group of order f, where $p = ef + 1$, M consisting of the multiplicative group of eth power residues modulo p. This may be generalized so that G is the additive group of order $p^r = q$ in the field $\mathrm{GF}(p^r)$ and M is the multiplicative group of eth powers in $\mathrm{GF}(p^r)$, where $q = ef + 1$. As we observed previously we cannot have f even for v odd, since this would make -1 a multiplier. Hence, we need consider only cases with e even (except for the possibility $q = 2^r$).

Here, $M(G)$ is the group of mappings of $\mathrm{GF}(q)$ into itself:

$$x \to c^e x + d, \qquad c \neq 0, \qquad c, d \in GF(q), \tag{11.6.10}$$

where $q = 1 + ef$. G is the additive group:

$$A(b): x \to x + b, \tag{11.6.11}$$

and the group M of multipliers is

$$M(c^e): x \to c^e x, \qquad c \neq 0, \tag{11.6.12}$$

We easily verify that

$$M(c^e)^{-1}A(b)M(c^e) = A(c^e b). \tag{11.6.13}$$

There is a correspondence, determined by (11.6.11),

$$b \rightleftharpoons A(b): x \to x + b, \qquad b \in GF(q), \tag{11.6.14}$$

such that

$$A(b_1)A(b_2) = A(b_1 + b_2), \tag{11.6.15}$$

and when there is no ambiguity, we shall identify the elements b and $A(b)$. From (11.6.13) we see that b and $c^e b$ are conjugate in $M(G)$. Let g be a primitive root of $\mathrm{GF}(q)$. Then, from (11.6.13), g^u and g^v are conjugate in $M(G)$ if and only if there is a c such that $g^u = g^v c^e$, and if $c = g^j$, then $g^u = g^v g^{je} = g^{v+je}$ or, otherwise expressed, if and only if $u \equiv v (\mathrm{mod}\ e)$, since e is a divisor of $q - 1$, the order of g.

The conjugate classes of G in $M(G)$ are therefore the zero element, which we write as z, and the classes K_i of the elements g^u with $u \equiv i$ modulo e, for $i = 0, 1, \ldots, e - 1$. We write this at length as

$$z; \quad K_i: g^u, u \equiv i (\mathrm{mod}\ e), \qquad i = 0, \ldots, e - 1. \tag{11.6.16}$$

Now $M(G)$ is a normal subgroup of $L(G)$, the full group of linear transformations

$$x \to mx + b, \qquad m \neq 0, \qquad b \in GF(q), \tag{11.6.17}$$

In $L(G)$ the multiplication by the primitive root g is

$$\alpha(g): x \to gx. \tag{11.6.18}$$

Now we find that

$$\alpha(g)^{-1}A(g^u)\alpha(g) = A(g^{u+1}). \qquad (11.6.19)$$

Thus, $\alpha = \alpha(g)$ induces an automorphism in $M(G)$ such that

$$\alpha^{-1}K_i\alpha = K_{i+1}, \qquad i = 0, \ldots, e-1, \qquad K_e = K_0. \qquad (11.6.20)$$

We now turn our attentions to the group ring $M(Z)$ of $M(G)$ over the integers Z. Here, z is the identity of $M(G)$, which is of course the zero element of G if G is written additively. Then in $M(Z)$ let us represent the elements as $z, x_0, x_1, \ldots, x_{q-2}$, where x_i is given by

$$x_i = A(g^i). \qquad (11.6.21)$$

Let us also write

$$C_i = \sum x, \qquad x \in K_i, \qquad i = 0, \ldots, e-1. \qquad (11.6.22)$$

The class sums C_i and z form the basis for a subring of $M(Z)$ and we have

$$C_iC_j = a_{ij}z + \sum_{k=0}^{e-1} c_{ijk}C_k, \qquad i, j = 0, \ldots, e-1, \qquad (11.6.23)$$

where the a_{ij} and c_{ijk} are nonnegative integers. Here, a_{ij} is the number of solutions of

$$g^u + g^v = z = 0, \qquad u \equiv i, \qquad v \equiv j (\text{mod } e), \qquad (11.6.24)$$

and c_{ijk} is the number of solutions of

$$g^u + g^v = g^w, \quad u \equiv i, \quad v \equiv j(\text{mod } e)\, w \text{ fixed}, \quad w \equiv k (\text{mod } e). \quad (11.6.25)$$

The relations (11.6.24) and (11.6.25) are a consequence of the fact that in any finite group, the equation $ab = c$, $a \in K_i$, $b \in K_2$, $c \in K_k$, the K's being classes of conjugate elements, holds under conjugation by an arbitrary element x, whence $(x^{-1}ax)(x^{-1}bx) = x^{-1}cx$. Hence, for fixed c and $x^{-1}cx$, $ab = c$ and $a'b' = x^{-1}cx$ have the same number of solutions with $a, a' \in K_i, b, b' \in K_j$. The same is true, of course, if we fix b or s in (11.6.25).

Since $-1 = g^{(q-1)/2}$, we can easily find the value of a_{ij}, the number of solutions of (11.6.24). We observe that as $q - 1 = ef$, it follows that $-1 \in K_0$ if f is even and $-1 \in K_{e/2}$ if f is odd. Here,

$$g^v = -g^u = g^{u+g(q-1)/2}$$

in (11.6.24), and so $a_{i,i} = f$ if f is even:

$$\begin{aligned} a_{i,i+e/2} &= f, \qquad \text{if } f \text{ is odd}, \\ a_{i,j} &= 0, \qquad \text{otherwise}. \end{aligned} \qquad (11.6.26)$$

If we multiply (11.6.25) by g^{-u}, we have

$$1 + g^{v-u} = g^{w-u} \qquad (11.6.27)$$

or

$$1 + g^{v'} = g^{w'}, \qquad v' = j - i, \qquad w' = k - i (\text{mod } e). \qquad (11.6.28)$$

As each class K_i, $i = 0, \ldots, e-1$ has j elements, this yields the basic relation

$$c_{i,j,k} = c_{0,j-i,k-i}. \tag{11.6.29}$$

In the theory of cyclotomy the number of solutions of

$$1 + g^s = g^t, \qquad s \equiv i, \qquad t \equiv j (\text{mod } e) \tag{11.6.30}$$

is denoted by (i, j) and so we have

$$(i, j) = c_{0,i,j} = c_{s,i+s,j+s}. \tag{11.6.31}$$

Multiplying (11.6.30) by g^{-s} gives

$$g^{-s} + 1 = g^{t-s}, \qquad -s \equiv -i, \qquad t - s \equiv j - i (\text{mod } e), \tag{11.6.32}$$

and transposing,

$$g^s = -1 + g^t, \tag{11.6.33}$$

whence

$$g^{s+(q-1)/2} = 1 + g^{t+(q-1)/2}. \tag{11.6.34}$$

Here, (11.6.32) and (11.6.34) give the rules

$$\begin{aligned} &(i, j) = (-i, j - i), \\ &(j, i) = (i, j), \qquad f \text{ even}, \\ &(j, i) = \left(i + \frac{e}{2},\ j + \frac{e}{2}\right), \qquad f \text{ odd}. \end{aligned} \tag{11.6.35}$$

Counting elements on both sides of (11.6.23),

$$f^2 = a_{ij} + f \sum_{k=0}^{e-1} c_{ijk}. \tag{11.6.36}$$

We may rewrite (11.6.26) in the form

$$a_{ij} = n_{j-i} f, \tag{11.6.37}$$

for which

$$\begin{aligned} n_0 &= 1, \qquad f \text{ even}, \\ \frac{ne}{2} &= 1, \qquad f \text{ odd}, \\ n_i &= 0, \qquad \text{otherwise}. \end{aligned}$$

With this notation, (11.6.36) gives

$$\sum_{k=0}^{e-1} (j, k) = f - n_j, \qquad j = 0, \ldots, e - 1. \tag{11.6.38}$$

The relations (11.6.35) and (11.6.38) are sufficient to determine the (i, j) when $e = 2$.

Case $e = 2$, f odd: ■ We find

$$\begin{aligned}(0, 0) + (0, 1) &= f, \\ (1, 0) + (1, 1) &= f - 1, \\ (0, 0) = (1, 1) &= (1, 0).\end{aligned} \tag{11.6.39}$$

Hence,

$$(0, 0) = (1, 1) = (1, 0) = \frac{f-1}{2}, \qquad (0, 1) = \frac{f+1}{2}. \tag{11.6.40}$$

This yields the difference sets of type Q, for here, in the notation of (11.5.14), we have $\theta(d) = C_0$, the squares in GF(q)(mod p the quadratic residues) and $\theta(d^{-1}) = C_1$. But (11.6.39) and (11.6.40) give

$$\begin{aligned}\theta(d)\theta(d^{-1}) = C_0C_1 &= fz + \frac{f-1}{2}(C_0 + C_1) \\ &= \frac{f+1}{2}z + \frac{f-1}{2}T \\ &= (k - \lambda)z + \lambda T,\end{aligned} \tag{11.6.41}$$

where $v = q = 4t - 1$, $k = 2t - 1$, $\lambda = t - 1$.

Case $e = 2$, f even: ■ We find

$$\begin{aligned}(0, 0) + (0, 1) &= f - 1, \\ (1, 0) + (1, 1) &= f, \\ (1, 1) = (1, 0) &= (0, 1).\end{aligned} \tag{11.6.42}$$

Hence,

$$(1, 1) = (1, 0) = (0, 1) = \frac{f}{2}, \qquad (0, 0) = \frac{f}{2} - 1. \tag{11.6.43}$$

This case does not yield any difference sets.

Case $e = 3$: ■ Writing (i, j) in row i and column j, we have

$$\begin{matrix}(0, 0) & (0, 1) & (0, 2) \\ (0, 1) & (0, 2) & (1, 2). \\ (0, 2) & (1, 2) & (0, 1)\end{matrix} \tag{11.6.44}$$

This array expresses the equalities

$$\begin{aligned}a &= (0, 0), \\ b &= (0, 1) = (1, 0) = (2, 2), \\ c &= (0, 2) = (2, 0) = (1, 1), \\ d &= (1, 2) = (2, 1).\end{aligned} \tag{11.6.45}$$

When $e = 3$, $q = 3f + 1$, and so $(j, i) = (i, j)$ for f even. But if f is odd, then $q = 2^{2s}$; then -1 and $+1$ are the same and (11.6.35) still gives $(j, i) = (i, j)$. The array (11.6.44) expresses the multiplication table of the classes in G:

$$\begin{aligned} C_0^2 &= fz + aC_0 + bC_1 + cC_2, \\ C_0C_1 &= \qquad bC_0 + cC_1 + dC_2, \\ C_0C_2 &= \qquad cC_0 + dC_1 + bC_2. \end{aligned} \tag{11.6.46}$$

From (11.6.38) we obtain the linear relations

$$\begin{aligned} a + b + c &= f - 1, \\ b + c + d &= f. \end{aligned} \tag{11.6.47}$$

From (11.6.31) we note that all products C_iC_j are determined; for example, $C_1C_2 = dC_0 + bC_1 + cC_2$. From the associative law, $(C_0C_1)C_2 = C_0(C_1C_2)$, we find

$$\begin{aligned} dfz &+ (bc + cd + bd)C_0 + (bd + bc + cd)C_1 \\ &+ (b^2 + c^2 + ad)C_2 \\ &= dfz + (ad + b^2 + c^2)C_0 + (bd + bc + cd)C_1 \\ &\quad + (cd + bd + bc)C_2. \end{aligned} \tag{11.6.48}$$

On equating coefficients we have

$$ad + b^2 + c^2 = bc + cd + bd,$$

into which we put $d = a + 1$, $c = f - 1 - a - b$ from (11.6.46) to obtain

$$3a^2 + 3ab + 3b^2 - (3f - 5)a - (3f - 3)b = -f^2 + 3f - 2. \tag{11.6.49}$$

Multiplying this by 36 and combining gives

$$(9a - 3f + 7)^2 + 27(f - 1 - a - 2b)^2 = 12f + 4 = 4q. \tag{11.6.50}$$

Thus, the numbers of (11.6.44) are determined by a solution of the Diophantine equation

$$L^2 + 27M^2 = 4q, \tag{11.6.51}$$

where the sign of L is determined by $L \equiv 1 \pmod 3$. The sign of M will depend on the choice of primitive root. We find

$$\begin{aligned} 9a &= 9(0, 0) = q - 8 + L, \\ 18b &= 18(0, 1) = 2q - 4 - L - 9M, \\ 18c &= 18(0, 2) = 2q - 4 - L + 9M, \\ 9d &= 9(1, 2) = q + 1 + L. \end{aligned} \tag{11.6.52}$$

For a difference set here, we must have $k = (q - 1)/3$, and $a = b = c$ in the first equation (11.6.46). But this yields $M = 0$, $L = 4$, $q = 4 = v$ and, trivially, $k = 1$.

With $q = p$ a prime, it is known that (11.6.51) determines L and M uniquely apart from sign. If $q = p^r$, $p \equiv 2 \pmod 3$, then r must be even, $r = 2s$, and we must have $M = 0$ and $L = \pm 2q^s$ in (11.6.51). If $q = p^r$, $p \equiv 1 \pmod 3)$, $r > 1$, the solution of (11.6.51) is not unique. In this case it is not clear which solution of (11.6.51) should be used in (11.6.52), but a little calculation seems to indicate that if $4p = u^2 + 27v^2$ is the unique solution for p, then for $q = p^r$ the appropriate solution is given by

$$\frac{L + M\sqrt{-27}}{2} = \left(\frac{u + v\sqrt{-27}}{2}\right)^r.$$

This solution is characterized by the property that L and M are not multiples of p. This has recently been shown to be true by the author [2], relying on results of Mitchell [1].

The relation

$$C_0^2 = fz + aC_0 + bC_1 + cC_2 \tag{11.6.53}$$

gives the number of solutions of

$$x + y = t_i, \qquad x, y \in K_0, \qquad t_i \in K_i, \qquad t_i \text{ fixed}. \tag{11.6.54}$$

Here, if (x, y) is a solution, then (y, x) is a solution, and the solutions occur in pairs except when $y = x$ and $2x = t_i$, in which case $2 \in K_i$. Thus, $2 \in K_0$ if a is odd, $2 \in K_1$ if b is odd and $2 \in K_2$ if c is odd.

Similarly,

$$C_0^3 = afz + (a^2 + b^2 + c^2 + f)C_0 + (ab + bc + cd)C_1 + (ac + bd + bc)C_2 \tag{11.6.55}$$

gives the number of solutions of

$$x + y + z = t_i, \qquad x, y, z \in K_0, \qquad t_i \in K_i, \qquad t_i \text{ fixed}. \tag{11.6.56}$$

Permuting x, y, z in (11.6.56), we have six solutions if all three of x, y, z are different, three solutions if two are equal and the third different, and one solution if $x = y = z$, in which case $3x = t_i$ and $3 \in K_i$. Thus, the coefficient of C_i on the right-hand side of (11.6.55) is congruent to 1 modulo 3 if and only if $3 \in K_i$. We calculate, noting that $4q = L^2 + 27M^2$,

$$\begin{aligned} A_0 &= a^2 + b^2 + c^2 + f = \frac{q^2 + 3q + 15 - 4L}{27}. \\ A_1 &= ab + bc + cd \quad = \frac{2q^2 - 12q + 12 + L + 27M}{54}. \\ A_2 &= ac + bd + bc \quad = \frac{2q^2 - 12q + 12 + L - 27M}{54}. \end{aligned} \tag{11.6.57}$$

To find the A_i modulo 3, we must know the values of the numerators on the right modulo 81. Now $A_i \equiv 16A_i \pmod 3$. Here use $4q = L^2 + 27M^2$ and, as

$L \equiv 1 \pmod 3$, put $L = 1 + 3t$. Then modulo 3,

$$A_0 \equiv 16A_0 \equiv \frac{(L^2 + 27M^2) + 12(L^2 + 27M^2) + 240 - 64L}{27}$$
$$\equiv \frac{L^4 + 12L^2 - 64L + 240 + 54L^2M^2}{27}$$
$$\equiv \frac{27 - 108(t - t^3) - 27M^2}{27}$$
$$\equiv 1 - (t - t^3) - M^2 \pmod 3. \tag{11.6.58}$$

But $t - t^3 \equiv 0 \pmod 3$ for any integer t, and so

$$A_0 \equiv 1 - M^2 \pmod 3. \tag{11.6.59}$$

This gives $A_0 \equiv 1 \pmod 3$ if $M \equiv 0 \pmod 3$; otherwise, $A_0 \equiv 0 \pmod 3$. Thus, 3 is a cubic residue if and only if $M \equiv 0 \pmod 3$. Similarly, modulo 3,

$$A_1 \equiv 16A_1 \equiv \frac{L^4 - 24L^2 + 8L + 96 + 54L^2M^2 + 216M}{27}$$
$$\equiv -t + t^3 - M^2 - M$$
$$\equiv -M^2 - M \pmod 3. \tag{11.6.60}$$

Hence, if $M \equiv 0$ or $2 \pmod 3$ $A_1 \equiv 0 \pmod 3$, and if $M \equiv 1 \pmod 3$, $A_1 \equiv 1 \pmod 3$. Hence, $3 \in K_1$ if and only if $M \equiv 1 \pmod 3$. Similarly,

$$A_2 \equiv -M^2 + M \pmod 3,$$

and $3 \in K_2$ if and only if $M \equiv 2 \pmod 3$.

We combine these results on the cubic character of 2 and 3 in:

THEOREM 11.6.1. *Let $q = p^r \equiv 1 \pmod 3$, p a prime. Let g be a primitive root in* GF(q), $4q = L^2 + 27M^2$, *with* $L \equiv 1 \pmod 3$ *and* M *satisfying relations* (11.6.52). *Let* $g^m = 2$, $g^{m'} = 3$. *Then* $m \equiv 0 \pmod 3$ *if* $L \equiv 0 \pmod 2$. $m \equiv 1 \pmod 3$ *if* $L + 9M \equiv 0 \pmod 4$, $m \equiv 2 \pmod 3$ *if* $L - 9M \equiv 0 \pmod 4$. *Also,* $m' \equiv M \pmod 3$.

We used the associativity of multiplication in the group ring $G(Z)$ to obtain relations on the cyclotomic coefficients (i, j) in the equation (11.6.48). There is another, somewhat more arithmetical, approach based on the group ring $G(\zeta)$, where ζ is a primitive root of $x^{q-1} = 1$. In $G(\zeta)$ the group elements are z, the zero element, and $x_0, \ldots, x_{q-2}$, with $x_i = A(g^i)$ as in (11.6.21), and the coefficients are the ring of polynomials in ζ over the integers.

In $G(\zeta)$

$$x_t x_{t+j} = A(g^t + g^{t+j}) = A(g^t(1 + g^j)).$$

Here, if $j = (q-1)/2$, $1 + g^j = 1 - 1 = 0$ in GF(q), and so $x_t x_{t+j} = z$. If $j \neq (q-1)/2$, then $1 + g^j = g^s$ for an appropriate s, and so

$$x_t x_{t+j} = A(g^t g^s) = A(g^{t+s}) = x_{t+s}.$$

Writing $T = z + x_0 + x_1 + \cdots + x_{q-2}$, we have

$$\begin{aligned}\sum_{t=0}^{q-2} x_t x_{t+j} &= \sum_{t=0}^{q-2} z = (q-1)z, \qquad \text{if } j = \frac{q-1}{2}, \\ &= \sum_{t=0}^{q-2} x_{t+s} = T - z, \qquad \text{if } j \neq \frac{q-1}{2}.\end{aligned} \tag{11.6.61}$$

Let $\alpha \neq 1$ be a root of $x^{q-1} = 1$. We define the function $F(\alpha)$ by the rule

$$F(\alpha) = \sum_{k=0}^{q-2} \alpha^k x_k. \tag{11.6.62}$$

We now find

$$F(\alpha)F(\alpha^{-1}) = \sum_{k=0}^{q-2} \alpha^k x_k \sum_{t=0}^{q-2} \alpha^{-t} x_t, \tag{11.6.63}$$

and putting $k - t \equiv j$,

$$F(\alpha)F(\alpha^{-1}) = \sum_{j=0}^{q-2} \alpha^j \sum_{t=0}^{q-2} x_t x_{t+j}. \tag{11.6.64}$$

Using (11.6.61), this becomes

$$F(\alpha)F(\alpha^{-1}) = \alpha^{(q-1)/2}(q-1)z + \sum_{j \neq (q-1)/2} \alpha^j (T - z). \tag{11.6.65}$$

But since

$$\sum_{j=0}^{q-2} \alpha^j = 0$$

this simplifies, and we have

$$\begin{aligned}F(\alpha)F(\alpha^{-1}) &= \alpha^{(q-1)/2}(q-1)z - \alpha^{(q-1)/2}(T - z) \\ &= \alpha^{(q-1)/2}(qz - T).\end{aligned} \tag{11.6.66}$$

With $q = 1 + ef$, the classes of eth power residues are, by (11.6.22), such that

$$C_i = \sum_j x_j, \qquad j \equiv i (\text{mod } e). \tag{11.6.67}$$

Hence, if $\beta^e = 1$, as $\beta^r = \beta^s$ if $r \equiv 2 (\text{mod } e)$, then with C_i determined by (11.6.22),

$$\begin{aligned}F(\beta) &= \sum_{k=0}^{q-2} \beta^k x_k = \sum_{i=0}^{e-1} \beta^i \sum_{j \equiv i (\text{mod } e)} x_j \\ &= \sum_{i=0}^{e-1} \beta^i C_i.\end{aligned} \tag{11.6.68}$$

Now let β be a primitive eth root of unity and m, n be integers $m \neq 0 (\text{mod } e)$, $n \neq 0 (\text{mod } e)$. We also assume $m + n \neq 0 (\text{mod } e)$. Then

$$F(\beta^m)F(\beta^n) = \sum_{t=0}^{e-1} \sum_{j=0}^{e-1} \beta^{mj}\beta^{nt}C_t C_j \tag{11.6.69}$$

and, putting $t = j + k$,

$$F(\beta^m)F(\beta^n) = \sum_{k=0}^{e-1} \beta^{nk} \sum_{j=0}^{e-1} \beta^{j(m+n)} C_j C_{j+k}$$

$$= \sum_{k=0}^{e-1} \beta^{nk} \sum_{j=0}^{e-1} \beta^{(m+n)j} \sum_{h=0}^{e-1} [n_k fz + (k, h)C_{j+h}], \tag{11.6.70}$$

using the multiplication rules (11.6.23) and (11.6.31), and (11.6.37). As we have assumed that $m + n \not\equiv 0 (\text{mod } e)$, we have

$$\sum_{j=0}^{e-1} \beta^{(m+n)j} n_k fz = 0. \tag{11.6.71}$$

Hence, (11.6.70) becomes

$$F(\beta^m)F(\beta^n) = \sum_{k=0}^{e-1} \sum_{h=0}^{e-1} \beta^{nk}\beta^{-(m+n)h}(k, h) \sum_{j=0}^{e-1} \beta^{(m+n)(j+h)} C_{j+h}. \tag{11.6.72}$$

If we define the function $R(m, n)$—which is an algebraic integer in $K(\beta)$—by

$$R(m, n) = \sum_{k=0}^{e-1} \sum_{h=0}^{e-1} \beta^{nk}\beta^{-(m+n)h}(k, h), \tag{11.6.73}$$

then (11.6.72) becomes

$$F(\beta^m)F(\beta^n) = R(m, n)F(\beta^{m+n}). \tag{11.6.74}$$

The excluded value $m + n \not\equiv 0 (\text{mod } e)$ corresponds to $m = -n$, and for this we may evaluate $F(\beta^m)F(\beta^{-m})$ from (11.6.66).

Let us state the main results as

THEOREM 11.6.2. *Let G be the additive group $A(q)$ of* GF(q), $q = p^r$ *with z the zero element and other elements $x_0, \ldots, x_{q-2}$, where $x_i = A(g^i)$, g a primitive root in* GF(q). *If $\alpha \neq 1$ is a root of $x^{q-1} = 1$, let us put*

1. $$F(\alpha) = \sum_{k=0}^{q-2} \alpha^k x_k.$$

If we write $T = z + x_0 + x_1 + \cdots + x_{q-2}$, then

2. $$F(\alpha)F(\alpha^{-1}) = \alpha^{(q-1)/2}(qz - T).$$

If β is a primitive solution of $x^e = 1$ and $m \not\equiv 0 (\text{mod } e)$, where $q = 1 + ef$, then

3. $$F(\beta^m) = \sum_{i=0}^{e-1} \beta^{mi} C_i,$$

where C_i is the class of (11.6.22). *If we define* $R(m, n)$ *by*

4. $$R(m, n) = \sum_{k=0}^{e-1} \sum_{h=0}^{e-1} \beta^{nk} \beta^{-(m+n)h}(k, h),$$

when $m, n, m + n \not\equiv 0 (\text{mod } e)$, *then*

5. $$F(\beta^m)F(\beta^n) = R(m, n)F(\beta^{m+n}).$$

Also we have

6. $$R(m, n)R(-m, -n) = q.$$

Proof: ∎ All parts of the theorem except property 6 have already been proved. In property 5, replace m and n by $-m$ and $-n$. This gives

$$F(\beta^{-m})F(\beta^{-n}) = R(-m, -n)F(\beta^{-m-n}) \tag{11.6.75}$$

Multiplying this by property 5, and noting that the ring $G(\zeta)$ is commutative, we find

$$F(\beta^m)F(\beta^{-m})F(\beta^n)F(\beta^{-n}) = R(m, n)R(-m, -n)F(\beta^{m+n})F(\beta^{-m-n}). \tag{11.6.76}$$

Now apply property 2 with $\alpha = \beta^m$, β^n, and β^{m+n}. This yields

$$\beta^{m(q-1)/2}(qz - T)\beta^{n(q-1)/2}(qz - T)$$
$$= R(m, n)R(-m, -n)\beta^{(m+n)(q-1)/2}(qz - T). \tag{11.6.77}$$

Here, as

$$\begin{aligned}(qz - T)^2 &= q^2z^2 - 2qzT + T^2 \\ &= q^2z - 2qT + qT \\ &= q(qz - T),\end{aligned} \tag{11.6.78}$$

from (11.6.77), and comparing coefficients, we find

$$R(m, n)R(-m, -n) = q. \tag{11.6.79}$$

THEOREM 11.6.3. *If* $q = p^r$ *is odd and if* m *is determined by* $g^m = 2$, *then with* $\alpha \neq 1$, $\alpha^{q-1} = 1$,

$$F(-1)F(\alpha^2) = \alpha^{2m}F(\alpha)F(-\alpha).$$

Proof: ∎ Following Dickson [1], we show that α^{2m+i} has, for every $i = 0, \ldots$ $q - 2$, the same coefficient on the left and the right of the equation

$$F(-1)F(\alpha^2) = \alpha^{2m}F(\alpha)F(-\alpha). \tag{11.6.80}$$

On the right-hand side the coefficient of α^{2m+i} is

$$\sum_j (-1)^j x_j x_{i-j}. \tag{11.6.81}$$

Here, if i is odd, the terms with $j = J$ and $j = i - J$ yield $(-1)^J x_J x_{i-J} + (-1)^{i-J} x_{i-J} x_J = 0$, and the sum is zero. Thus, if i is odd, the coefficient of α^{2m+i} on the right is zero. Clearly, the left-hand side yields only even powers of α. Now suppose i even, $i = 2t$. In $F(\alpha^2)$, both x_{m+t} and $x_{m+t+(q-1)/2}$ have α^{2m+t} as their coefficient, and so the coefficient of $\alpha^{2m+i} = \alpha^{2m+2t}$ is

$$F(-1)(x_{m+t} + x_{m+t+(q-1)/2}). \tag{11.6.82}$$

But here

$$x_{m+t} = A(g^{m+t}) = A(2g^t)$$

$$x_{m+t+(q-1)/2} = A(-2g^t).$$

Hence,

$$F(-1) = \sum_{k=0}^{q-2}(-1)^k x_k,$$

and (11.6.82) becomes

$$\sum_{k=0}^{q-2}(-1)^k(A(g^k + 2g^t) + A(g^k - 2g^t)), \tag{11.6.83}$$

and for $i = 2t$, (11.6.81) becomes

$$\sum_{j=0}^{q-2}(-1)^j A(g^j + g^{2t-j}). \tag{11.6.84}$$

We now show that for any c, $A(c)$ occurs with the same coefficient in (11.6.83) and (11.6.84). If $c = +2g^t$, we have $g^j = g^t$ in (11.6.84), $g^k = 4g^t$ in the second sum of (11.6.83), and no g^k in the first sum. Here, k and j have the same parity. If $c = -2g^t$, we have $g^j = -g^t$ in (11.6.84), $g^k = -4g^t$ in the first sum of (11.6.83), and j and k have the same parity. If $c \neq \pm 2g^t$ is a value with $c = g^J + g^{2t-J}$, then there are two terms in (11.6.84) with $j = J$ and $j = 2t - J$, both of the same parity, and we note that

$$c^2 - 4g^{2t} = (g^J - g^{2t-J})^2 = w^2$$

is a nonzero square. Here, $g^k = g^{-J}(g^J - g^t)^2$ gives c in the first term of (11.6.83) and $g^k = g^{-J}(g^J + g^t)^2$ gives c in the second sum, both k's with the same parity as J. If c is not a value taken on in (11.6.84), then $c^2 - 4g^{2t}$ is not a square, and then for $g^{k_1} + 2g^t = c$ and $g^{k_2} - 2g^t = c$ in the first and second sums of (11.6.83), we have

$$g^{k_1+k_2} = c^2 - 4g^{2t},$$

and so k_1 and k_2 are of opposite parity and the two terms cancel. Thus, for every power of α, the left- and right-hand sides of (11.6.80) have the same coefficient, and the identity is proved.

Case $e = 4$: ■ Here $\beta = i$ and

$$R(1,1) = (0,0) - (0,1) + (0,2) - (0,3) - (2,0) + (2,1) - (2,2) + (2,3) + 2i((1,0) - (1,1) + (1,2) - (1,3)). \tag{11.6.85}$$

Subcase $e = 4, f$ *even:* ■ The table for (i, j) is, with (i, j) in the ith row and jth column, $i, j = 0, 1, 2, 3$.

$$\begin{matrix} (0,0) & (0,1) & (0,2) & (0,3) \\ (0,1) & (0,3) & (1,2) & (1,2) \\ (0,2) & (1,2) & (0,2) & (1,2) \\ (0,3) & (1,2) & (1,2) & (0,1) \end{matrix} \qquad (11.6.86)$$

The linear relations (11.6.38) become

$$\begin{aligned} (0,0) + (0,1) + (0,2) + (0,3) &= f - 1, \\ (0,1) + (0,3) + 2(1,2) &= f, \\ (0,2) + (1,2) &= \frac{f}{2}. \end{aligned} \qquad (11.6.87)$$

and we find, substituting in (11.6.85),

$$\begin{aligned} R(1,1) &= -x + 2iy, \\ x &= 2f + 1 - 8(1,2), \\ y &= (0,1) - (0,3). \end{aligned} \qquad (11.6.88)$$

Thus, from (11.6.79),

$$q = x^2 + 4y^2, \qquad (11.6.89)$$

with the sign of x chosen so that $x \equiv 1 \pmod 4$. This, with (11.6.81), yields

$$\begin{aligned} 16(0,0) &= q - 11 - 6x, \\ 16(0,1) &= q - 3 + 2x + 8y, \\ 16(0,2) &= q - 3 + 2x, \\ 16(0,3) &= q - 3 + 2x - 8y, \\ 16(1,2) &= q + 1 - 2x. \end{aligned} \qquad (11.6.90)$$

Subcase $e = 4, f$ *odd:* ■ The table reads

$$\begin{matrix} (0,0) & (0,1) & (0,2) & (0,3) \\ (1,0) & (1,0) & (0,3) & (0,1) \\ (0,0) & (1,0) & (0,0) & (1,0) \\ (1,0) & (0,3) & (0,1) & (1,0) \end{matrix} \qquad (11.6.91)$$

and the linear relations are

$$\begin{aligned} (0,0) + (0,1) + (0,2) + (0,3) &= f \\ 2(1,0) + (0,3) + (0,1) &= f \\ (0,0) + (1,0) &= \frac{f-1}{2}. \end{aligned} \qquad (11.6.92)$$

Here,

$$R(1, 1) = -x + 2iy,$$
$$x = 2f - 1 - 8(1, 0), \tag{11.6.93}$$
$$y = (0, 3) - (0, 1),$$

and, by (11.6.79),

$$q = x^2 + 4y^2, \tag{11.6.94}$$

with the sign of x chosen so that $x \equiv 1 (\text{mod } 4)$, and we find

$$\begin{aligned} 16(0, 0) &= q - 7 + 2x, \\ 16(0, 1) &= q + 1 + 2x - 8y, \\ 16(0, 2) &= q + 1 - 6x, \\ 16(0, 3) &= q + 1 + 2x + 8y, \\ 16(1, 0) &= q - 3 - 2x. \end{aligned} \tag{11.6.95}$$

General e: ■ If

$$C_0^2 = n_0 fz + (0, 0)C_0 + (0, 1)C_1 + \cdots + (0, e - 1)C_{e-1}, \tag{11.6.96}$$

the number $(0, 0)$ gives the number of solutions of $u + v = 1$ in GF(q), with $u, v \in K_0$, and this number is even unless there is a solution $2u = 1$, in which case $2 \in K_0$. We observed this when proving Theorem 11.6.1.

From (11.6.40) and (11.6.43) we conclude that 2 is a quadratic residue in GF(q) if $q = 1, 7 (\text{mod } 8)$, and 2 is a quadratic nonresidue if $q \equiv 3, 5 (\text{mod } 8)$. We now ask for which $q \equiv 1 (\text{mod } 4)$ is 2 a quartic residue. If $q \equiv 5 (\text{mod } 8)$, then 2, being a quadratic nonresidue, is certainly not a quartic residue. But if $q \equiv 1 (\text{mod } 8)$, the values in (11.6.90) apply. Thus, 2 is a quartic residue of $q \equiv 1 (\text{mod } 8)$ if and only if

$$q - 11 - 6x \equiv 16 (\text{mod } 32), \qquad x \equiv 1 (\text{mod } 4), \tag{11.6.97}$$

Putting $x = 1 + 4t$ and $q = x^2 + 4y^2$, (11.6.97) becomes

$$1 + 8t + 16t^2 + 4y^2 - 11 - 6 - 24t \equiv 16 (\text{mod } 32), \tag{11.6.98}$$

and as $t + t^2 \equiv 0 (\text{mod } 2)$ for any t, this becomes

$$4y^2 \equiv 0 (\text{mod } 32) \tag{11.6.99}$$

or

$$y \equiv 0 (\text{mod } 4) \tag{11.6.100}$$

as the necessary and sufficient condition that 2 be a quartic residue of $q \equiv 1 (\text{mod } 8)$. We state our results as

THEOREM 11.6.4. *Let $q = p^r = 1 + 4f$, and suppose $q = x^2 + 4y^2$, with the sign of x chosen so that $x \equiv 1 (\text{mod } 4)$. Then for f even, the cyclotomic numbers (i, j) are given by* (11.6.90) *and* (11.6.86); *for f odd, by* (11.6.91) *and* (11.6.95).

Now 2 is a quadratic residue of $q = p^r$ *in general if* $q \equiv 1, 7 \pmod 8$ *and is a quadratic nonresidue if* $q \equiv 3, 5 \pmod 8$. *If* $q \equiv 1 \pmod 8$, *then 2 is a quartic residue if and only if* $y \equiv 0 \pmod 4$ *in* $q = x^2 + 4y^2$.

Remark: ■ If q is a prime, the representation $q = x^2 + 4y^2$ is known to be unique. If $q = p^r, p \equiv 3 \pmod 4$, then r is even, $r = 2s$, and the unique solution is $x = \pm p^s$, $y = 0$. If $q = p^r$, $r > 1$, $p \equiv 1 \pmod 4$, the representation is not unique, and so the values in (11.6.90) and (11.6.95) are not completely determined by the theorem.

To obtain difference sets with the quartic residues as multipliers, D is made up of one or more of the classes z, k_0, k_1, k_2, k_3. As -1 may not be a multiplier, we must have f odd.

THEOREM 11.6.5. *If* $q = p^r \equiv 5 \pmod 8$, *the quartic residues in* GF(q) *form a difference set if q is of the form* $q = 1 + 4y^2$; *the quartic residues and zero form a difference set if q is of the form* $q = 9 + 4y^2$. *These are the types B and* B_0 *of the families of difference sets listed at the beginning of this section. No additional combinations of quartic residues essentially different from these form a difference set.*

Proof: ■ We have f odd and use the values in (11.6.91) and (11.6.95). If a single class is included in D, it may be taken as K_0. Thus, with $\theta(d) = C_0$, $\theta(d^{-1}) = C_2$ and (11.6.91) gives

$$\begin{aligned} \theta(d)\theta(d^{-1}) &= C_0C_2 = fz + aC_0 + bC_1 + aC_2 + bC_3, \\ 16a &= 16(0, 0) = q - 7 + 2x, \\ 16b &= 16(1, 0) = q - 3 - 2x. \end{aligned} \tag{11.6.101}$$

To give a difference set we must have $\theta(d)\theta(d)^{-1} = (k - \lambda)z + \lambda T$, whence this holds in (11.6.91) if and only if $a = b$, whence if and only if $x = 1$. In this case, q is of the form $q = 1 + 4y^2$. If we have a single class and zero, we may take $\theta(d) = z + C_0$. In this case

$$\begin{aligned} \theta(d)\theta(d^{-1}) &= (C_0 + z)(C_2 + z) \\ &= (f + 1)z + (a + 1)C_0 + bC_1 \\ &\qquad + (a + 1)C_2 + bC_3. \end{aligned} \tag{11.6.102}$$

Here we have a difference set if and only if $a + 1 = b$, or $x = -3$, and in this case, $q = 9 + 4y^2$. For $q = p = 1 + 4y^2$, we have type B, and for $q = p = 9 + 4y^2$, we have B_0. The condition that y be odd is necessary in order that f be odd. The types B and B_0 are listed for primes only. For prime powers $q = p^r$, we must have r odd and $p \equiv 1 \pmod 4$. It can be shown that $p^r = 1 + 4y^2$ and $p^r = 9 + 4y^2$ do not have solutions with r odd and greater than 1.

THEOREM 11.6.6. *If no one of m, n, $m + n$ is divisible by e, then*

$$R(m, n) = R(n, m) = (-1)^{nf} R(-m - n, n).$$

Proof: ■ From Theorem 11.6.2, if no one of m, n, $m + n$ is divisible by e, then

$$F(\beta^m)F(\beta^n) = R(m, n)F(\beta^{m+n}). \tag{11.6.103}$$

From this it is immediate that $R(n, m) = R(m, n)$. Multiply by $F(\beta^{-m-n})$ and apply (11.6.66):

$$F(\beta^m)F(\beta^{-m-n})F(\beta^n) = R(m, n)\beta^{(m+n)(q-1)/2}(qz - T). \tag{11.6.104}$$

Now on the left use (11.6.103) with m replaced by $-m - n$ and we have

$$F(\beta^m)R(-m - n, n)F(\beta^{-m}) = R(m, n)\beta^{(m+n)(q-1)/2}(qz - T). \tag{11.6.105}$$

In this apply (11.6.66) and we have

$$\beta^{m(q-1)/2}(qz - T)R(-m - n, n) = R(m, n)\beta^{(m+n)(q-1)/2}(qz - T). \tag{11.6.106}$$

Comparing coefficients, we have

$$R(m, n) = \beta^{-n(q-1)/2}R(-m - n, n). \tag{11.6.107}$$

As β is a primitive eth root of unity and $q - 1 = ef$, $\beta^{(q-1)/2} = (-1)^f$, since, if e is even, $\beta^{e/2} = -1$, but if e is odd, f is even and $\beta^{(q-1)/2} = 1 = (-1)^f$. This establishes the relation of the theorem.

Case $e = 6$: ■ For $e = 6$, the cubic character of 2 is relevant and the relation $F(-1)F(\alpha^2) = \alpha^{2m}F(\alpha)F(-\alpha)$, where $g^m = 2$ in GF(q), established in Theorem 11.6.3, shows how this affects the values of the $R(m, n)$. For $e = 6$, we take $\beta = (1 + \sqrt{-2})/2$.

In general, replacing β by β^j, where j is prime to e, replaces $R(m, n)$ by a conjugate, which is $R(jm, jn)$. The equation (11.6.79) yields the same relation on the (h, k) for $R(m, n)$ and a conjugate. For this reason Dickson introduces the term *reduced* for a minimal set of $R(m, n)$ such that all others may be derived from these by the relations of Theorem 11.6.6 and conjugacy. For $e = 6$, a set of reduced $R(m, n)$ consists of $R(1, 1)$, $R(1, 2)$ and $R(2, 2)$.

Taking $\alpha = \beta$, as $\beta^3 = -1$, (11.6.80) gives

$$F(\beta^3)F(\beta^2) = \beta^{2m}F(\beta)F(\beta^4) \tag{11.6.108}$$

or

$$R(3, 2)F(\beta^5) = \beta^{2m}R(1, 4)F(\beta^5), \tag{11.6.109}$$

whence

$$R(3, 2) = \beta^{2m}R(1, 4). \tag{11.6.110}$$

Let us evaluate

$$[F(\beta^2)F(\beta^2)]F(\beta) = F(\beta^2)[F(\beta^2)F(\beta)],$$
$$R(2, 2)F(\beta^4)F(\beta) = F(\beta^2)R(2, 1)F(\beta^3), \tag{11.6.111}$$
$$R(2, 2)R(4, 1)F(\beta^5) = R(2, 1)R(2, 3)F(\beta^5).$$

Hence,

$$R(2, 2)R(4, 1) = R(2, 1)R(2, 3). \tag{11.6.112}$$

From Theorem 11.6.6,

$$\begin{aligned} R(2, 3) &= (-1)^f R(1, 2), \qquad R(2, 1) = R(1, 2), \\ R(4, 1) &= R(1, 4), \qquad R(1, 1) = (-1)^f R(4, 1). \end{aligned} \tag{11.6.113}$$

Using (11.6.113), equation (11.6.110) becomes

$$(-1)^f R(1, 2) = \beta^{2m}(-1)^f R(1, 1) \tag{11.6.114}$$

and (11.6.112) becomes

$$R(2, 2)(-1)^f R(1, 1) = (-1)^f R(1, 2)^2. \tag{11.6.115}$$

Combining these we obtain

$$\begin{aligned} R(1, 1) &= (-1)^f \beta^{2m} R(2, 2), \\ R(1, 2) &= \beta^{4m} R(2, 2). \end{aligned} \tag{11.6.116}$$

Dickson erroneously states that the second of these relations may be derived by putting $\alpha = \beta$ in (11.8.80).

Now $R(2, 2)$ was originally defined by (11.6.74) from

$$F(\beta^2)F(\beta^2) = R(2, 2)F(\beta^4). \tag{11.6.117}$$

But with $\beta^2 = w$, w is a primitive cube root of unity, whence writing $R(2, 2) = R_6(2, 2)$ to indicate that $e = 6$, and by replacing β^2 by w in (11.6.107), we have

$$R_6(2, 2) = R_3(1, 1). \tag{11.6.118}$$

But with a, b, c, d as in (11.6.46),

$$R_3(1, 1) = (a + 2d) + 3bw + 3cw^2. \tag{11.6.119}$$

As $w = (-1 + \sqrt{-3})/2$,

$$\begin{aligned} R_6(2, 2) = R_3(1, 1) &= \frac{(2a + 4d - 3b - 3c) + 3\sqrt{-3}(b - c)}{2} \\ &= \frac{(L - 3\sqrt{-3}\, M)}{2}, \end{aligned} \tag{11.6.120}$$

where L and M are the same values as those appearing in (11.6.51) and (11.6.52).

We now have sufficient information to determine the cyclotomic numbers for $e = 6$. For difference sets we need only the values with f odd.

Subcase $e = 6, f$ *odd:* ■ The table of the (i, j) reads

$$\begin{matrix} (0,0) & (0,1) & (0,2) & (0,3) & (0,4) & (0,5) \\ (1,0) & (2,0) & (1,2) & (0,4) & (0,2) & (1,2) \\ (2,0) & (2,1) & (1,0) & (0,5) & (1,2) & (0,1) \\ (0,0) & (1,0) & (2,0) & (0,0) & (1,0) & (2,0) \\ (1,0) & (0,5) & (1,2) & (0,1) & (2,0) & (2,1) \\ (2,0) & (1,2) & (0,4) & (0,2) & (1,2) & (1,0) \end{matrix} \tag{11.6.121}$$

If we write

$$\begin{aligned} R(2,2) &= \frac{L - 3M\sqrt{-3}}{2}, \\ R(1,1) &= \frac{E + F\sqrt{-3}}{2}, \\ R(1,2) &= -A + B\sqrt{-3}, \end{aligned} \tag{11.6.122}$$

we have the linear relations

$$\begin{aligned} (0,0) + (0,1) + (0,2) + (0,3) + (0,4) + (0,5) &= f, \\ (1,0) + (2,0) + (1,2) + (0,4) + (0,2) + (1,2) &= f, \\ (2,0) + (2,1) + (1,0) + (0,5) + (1,2) + (0,1) &= f, \\ (0,0) + (1,0) + (2,0) &= \frac{f-1}{2}, \\ (1,0) - (2,0) + (0,2) - (0,4) &= B, \\ (0,1) + (0,4) - (0,2) - (0,5) + 2(1,0) - 2(2,0) &= -M, \\ (0,4) - (0,2) + 3(0,5) - 3(0,1) + 2(1,0) - 2(2,0) &= F, \\ (0,0) + (0,2) - (0,3) + (0,4) - (1,0) - (2,0) + (1,2) - (2,1) &= -A, \\ 6(0,0) - 3(0,1) - 3(0,2) + 2(0,3) - 3(0,4) - 3(0,5) & \\ - 6(1,0) - 6(2,0) + 12(1,2) + 4(2,1) &= L, \\ 2(0,0) + 3(0,1) - (0,2) - 2(0,3) - (0,4) + 3(0,5) & \\ - 2(1,0) - 2(2,0) + 4(1,2) - 4(2,1) &= -E. \end{aligned} \tag{11.6.123}$$

The first four of these arise from (11.6.38), and the remainder from the definition of $R(m, n)$ in (11.6.73), using the forms of (11.6.122). The solution depends on the cubic character of 2, which determines the relations (11.6.116). There are three cases, depending on the value of m modulo 3:

$$q = A^2 + 3B^2, \quad 4q = L^2 + 27M^2 = E^2 + 3F^2, \quad g^m = 2. \qquad (11.6.124)$$

	Two-cubic residue $L = -2A$, $-3M = 2B$, $E = 2A$, $F = -2B$	$m \equiv 1 \pmod 3$ $L = A - 3B$, $3M = A + B$, $E = -A - 3B$, $F = -A + B$	$m \equiv 2 \pmod 3$ $L = A + 3B$, $3M = -A + B$, $E = -A + 3B$, $F = A + B$
36(0, 0)	$q - 11 - 8A$	$q - 11 - 2A$	$q - 11 - 2A$
36(0, 1)	$q + 1 - 2A + 12B$	$q + 1 + 4A$	$q + 1 - 2A - 12B$
36(0, 2)	$q + 1 - 2A + 12B$	$q + 1 - 2A + 12B$	$q + 1 - 8A + 12B$
36(0, 3)	$q + 1 + 16A$	$q + 1 + 10A - 12B$	$q + 1 + 10A + 12B$
36(0, 4)	$q + 1 - 2A - 12B$	$q + 1 - 8A - 12B$	$q + 1 - 2A - 12B$
36(0, 5)	$q + 1 - 2A - 12B$	$q + 1 - 2A + 12B$	$q + 1 + 4A$
36(1, 0)	$q - 5 + 4A + 6B$	$q - 5 - 2A + 6B$	$q - 5 + 4A + 6B$
36(2, 0)	$q - 5 + 4A - 6B$	$q - 5 + 4A - 6B$	$q - 5 - 2A - 6B$
36(1, 2)	$q + 1 - 2A$	$q + 1 + 4A$	$q + 1 + 4A$
36(2, 1)	$q + 1 - 2A$	$q + 1 - 8A - 12B$	$q + 1 - 8A + 12B$

THEOREM 11.6.7. *For $q = p^r \equiv 7 \pmod{12}$, the quadratic residues form a difference set, with $v = q = 4t - 1$, $k = 2t - 1$, $\lambda = t - 1$. If $q = A^2 + 27$, then also the cubic residues and the sextic class, including the residue 3, form a difference set with the same parameters. This is the type H_6. A difference set that has the sextic residues as multipliers is equivalent to one of these types.*

Proof: ■ If D is a difference set in GF(q) that has the sextic residues as multipliers, then $\theta(d)$ is the sum of one or more of the C_i and possibly z. If $\theta(d)$ has only one class, we may take this to be C_0; if $\theta(d) = C_0$, then $\theta(d^{-1}) = C_3$, and

$$\theta(d)\theta(d^{-1}) = f + (0,0)C_0 + (1,0)C_1 + (2,0)C_2 + (0,0)C_3 + (1,0)C_4 + (2,0)C_5, \qquad (11.6.125)$$

with $(0,0) = (1,0) = (2,0)$ if $\theta(d)$ is to be a difference set. But this requires $B = 0$, $12A = -6$ for 2 a cubic residue, and this is impossible. For $m \equiv 1 \pmod 3$, this requires $A = -2$, $B = -1$ whence $q = 7$, and the "difference set" consists of a single residue. Similarly for $m \equiv 2 \pmod 3$, we find $A = -2$, $B = 1$, and $q = 7$; $\theta(d) = z + C_0$ leads to $(0,0) + 1 = (1,0) = (2,0)$. For 2 a cubic residue, this gives $B = 0$, $12A = 30$, which is impossible. For $m \equiv \pm 1 \pmod 3$, we find $A = 10$, $B = \pm 5$, and $q = 175$, again impossible.

For $\theta(d) = C_0 + C_1 + C_3$, we find

$$\theta(d)\theta(d^{-1}) = 3f + RC_0 + SC_1 + TC_2 + RC_3 + SC_4 + TC_5, \qquad (11.6.126)$$

where, according to the value of m modulo 3, we find

	$m \equiv 0$	$m \equiv 1$	$m \equiv 2$
$36R =$	$9q - 45 + 6B,$	$9q - 45 - 18B,$	$9q - 45 + 6A - 6B,$
$36S =$	$9q - 27,$	$9q - 27 - 6A + 12B,$	$9q - 27 - 6A - 12B,$
$36T =$	$9q - 9 - 6B,$	$9q - 9 + 6A + 6B,$	$9q - 9 + 18B.$

(11.6.127)

For this to be a difference set, we must have $R = S = T$. When 2 is a cubic residue and $m \equiv 0 \pmod 3$, this gives $B = 3$. For $m \equiv 1 \pmod 3$, we find $A = -2, B = -1$, and so $q = 7$. For $m \equiv 2$(mod), we find $A = 2, B = -1$, and so $q = 7$. With $q = 7$, the residues 1, 5, 6 form a difference set equivalent to the quadratic residues 1, 2, 4. In the first case, with $B = 3$, we have $q = A^2 + 27$. This leads to the type H_6. Here, $4q = (2A)^2 + 27(2^2)$, and from this representation Theorem 11.6.1 tells us that 2 is certainly a cubic residue if $q = p$ is prime, and also if this is the appropriate representation for g a prime power. As $-3M = 2B$ for 2 a cubic residue, $B = 3$ gives $M = -2$, and by Theorem 11.6.1, the index of the residue 3 is congruent to 1 modulo 3. Also, for $q \equiv 7 \pmod{12}$, 3 is a quadratic nonresidue. Hence, the residue 3 is in the sextic class K_1, for the choice of primitive root yielding the values in (11.6.113). Thus, our difference set D consists of the cubic residues and the sextic class, including the residue 3. Of course the quadratic residues give the difference set of type Q. No other combinations of sextic residues give additional difference sets, though of course, with a change of primitive root and $B = -3$, we obtain type H_6 again with $\theta(d) = C_0 + C_3 + C_5$.

Case $e = 8, f$ *odd:* ■ Here we have

$$q = x^2 + 4y^2 = a^2 + 2b^2, \qquad x \equiv a \equiv 1 \pmod 4. \tag{11.6.128}$$

The table reads

(0, 0)	(0, 1)	(0, 2)	(0, 3)	(0, 4)	(0, 5)	(0, 6)	(0, 7)
(1, 0)	(1, 1)	(1, 2)	(1, 3)	(0, 5)	(1, 3)	(0, 3)	(1, 7)
(2, 0)	(2, 1)	(2, 0)	(1, 7)	(0, 6)	(1, 3)	(0, 2)	(1, 2)
(1, 1)	(2, 1)	(2, 1)	(1, 0)	(0, 7)	(1, 7)	(1, 2)	(0, 1)
(0, 0)	(1, 0)	(2, 0)	(1, 1)	(0, 0)	(1, 0)	(2, 0)	(1, 1)'
(1, 0)	(0, 7)	(1, 7)	(1, 2)	(0, 1)	(1, 1)	(2, 1)	(2, 1)
(2, 0)	(1, 7)	(0, 6)	(1, 3)	(0, 2)	(1, 2)	(2, 0)	(2, 1)
(1, 1)	(1, 2)	(1, 3)	(0, 5)	(0, 3)	(0, 3)	(1, 3)	(1, 0)

where the following two sets of values apply:

	If 2 is a quartic residue	If 2 is not a quartic residue
64(0, 0)	$q - 15 - 2x$	$q - 15 - 10x - 8a$
64(0, 1)	$q + 1 + 2x - 4a + 16y$	$q + 1 + 2x - 4a - 16b$
64(0, 2)	$q + 1 + 6x + 8a - 16y$	$q + 1 - 2x + 16y$
64(0, 3)	$q + 1 + 2x - 4a - 16y$	$q + 1 + 2x - 4a - 16b$
64(0, 4)	$q + 1 - 18x$	$q + 1 + 6x + 24a$
64(0, 5)	$q + 1 + 2x - 4a + 16y$	$q + 1 + 2x - 4a + 16b$
64(0, 6)	$q + 1 + 6x + 8a + 16y$	$q + 1 - 2x - 16y$
64(0, 7)	$q + 1 + 2x - 4a - 16y$	$q + 1 + 2x - 4a + 16b$
64(1, 0)	$q - 7 + 2x + 4a$	$q - 7 + 2x + 4a + 16y$
64(1, 1)	$q - 7 + 2x + 4a$	$q - 7 + 2x + 4a - 16y$
64(1, 2)	$q + 1 - 6x + 4a + 16b$	$q + 1 + 2x - 4a$
64(1, 3)	$q + 1 + 2x - 4a$	$q + 1 - 6x + 4a$
64(1, 7)	$q + 1 - 6x + 4a - 16b$	$q + 1 + 2x - 4a$
64(2, 0)	$q - 7 - 2x - 8a$	$q - 7 + 6x$
64(2, 1)	$q + 1 + 2x - 4a$	$q + 1 - 6x + 4a$

With these values we find that $\theta(d) = C_0$ yields a difference set if $x = 1$, $a = -3$, yielding the difference sets of type O. Also $\theta(d) = C_0 + z$ yields a difference set if $x = +21$, $a = -7$, the type O_0.

We now proceed to difference sets of type T. Let p^r and q^s be prime powers such that $p^r + 2 = q^s$. (This includes $5^2 + 2 = 3^3$ but the writer knows of no other case in which both $r > 1$ and $s > 1$.) Consider the system (a, b) of ordered pairs, $a \in GF(p^r)$, $b \in GF(q^s)$, combined according to the rules

$$(a, b) + (c, d) = (a + c, b + d) \qquad \text{and} \qquad (a, b)(c, d) = (ac, bd).$$

This is called the *direct sum* of $GF(p^r)$ and $GF(q^s)$, and is written

$$S = GF(p^r) \oplus GF(q^s).$$

If p and q are primes, $GF(p) \oplus GF(q)$ is the system of residues modulo pq, where $e \equiv (1, 0)$ and $f \equiv (0, 1)$ are residues modulo pq such that $e \equiv 1 (\text{mod } p)$, $e \equiv 0 (\text{mod } q)$, and $f \equiv 0 (\text{mod } p)$, $f \equiv 1 (\text{mod } q)$. In S we form the set D of the following pairs:

1. (c, d), where both c and d are nonzero squares.
2. (g, h), where both g and h are nonsquares.
3. Pairs $(u, 0)$.

Since (excluding $p^r = 2$, $q^s = 4$) both p^r and q^s are odd, there are

$$(p^r - 1)(q^s - 1)/4$$

pairs of type 1 and the same number of type 2, and also p^r of type 3. Then D contains $k = [(p^rq^s - p^r - q^s + 1)/2] + p^r$ elements, and with $v = p^rq^s$, noting that $q^s = p^r + 2$, we have $k = (v - 1)/2$. As p^r and q^s are odd and differ by 2, one of them is congruent to 1(mod 4) and the other to 3(mod 4). Thus, $v = p^rq^s \equiv -1 \pmod 4$, and if we write $v = 4t - 1$, then $k = 2t - 1$. Here, D is a system in S with all pairs of types 1 and 2 as multipliers M. Furthermore, $(-1, -1)$ is not in D, since -1 is a quadratic residue in one of GF(p^r), GF(q^s), and a nonresidue in the other. If $(x, y) - (z, w) = (a, b)$ with $(x, y), (z, w) \in D$ and $(m_1, m_2) \in M$, then multiplying by (m_1, m_2),

$$(m_1x, m_2y) - (m_1z, m_2w) = (m_1a, m_2b).$$

Hence, elements (a, b) and (m_1a, m_2b) occur equally often as differences of elements of D. Hence, we find the elements of S in classes, equivalent under multiplication from M, as

$$\begin{array}{ll} (a, b), & a, b \text{ same quadratic type}. \\ (c, d), & c, d \text{ opposite quadratic type}. \\ (e, 0), & e \neq 0. \\ (0, f), & f \neq 0. \\ (0, 0). & \end{array}$$

If $(x, y) - (z, w) = (a, b)$ with $(x, y), (z, w) \in D$, then

$$(z, w) - (x, y) = (-a, -b).$$

As $(-1, -1)$ is not in D, then if a, b are of the same quadratic character, the $-a, -b$ are of opposite quadratic character. Thus, all elements (a, b), $a \neq 0$, $b \neq 0$, of S occur the same number of times (say λ_1) as differences of elements of D. A difference of the form $(e, 0)$, $e \neq 0$, is given by

$$(a_1, b) - (a_0, b) = (a_1 - a_2, 0), \qquad b \neq 0,$$

for each of the $(q^s - 1)$ b's and for a fixed b with a_1 and a_2 distinct elements whose quadratic character is the same as that of b. This gives $(q^s - 1)(p^r - 1)(p^r - 3)/4$ such differences.

We also have from elements of type $(u, 0)$ in D,

$$(u_1, 0) - (u_0, 0) = (u_1 - u_2, 0),$$

yielding $p^r(p^r - 1)$ differences $(e, 0)$, $e \neq 0$. Since for each $e \neq 0$ we have $(e, 0)$ the same number of times as a difference, then from each of $p^r - 1$ elements $(e, 0)$, $e \neq 0$, arises

$$\lambda = \frac{(q^s - 1)(p^r - 3)}{4} + p^r = \frac{p^rq^s - 3}{4},$$

using $q^s = p^r + 2$, and with $v = p^rq^s = 4t - 1$, we find $\lambda = (v - 3)/4 = t - 1$.

A difference $(0,f)$ $f \neq 0$ arises in the form $(a, b_1) - (a, b_2) = (0, b_1 - b_2)$ and in the form

$$(a, b) - (a, 0) = (0, b) \qquad (a, 0) - (a, b) = (a, -b).$$

In the first way we have $(p^r - 1)(q^s - 1)(q^s - 3)/4$ such differences, and for the others, $2(p^r - 1)(q^s - 1)/2$. Thus, for each of $(q^s - 1)$ f's, $f \neq 0$, we have the difference $(0,f)$ in

$$\frac{(p^r - 1)(q^s - 3)}{4} + p^r - 1 = \frac{p^r q^s - 3}{4} = \frac{v - 3}{4} = t - 1$$

ways, using $q^s = p^r + 2$.

We have the differences $(e, 0)$, $e \neq 0$ and $(0,f)$, $f \neq 0$, each occurring $\lambda = t - 1$ times, and have a total of $k(k - 1)$ differences. Thus, as each of the $(p^r - 1)(q^s - 1)$ elements (a, b) occurs λ_1 times, we have

$$k(k - 1) = (p^r - 1)(q^s - 1)\lambda_1 + (p^r - 1)(t - 1) + (q^s - 1)(t - 1)$$

and as $k = 2t - 1$, $v = p^r q^s = 4t - 1$, this gives

$$\begin{aligned}(p^r - 1)(q^s - 1)\lambda_1 &= (t - 1)(4t - 2 - p^r - q^s + 2)\\ &= (t - 1)(p^r q^s - p^r - q^s + 1),\end{aligned}$$

whence $\lambda_1 = t - 1$.

Thus, every nonzero difference (a, b) in S arises $\lambda = t - 1$ times and D yields the difference set of type T with

$$v = p^r q^s = 4t - 1, \qquad q^s = p^r + 2, \qquad k = 2t - 1, \qquad \lambda = t - 1.$$

The derivation of type W_4 is conceptually similar to the derivation of type T. For the proof the reader is referred to Whiteman's original paper [1].

12

Finite Geometries

12.1 Foundations

From the standpoint of this book, a geometry is a particular kind of incidence system. The undefined elements are "points" and "lines" and the basic relation is the incidence relation $P \in L$, read "the point P is on the line L." We also say, "L contains P." Veblen and Young [1] regard points as the undefined elements, and lines as certain distinguished subsets of points. According to the axioms, two different lines cannot contain the same points, and so the distinction between the viewpoints is a minor one. A finite geometry is one that contains a finite number of points.

AXIOMS FOR PROJECTIVE GEOMETRY

PG 1. *There is one and only one line containing two distinct points.*

PG 2. *If A, B, C are three points not on a line, and if $D \neq A$ is a point on the line through A and B, and if $E \neq A$ is a point on the line through A and C, then there is a point F on a line with D and E and also on a line with B and C.*

PG 3. *Every line contains at least three distinct points.*

Veblen and Young [1] define "dimension" of a projective space recursively, saying that a point is a space of dimension zero, a line of dimension 1; and recursively, if X_{n-1} is a space of dimension $n - 1$, P a point not contained in X_{n-1}, the set of all points in all lines PB, B a point of X_{n-1}, is a space X_n of dimension n. Using these definitions, it is desirable to assign dimension -1 to the void space that contains no points. They prove that a space X_n is determined by any X_{n-1} and $P \notin X_{n-1}$, both of which are in X_n. The subspaces of a projective space X form a lattice in which, if A and B are subspaces of X, the intersection $A \cap B$ is the subspace of all points belonging to both A and B, and the union $A \cup B$ consists of all points on lines joining a point P of A and a point Q of B. It can be shown that by this definition, $A \cup B$ is a subspace. The dimension $d(A)$ of a subspace satisfies the following law:

$$d(A \cup B) + d(A \cap B) = d(A) + d(B). \qquad (12.1.1)$$

From this and the property that if $R \subseteq S$ and $d(R) = d(S)$, then $R = S$, we have the modular law for subspaces:

MODULAR LAW. *If* $A \supseteq B$, *then* $A \cap (B \cup C) = B \cup (A \cap C)$.

A *configuration* is a finite set of points and lines with specified incidences. Two important theorems deal with configurations.

THEOREM 12.1.1 (THEOREM OF DESARGUES). *If* O, A_1, B_1, C_1, A_2, B_2, C_2 *are distinct points and if* OA_1A_2, OB_1B_2, OC_1C_2 *are distinct lines, then* A_1B_1 *and*

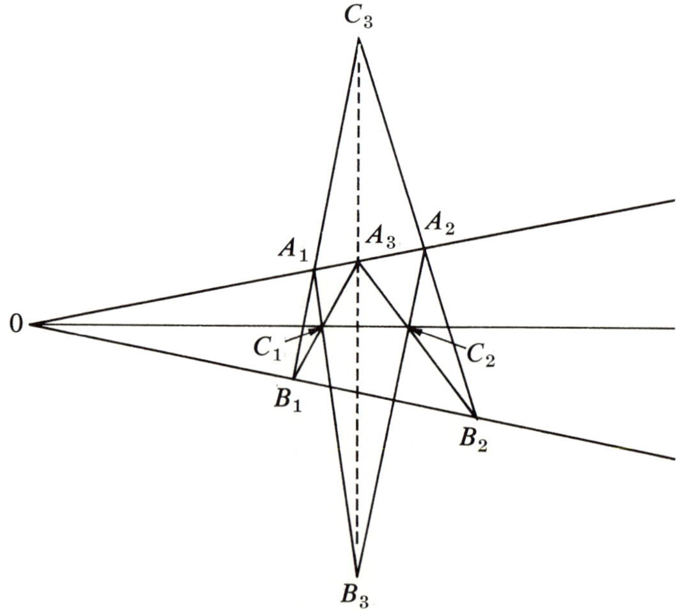

FIGURE 12.1. Theorem of Desargues.

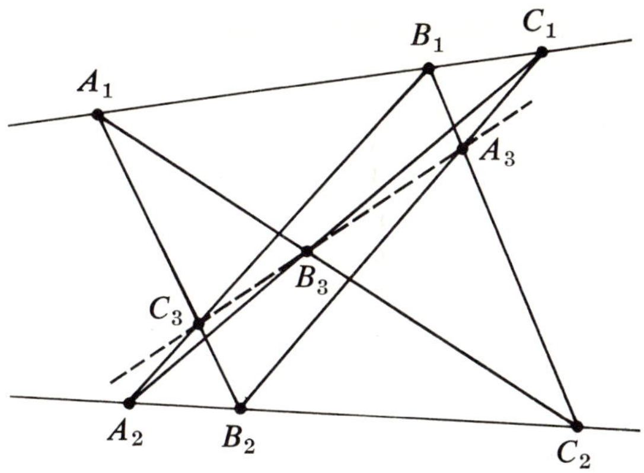

FIGURE 12.2. Theorem of Pappus.

A_2B_2 meet in a point C_3, A_1C_1 and A_2C_2 meet in a point B_3, B_1C_1 and B_2C_2 meet in a point A_3, and the points A_3, B_3, C_3 lie on a line.

THEOREM 12.1.2 (THEOREM OF PAPPUS). *If A_1, B_1, C_1 are points of a line and A_2, B_2, C_2 are points of another line in the same plane, and if A_1B_2 and A_2B_1 intersect in C_3, A_1C_2 and A_2C_1 intersect in B_3, and B_1C_2 and B_2C_1 intersect in A_3, then A_3, B_3, C_3 lie on a line.*

No attempt will be made here to prove Theorems 12.1.1 and 12.1.2. The reader can find proofs in Veblen and Young [1] or in the more modern book by Blumenthal [1].

THEOREM 12.1.3. *The theorem of Desargues is valid in any projective space of dimension 3 or more. If the theorem of Desargues is valid in a projective plane π, then π may be embedded in a projective space S_n of any finite dimension $n \geq 3$, and for a given n, S_n is unique up to isomorphism.*

There exist projective planes in which the theorem of Desargues is not valid. We speak of a Desarguesian or non-Desarguesian plane, depending on whether or not the theorem of Desargues is assumed to hold.

THEOREM 12.1.4 (INTRODUCTION OF COORDINATES). *A Desarguesian plane S_2, or projective space S_n, $n \geq 3$, may be coordinated by a division ring R so that a point P is represented by the left scalar multiples of a nonzero $n + 1$ vector over R thus:*

$$P = u(x_1, \ldots, x_{n+1}) = (ux_1, \ldots, ux_{n+1}), \qquad (x_1, \ldots, x_{n+1}) \neq (0, \ldots, 0).$$

If P_1 and $P_2 = u_2(y_1, \ldots, y_{n+1})$ are distinct points, then the nonzero vectors uQ in

$$u_1P_1 + u_2P_2 = (u_1x_1 + u_2y_1, \ldots, u_1x_{n+1} + u_2y_{n+1})$$

represent the points on the line joining P_1 and P_2. The division ring is uniquely determined by S_n. The vectors $(1, 0, \ldots, 0), (0, 1, \ldots, 0), \ldots, (0, 0, \ldots, 1)$, *and* $(1, 1, \ldots, 1)$ *may be taken as coordinates for any* $(n + 2)$ *points* $P_1, \ldots, P_{n+2}$ *such that no subset of* $n + 1$ *of these lies in an* S_{n-1}. *Coordinates for all other points are then uniquely determined. Conversely, given any division ring R, the points and lines as given above form a Desarguesian projective space* S_n.

THEOREM 12.1.5. *The theorem of Pappus implies the theorem of Desargues and the further property that multiplication is commutative, whence the coordinating ring R is a field F. Conversely, in a projective space coordinatized by a field F, the theorem of Pappus is valid.*

From Theorem 12.1.4 the number of points on a line is infinite if the ring R is infinite. Thus, for a finite geometry, the ring R is finite. From the theorem of Wedderburn, it follows that a finite Desarguesian geometry also satisfies the theorem of Pappus.

THEOREM 12.1.6 (WEDDERBURN). *A finite division ring is a finite field.*

A very brief proof of the Wedderburn theorem is given by Witt [1].

In a projective space S_n coordinatized by a division ring R, the set of points $P = u(x_1, \ldots, x_{n+1})$ satisfying a linear equation not identically zero,

$$x_1a_1 + x_2a_2 + \cdots + x_{n+1}a_{n+1} = 0, \tag{12.1.2}$$

is an S_{n-1} and is called a *hyperplane*. Every S_{n-1} in S_n may be defined in this way. Thus, the hyperplanes may be associated with vectors $[a_1, \ldots, a_{n+1}] \neq 0$ and their right multiples $[a_1v, \ldots, a_{n+1}v]$. Sometimes it is convenient to think of a projective geometry as determined by points $P = u(x_1, \ldots, x_{n+1})$ and hyperplanes $S_{n-1} = [a_1, \ldots, a_{n+1}]v$, with an incidence $P \in S_{n-1}$ if and only if (12.1.2) holds.

We may form a ring R^* "dual" to R by taking the elements of R as the elements of R^*, keeping the same addition but defining multiplication $x \circ y$ in R^* by the rule

$$x \circ y = yx \qquad \text{in } R. \tag{12.1.3}$$

This yields a correspondence

$$\begin{aligned} [a_1, \ldots, a_{n+1}]v &\to v \circ (a_1, \ldots, a_{n+1}), \\ u(x_1, \ldots, x_{n+1}) &\to [x_1, \ldots, x_{n+1}] \circ u, \end{aligned} \tag{12.1.4}$$

or

$$P \to P^*, \qquad K \to K^*, \tag{12.1.5}$$

where P is a point, K a hyperplane of $S_n(R)$, and P^* is a hyperplane, K^* a point of $S_n(R^*)$, and

$$K^* \in P^*, \qquad \text{if and only if } P \in K. \tag{12.1.6}$$

When the hyperplanes of a space are considered as the blocks of a design, the points being the objects, then this duality is the same as that observed earlier in (10.2.13).

The *affine* space A_n of dimension n over a division R has as points

$$P = (x_1, \ldots, x_n), \qquad x_i \in R, \tag{12.1.7}$$

and if P, $Q = (y_1, \ldots, y_n)$ are distinct points, the line joining P and Q consists of all points T such that

$$T = t(x_1, \ldots, x_n) + (1 - t)(y_1, \ldots, y_n), \qquad t \in R. \tag{12.1.8}$$

An affine space A may be derived from a projective space S_n by deleting a hyperplane, which is then considered the hyperplane at infinity. This may be easily described in terms of the coordinates, where with all projective points $P = u(x_1, \ldots, x_n, x_{n+1})$, we delete the hyperplane $x_{n+1} = 0$ and fix the factor u by taking $ux_{n+1} = 1$ and $P = (y_1, \ldots, y_n, 1)$, or $(y_1, \ldots, y_n)$ in affine form.

12.2 Finite Geometries as Block Designs

Let S_n be a projective geometry of dimension n over the finite field $F_q = \mathrm{GF}(q)$ with $q = p^r$ elements, where p is a prime. We sometimes write $S_n = \mathrm{PG}(n, q)$ for this. From Theorem 12.1.4 we see that the $(q^{n+1} - 1)$ vectors

$$(x_1, \ldots, x_{n+1}) \neq 0, \qquad x_i \in GF(q)$$

fall into sets of $q - 1$, representing a point. Hence, S_n contains $(q^{n+1} - 1)/(q - 1)$ points. An S_{n-1} in S_n contains $(q^n - 1)/(q - 1)$ points. From the dimension law (12.1.1), since two different S_{n-1}'s have S_n as their union, it follows that the intersection of any two is an S_{n-2} with $(q^{n-1} - 1)/(q - 1)$ points. Hence, the hyperplanes in an S_n form a symmetric block design with

$$v = \frac{q^{n+1} - 1}{q - 1}, \qquad k = \frac{q^n - 1}{q - 1}, \qquad \lambda = \frac{q^{n-1} - 1}{q - 1}. \tag{12.2.1}$$

By Singer's theorem (Theorem 11.3.1), this design has a cyclic automorphism of order v and can be represented by a difference set. The full group of automorphisms of this design is of much larger order, since this is the full collineation group of S_n. The order of the full collineation group is

$$r(q^{n+1} - 1)(q^{n+1} - q) \cdots (q^{n+1} - q^n)/(q - 1).$$

This is the projective group of all nonsingular matrices of order $n + 1$ over $\mathrm{GF}(p^r)$, where matrices differing by a scalar factor ($q - 1$ such factors) are the same projectively, together with the group of automorphisms of $\mathrm{GF}(p^r)$, which is cyclic of order r.

We may form nonsymmetric designs from $\mathrm{PG}(n, q)$, taking points as objects and blocks as spaces S_s, $1 \leq s < n - 1$. There are $(q^{n+1} - 1)/(q - 1)$ points in S_n, and we may choose $(s + 1)$ independent representatives of points to determine an S_s in $(q^{n+1} - 1)(q^{n+1} - q)(q^{n+1} - q^2) \cdots (q^{n+1} - q^s)$ ways. But within S_s, these $(s + 1)$ independent representatives could have been chosen in $(q^{s+1} - 1)(q^{s+1} - q) \cdots (q^{s+1} - q^s)$ ways. Thus, the number of distinct subspaces S_s is the quotient of these two values and is the number b of blocks. Hence, for this design,

$$
\begin{aligned}
b &= \frac{(q^{n+1} - 1)(q^{n+1} - q) \cdots (q^{n+1} - q^s)}{(q^{s+1} - 1)(q^{s+1} - q) \cdots (q^{s+1} - q^s)}, \\
v &= \frac{q^{n+1} - 1}{q - 1}, \\
r &= \frac{(q^{n+1} - q)(q^{n+1} - q^2) \cdots (q^{n+1} - q^s)}{(q^{s+1} - q)(q^{s+1} - q^2) \cdots (q^{s+1} - q^s)}, \\
k &= \frac{q^{s+1} - 1}{q - 1}, \\
\lambda &= \frac{(q^{n+1} - q^2) \cdots (q^{n+1} - q^s)}{(q^{s+1} - q^2) \cdots (q^{s+1} - q^s)}.
\end{aligned}
\tag{12.2.2}
$$

Similarly, the subspaces of given dimension of an affine space form a block design. Over the finite field GF(q) $q = p^r$, the affine space A_n of dimension n contains q^n points. With the points as objects and spaces S_s, $1 \leq s \leq n - 1$, we have a block design with

$$
\begin{aligned}
b &= \frac{q^n(q^n - 1) \cdots (q^n - q^{s-1})}{q^s(q^s - 1) \cdots (q^s - q^{s-1})}, \\
v &= q^n, \\
r &= \frac{(q^n - 1)(q^n - q) \cdots (q^n - q^{s-1})}{(q^s - 1)(q^s - q) \cdots (q^s - q^{s-1})}, \\
k &= q^s, \\
\lambda &= \frac{(q^n - q) \cdots (q^n - q^{s-1})}{(q^s - q) \cdots (q^s - q^{s-1})}.
\end{aligned}
\tag{12.2.3}
$$

It should be noted here that a block design (even a symmetric one) with the parameters of a geometry is not necessarily a geometry.

For $v = 31$, $k = 15$, $\lambda = 7$, there are two difference sets:

$$
\begin{aligned}
B_0 &= 1, 2, 3, 4, 6, 8, 12, 15, 16, 17, 23, 24, 27, 29, 30 \pmod{31}, \\
B_0' &= 1, 2, 4, 5, 7, 8, 9, 10, 14, 16, 18, 19, 20, 25, 28 \pmod{31}.
\end{aligned}
\tag{12.2.4}
$$

Of these, the first consists of the hyperplanes in the four-dimensional space PG(2, 4). The second consists of the quadratic residues modulo 31. Here,

B_0, B_1, and B_2 have the three objects 9, 10, and 20 in common, while B_0, B_1, and B_4 have the four objects 5, 8, 10, 19 in common. But over GF(2) a projective line has three points; a plane, seven points. Hence, the second design is *not* PG(2, 4).

12.3 Finite Planes

For projective planes, the axioms of Section 12.1 may be put in a simpler form:

AXIOMS FOR PROJECTIVE PLANES

PP 1. *There is one and only one line containing two distinct points.*
PP 2. *There is one and only one point common to two distinct lines.*
PP 3. *There exist four points, no three of which are on a line.*

The axiom PG 2 can be interpreted as saying that two lines in the same plane have a point of intersection. Thus, PP 2 says that all lines are in the same plane. The axiom PP 3 readily yields axiom PG 3. The first two axioms, PP 1 and PP 2, exhibit clearly the duality between points and lines, and it is not difficult to show that PP 3 is equivalent to its dual:

PP 3. *There exist four lines, no three of which go through the same point.*

There are not many general results that can be proved about finite projective planes. An easy one is the following theorem.

THEOREM 12.3.1. *Let $n \geq 2$ be an integer. In a projective plane π any one of the following properties implies the rest:*

1. *One line contains exactly $(n + 1)$ points.*
2. *One point is on exactly $(n + 1)$ lines.*
3. *Every line contains exactly $(n + 1)$ points.*
4. *Every point is on exactly $(n + 1)$ lines.*
5. *There are exactly $(n^2 + n + 1)$ points in π.*
6. *There are exactly $(n^2 + n + 1)$ lines in π.*

Proof: ∎ Let A, B, C, D be four points, no three on a line, whose existence is given by PP 3. Then we have lines $L_1, \ldots, L_6$ containing A, B, C, D and three additional points X, Y, Z:

$$\begin{aligned} L_1&: \ A, B, X \ldots. \\ L_2&: \ A, C, Y \ldots. \\ L_3&: \ A, D, Z \ldots. \\ L_4&: \ B, C, Z \ldots. \\ L_5&: \ B, D, Y \ldots. \\ L_6&: \ C, D, X \ldots. \end{aligned} \tag{12.3.1}$$

From the axioms it is not difficult to check that the six lines $L_1, \ldots, L_6$ are distinct, and the points $A, \ldots, Z$ are distinct, where X, Y, Z are the points of intersection of the pairs of lines L_1, L_6; L_2, L_5; and L_3, L_4. Furthermore, we may also check that there are no further incidences between these seven points and these six lines; for example, we could not have $A \in L_4$, since then the distinct points A, B would lie on the two distinct lines L_1 and L_4.

Let us now turn to the proof of our theorem and assume property 1. Let L be a line with exactly $(n + 1)$ points on it, say, $Q_1, Q_2, \ldots, Q_{n+1}$. If P is a point not on L, the lines PQ_i, $i = 1, \ldots, n + 1$, are distinct, since if $PQ_i = PQ_j$, then $P \in Q_iQ_j = L$, contrary to assumption. Furthermore, every line through P intersects L and so must be one of the $(n + 1)$ lines PQ_i, $i = 1, \ldots, n + 1$. At least two of the points A, B, C, D of (12.3.1) are not on L, and so such points P exist. Now let P be a point on exactly $(n + 1)$ lines $K_1, \ldots, K_{n+1}$. If M is any line not through P, then M intersects $K_1, \ldots, K_{n+1}$ in points $Q_1, \ldots, Q_{n+1}$, which are all distinct because P is the only point on more than one of $K_1, \ldots, K_{n+1}$. If there were one more point Q_{n+2} on M, then PQ_{n+2} would be another line through P, since if PQ_{n+2} were some K_j, it would also contain Q_j and then

$$PQ_{n+2} = PQ_{n+2}Q_j = Q_{n+2}Q_j = M,$$

contrary to our assumption that $P \notin M$.

Now our original line L contained exactly $(n + 1)$ points, and so every point not on L is on exactly $(n + 1)$ lines, including at least two of A, B, C, D; say, A and B. Then every line not through A or not through B contains exactly $(n + 1)$ points. Hence, every line except possibly L_1: A, B, X contains exactly $(n + 1)$ points. Hence, L_2: A, C, Y contains exactly $(n + 1)$ points and the point Z not on L_2 is on exactly $(n + 1)$ lines; whence L_1, which does not contain Z, must also contain exactly $(n + 1)$ points. Thus property 1 implies property 3.

But now for any point P we may find a line not through it, and so by our previous argument there are exactly $(n + 1)$ lines through P, giving properties 2 and 4. Now let P_0 be a particular point and let $L_1, \ldots, L_{n+1}$ be the $(n + 1)$ lines through P_0. Each of these contains exactly n points different from P_0, and since P_0 is joined to any point by one of these lines, they account for all points of π, giving a total of $1 + (n + 1)n = n^2 + n + 1$ points in π, proving property 5. Similarly, if L_0 is a line containing points $P_1, \ldots, P_{n+1}$, each of these points is on exactly n additional lines, and we have $1 + (n + 1) = n^2 + n + 1$ lines in π, and property 6 is proved.

We have shown that property 1 implies the remaining five properties. Property 3 implies the others a fortiori. By duality, interchanging the roles of "point" and "line," property 2 and property 4 implies the others a fortiori. If property 5 holds, then the line L_1 of (12.3.1) contains $(m + 1)$ points for some integer $m \geq 2$, and so by our earlier argument, π contains $(m^2 + m + 1)$ points. But from $m^2 + m + 1 = n^2 + n + 1$ for positive integers m and n, we may conclude $m = n$, whence property 5 implies property 1 and so the remaining ones. Similarly, 6 implies 2 and so the others. Thus, all parts of our theorem are proved.

DEFINITION. *A finite projective plane is said to be of order n if a line contains exactly* $(n + 1)$ *points.*

We note that a plane is of order n if it has any of the six properties of Theorem 12.3.1. If we take the points as objects and the lines as blocks, a finite projective plane is a symmetric block design with parameters

$$v = n^2 + n + 1, \qquad k = n + 1, \qquad \lambda = 1. \tag{12.3.2}$$

Conversely, a block design with these parameters is a finite projective plane, since we readily verify that the axioms are satisfied. A finite projective plane coordinatized by GF(q), $q = p^r$, as given in Section 12.1, is of order $n = p^r$. These are the Desarguesian planes and they exist for every order that is a prime power. All finite planes known at this writing are of prime-power order, though non-Desarguesian planes exist for all prime-power orders p^r, $r \geq 2$, except for orders 4 and 8.

Application of Theorem 10.3.1 yields a necessary condition for the existence of a finite projective plane.

THEOREM 12.3.2 (BRUCK-RYSER). *A necessary condition for the existence of a finite projective plane of order n is that for* $n \equiv 1, 2 \pmod 4$, *integers x, y exist with* $n = x^2 + y^2$.

Proof: ■ We have $v = n^2 + n + 1$, $k = n + 1$, $\lambda = 1$, and $k - \lambda = n$, whence the order n of the plane agrees with the notation $n = k - \lambda$ used in Theorem 10.3.1. Here, since $n(n + 1) \equiv 0 \pmod 2$, v is certainly odd, and the condition takes the form of requiring the existence of integers x, y, z not all zero such that

$$z^2 = nx^2 + (-1)^{(v-1)/2}y^2. \tag{12.3.3}$$

Here, as $v = n^2 + n + 1$, if $n \equiv 0, 3 \pmod 4$, then $(v - 1)/2$ is even, and we have

$$z^2 - y^2 = nx^2. \tag{12.3.4}$$

If $n \equiv 3 \pmod 4$, we can take $x = 1$, $y = (n - 1)/2$, $z = (n + 1)/2$, to satisfy this; if $n \equiv 0 \pmod 4$, we can take $x = 1$, $y = (n - 4)/4$, $z = (n + 4)/4$. If $n \equiv 1, 2 \pmod 4$, then $(v - 1)/2$ is odd and (12.3.3) takes the form

$$z^2 + y^2 = nx^2. \tag{12.3.5}$$

It is well known (for example, see Hardy and Wright [1] p. 299) that (12.3.5) implies the existence of integers a, b such that $n = a^2 + b^2$, the statement of our theorem. This shows that infinitely many orders n are impossible, starting with $n = 6, 14, 21, 22, \ldots$. Since there is a plane of any prime-power order, the first order left in doubt is $n = 10$.

If $v = n^2 + n + 1$, $k = n + 1$, $\lambda = 1$ are the parameters for a finite projective plane regarded as a block design, the residual design, as defined in Section 10.1, is an affine plane and is the design with parameters

$$b = n^2 + n, \qquad v = n^2, \qquad k = n, \qquad r = n + 1, \qquad \lambda = 1. \quad (12.3.6)$$

Given a design with these parameters, we may always embed it in a symmetric design. We shall prove this, using a process equivalent to adjoining a line at infinity to an affine Euclidean plane.

THEOREM 12.3.3. *Given a block design D^* with parameters $b^* = n^2 + n$, $v^* = n^2$, $r^* = n + 1$, $k^* = n$, $\lambda^* = 1$, we may adjoin, in a unique way, a further block with $(n + 1)$ new objects, and add one of the new objects to each of the original blocks to form a symmetric design D with $v = b = n^2 + n + 1$, $r = k = n + 1$, $\lambda = 1$.*

Proof: ∎ Let D^* be given. Consider a given block B^* with objects $a_1, \ldots, a_n$. Let x be one of the remaining objects. Then the n blocks B_i: $x, a_i, \ldots, i = 1, \ldots, n$, are uniquely determined, since $\lambda = 1$, and are distinct because $x \notin B^*$ and no block except B^* contains more than one of $a_1, \ldots, a_n$. This accounts for n of the $(n + 1)$ blocks containing x. Thus, through x, there is exactly one block B_0: $x \cdots$ that does not contain any object of B^*. We call B_0 the "parallel" to B^* through x. What we have proved so far is the Euclidean parallel postulate for D^*. Hence, given any block B^*, there is a unique block parallel to B^* through any object not in B^*. Two distinct blocks B_u and B_v parallel to B^* are also parallel to each other, since if they had an object y in common, this would conflict with the uniqueness of the parallel to B^* through y. Hence, given any block B_1, the remaining $(n^2 - n)$ objects are contained in parallels $B_2, \ldots, B_n$. Here, $B_1, B_2, \ldots, B_n$ are a family of n parallel blocks, which together contain each of the n^2 objects exactly once. Hence, the $b = n^2 + n$ blocks of D^* can be divided in a unique way into $(n + 1)$ families $F_1, \ldots, F_{n+1}$ of parallel blocks, each F_i containing n blocks and every object of D^* appearing exactly once in a block of F_i. We now take new objects $y_1, \ldots, y_{n+1}$ and a new block B_∞ containing $y_1, \ldots, y_{n+1}$ and no other objects, and also adjoin the object y_i to every block of the family F_i but to no other block. It is now almost immediate that the new block B_∞ and the blocks of D^* with the adjoined y's form the blocks of a symmetric design with $v = b = n^2 + n + 1$, $r = k = n + 1$, $\lambda = 1$.

In the second example of Section 10.1, we have $b = 12$, $v = 9$, $r = 4$, $k = 3$, $\lambda = 1$, and the 12 blocks fall into four families: F_1: B_1, B_2, B_3; F_2: B_4, B_5, B_6; F_3: B_7, B_8, B_9; F_4: B_{10}, B_{11}, B_{12}. We may take new objects 10, 11, 12, 13 as the objects of a block B_{13} and adjoin 10 to the blocks of F_1, 11 to F_2, 12 to F_3

and 13 to F_4 to obtain the symmetric design with $b = v = 13$, $r = k = 4$, $\lambda = 1$, the finite projective plane of order 3.

From Theorem 12.3.3 an affine plane of order n exists if and only if a projective plane of order n exists, and an affine plane is the residual design of a projective plane. But if we take a projective plane, the affine planes obtained by deleting two different lines will be isomorphic if and only if there is an automorphism of the projective plane taking one of these lines into the other. This follows, since the proof of Theorem 12.3.3 shows that an isomorphism between affine planes is uniquely extendable to an isomorphism between the projective planes in which they are embedded.

The affine form of a finite plane of order n leads to a representation of the plane by a family of "mutually orthogonal" Latin squares of order n. Two Latin squares of order n containing the numbers 1 to n (once in each row and once in each column) are said to be orthogonal if, when superposed, the n^2 cells (one containing i from the first square and j from the second) are such that the n^2 ordered pairs (i, j) produced in this way are all distinct and so consist of all pairs (u, v) $u, v = 1, \ldots, n$, each exactly once. More than two Latin squares are called "mutually orthogonal" if any two are orthogonal.

Given an affine plane of order n with its lines (that is, blocks) divided into $(n + 1)$ families of parallels. We arbitrarily designate two of these families as F_r and F_c, the row and column families, and the remainder as $F_1, \ldots, F_{n-1}$. Each of the n^2 objects is a point. We number the lines of each family from 1 to n in an arbitrary fashion. The numbering in F_r is that of the rows of a square, in F_c that of the columns of a square. A point P is on one line of F_r and one of F_c, and so is associated with a particular cell of a square. Thus, we may write $P = (i, j)$ for that point in the ith row and jth column. Thus, F_r and F_c merely correspond to rows and columns. For each of $F_1, \ldots, F_{n-1}$, we construct an $n \times n$ Latin square in the following way: Let F_u be one of these families containing lines $L_1^u, \ldots, L_n^u$, where the subscripts give the arbitrary numbering. We construct a square S_u from F_u. We take an $n \times n$ square and insert the number v in the cell of the ith row and jth column if the point P associated with this cell (from being on the ith line of F_r and on the jth line of F_c) is on the vth line L_v^u of the family F_u. Since every point is on exactly one line of F_u, we have inserted exactly one number in each cell of the square S_u. A line of F_r (or F_c) intersects each line of F_u exactly once, and so a row (or column) of S_u contains each of $1, \ldots, n$ exactly once. Hence, S_u is a Latin square.

Now consider two squares S_u and S_w, $w \neq u$. Let $a(i, j)$ be the number in cell (i, j) of S_u and $b(i, j)$ that of S_w. If we should have

$$a = a(i_1, j_1) = a(i_2, j_2) \quad \text{and} \quad b = b(i_1, j_1) = b(i_2, j_2),$$

the points $P_1 = (i_1, j_1)$ and $P_2 = (i_2, j_2)$ both lie on line L_a^u of F_u and line L_b^w of F_w, contrary to the fact that $\lambda = 1$—that is, two distinct points lie on one and only one line. Thus, this cannot happen, and the ordered pairs $(a(i, j), b(i, j))i$,

$j = 1, \ldots, n$, are all distinct, and so the squares S_u and S_w are orthogonal. Thus, a finite plane of order n leads to a family of $(n - 1)$ mutually orthogonal Latin squares of order n.

Suppose we have a family of m mutually orthogonal Latin squares. The property of orthogonality is unaltered if a substitution is made on the numbers of any square. Hence, we may suppose the numbering so chosen that the first row of every square is $1, 2, \ldots, n$. Hence, the number appearing in row 2, column 1, of each square is one of the $n - 1$ numbers $2, \ldots, n$. If the number i were to occur twice in this position, then on superposition of these two squares we would have the pair (i, i) in cell $(2, 1)$ and also in cell $(1, i)$, conflicting with orthogonality. Hence, there are no repeats in cell $(2, 1)$ and so there are at most $(n - 1)$ mutually orthogonal squares of order n. Suppose $m = n - 1$. We now consider each cell (i, j) as a point $P(i, j)$ and form families of lines $F_r, F_c, F_1, \ldots, F_{n-1}$, where the $P(i, j)$ is on the ith line of F_r, the jth line of F_c, and is on the vth line of F_u, $u = 1, \ldots, n - 1$, if the entry in cell (i, j) of S_u is v. This gives us $(n + 1)$ families of n lines, each containing n points. The total number of points is n^2. The Latin property and orthogonality of $S_1, \ldots, S_{n-1}$ assures us that two distinct points do not both lie on distinct lines. As a particular point, if P_0 is on $(n + 1)$ lines, each of which contains $(n - 1)$ further points, it follows that P_0 is joined exactly once to every other point, and we have $\lambda = 1$. Thus, we have shown that a family of $(n - 1)$ mutually orthogonal Latin squares is equivalent to a finite plane of order n.

As an example of this equivalence, consider the design with $v = 21$, $k = 5$, $\lambda = 1$, the plane of order 4, which we may take as determined by the difference set B_0: 3, 6, 7, 12, 14(mod 21). We take B_0 as the line at infinity, deleting its points on the other lines. The projective pencils through 3, 6, 7, 12, 14 become the affine parallel families F_r, F_c, F_1, F_2, F_3, respectively. Thus, we have where $B_i = 3 + i, 6 + i, 7 + i, 12 + i, 14 + i$(mod 21); and B_i^* is the block B_i with 3, 6, 7, 12, or 14 deleted.

$F_r(3)$			$F_c(6)$		
1:	B_{10}^*:	13, 16, 17, 1.	1:	B_3^*:	9, 10, 15, 17.
2:	B_{12}^*:	15, 18, 19, 5.	2:	B_{13}^*:	16, 19, 20, 4.
3:	B_{17}^*:	20, 2, 8, 10.	3:	B_{15}^*:	18, 0, 1, 8.
4:	B_{18}^*:	0, 4, 9, 11.	4:	B_{20}^*:	2, 5, 11, 13.

(12.3.7)

$F_1(7)$			$F_2(12)$			$F_3(14)$		
4:	B_1^*:	4, 8, 13, 15.	1:	B_5^*:	8, 11, 17, 19.	2:	B_2^*:	5, 8, 9, 16.
2:	B_4^*:	10, 11, 16, 18.	4:	B_6^*:	9, 13, 18, 20.	4:	B_7^*:	10, 13, 19, 0.
1:	B_{14}^*:	17, 20, 0, 5.	2:	B_9^*:	15, 16, 0, 2.	3:	B_8^*:	11, 15, 20, 1.
3:	B_{16}^*:	19, 1, 2, 9.	3:	B_{19}^*:	1, 4, 5, 10.	1:	B_{11}^*:	17, 18, 2, 4.

In terms of cell coordinates, from F_r and F_c we have

$$\begin{array}{llll} 17 = (1, 1); & 16 = (1, 2); & 1 = (1, 3); & 13 = (1, 4); \\ 15 = (2, 1); & 19 = (2, 2); & 18 = (2, 3); & 5 = (2, 4); \\ 10 = (3, 1); & 20 = (3, 2); & 8 = (3, 3); & 2 = (3, 4); \\ 9 = (4, 1); & 4 = (4, 2); & 0 = (4, 3); & 11 = (4, 4). \end{array} \tag{12.3.8}$$

We have numbered the lines of F_1, F_2, F_3 so that the squares S_1, S_2, S_3 will have their first rows 1, 2, 3, 4. Following our construction, the squares are

$$\begin{array}{ccc} S_1 & S_2 & S_3 \\ 1, 2, 3, 4, & 1, 2, 3, 4, & 1, 2, 3, 4, \\ 4, 3, 2, 1, & 2, 1, 4, 3, & 3, 4, 1, 2, \\ 2, 1, 4, 3, & 3, 4, 1, 2, & 4, 3, 2, 1, \\ 3, 4, 1, 2, & 4, 3, 2, 1, & 2, 1, 4, 3. \end{array} \tag{12.3.9}$$

12.4 Some Types of Finite Planes

We can introduce a system of coordinates into any projective plane. This is most easily described for finite planes in terms of the division of the affine plane into the families of parallels $F_r, F_c, F_1, \ldots, F_{n-1}$. Our coordinates will be n letters $a_0, a_1, \ldots, a_{n-1}$, where two of them play the role of a zero and a unit, and we write $a_0 = 0$, $a_1 = 1$.

If $P_0, P_1, \ldots, P_{n-1}$ are the points of the first line of F_1 (in some fixed order), we write

$$P_0 = (0, 0), \quad P_1 = (1, 1), \ldots, \quad P_i = (a_i, a_i), \ldots, \quad P_{n-1} = (a_{n-1}, a_{n-1}). \tag{12.4.1}$$

The line of F_r containing P_i we designate as $x = a_i$, and the line of F_c containing P_j we designate as $y = a_j$. The point in common to $x = a_i$ and $y = a_j$ we represent by the ordered pair (a_i, a_j) and note that this representation is consistent with (12.4.1). In any family F_u there is a unique line containing $P_0 = (0, 0)$ and the intersection of this line with $x = 1$ is a point $(1, w)$. We use this value of w to renumber the families, writing $F_u = F_w$. This rule assigns the number $a_1 = 1$ to the family F_1 from (12.4.1). It also assigns the number $a_0 = 0$ to the family F_c, but does not assign a number to the family F_r. The numbering of the families is the assignment of a "slope" to a family of parallel lines, and we may assign slope "∞" to the lines $x = c$ of F_r, this being different from $a_0, \ldots, a_{n-1}$.

We now define a ternary operation $x \cdot m \circ b$ on the elements $a_0, a_1, \ldots, a_{n-1}$.

Ternary Rule: ■ $y = x \cdot m \circ b$ *if the point* (x, y) *is on the line of* F_m *containing the point* $(0, b)$.

It is easily checked that this operation is well defined for any x, m, and b chosen from $a_0, a_1, \ldots, a_{n-1}$. We define a multiplication xm and an addition $x + b$ as special cases of the ternary operation by the rules

$$xm = x \cdot m \circ 0, \qquad x + b = x \cdot 1 \circ b, \tag{12.4.2}$$

The ternary operation satisfies the following laws:

T1. $0 \cdot m \circ c = a \cdot 0 \circ c = c$.
T2. $1 \cdot m \circ 0 = m \cdot 1 \circ 0 = m$.
T3. Given a, m, c, there exists exactly one z such that $a \cdot m \circ z = c$.
T4. If $m_1 \neq m_2, b_1, b_2$ are given, there exists a unique x such that $x \cdot m_1 \circ b_1 = x \cdot m_2 \circ b_2$.
T5. If $a_1 \neq a_2$, c_1, c_2 are given, there exists a unique pair m, b such that $a_1 \cdot m \circ b = c_1$ and $a_2 \cdot m \circ b = c_2$.

Here, T1 and T2 are direct consequences of the definition. T3 says that there is a line in the family F_m through the point (a, c). T4 says that nonparallel lines have a unique intersection point. T5 says that two points not on a line $x = a$ are on a unique line in one of the other families. These properties are not independent, but are sufficient so that a ternary ring with these properties determines a projective plane. The necessary verifications are straightforward and will not be given here.

The coordinates and ternary ring are given here for an affine plane. To extend them to a projective plane, we introduce points at infinity (0), (1), $\ldots$, (a_{n-1}) and (∞), all lying on a single line at infinity L_∞, and adjoin (∞) to the lines of F_r and (a_i) to the lines of Fa_i, $i = 0, \ldots, n - 1$.

It may happen that the ternary operation can be expressed as a combination of multiplication and addition. We may have

$$x \cdot m \circ b = xm + b. \tag{12.4.3}$$

The algebraic relation (12.4.3) may be shown to be equivalent to the validity of the theorem of Desargues in certain cases with L_∞ the axis of perspectivity and the point (∞) as the center of perspectivity.

An important type of ternary ring is called a Veblen-Wedderburn system. Such a system has a well-defined addition $a + b$ and multiplication ab together with distinguished elements 0 and 1, and satisfies:

VW 1. Addition is an Abelian group.
VW 2. In $xy = z$, if any two of x, y, z are given and different from zero.
VW 3. $1x = x1 = x$, $0x = x0 = 0$, $x + 0 = 0 + x = x$.
VW 4. $(a + b)m = am + bm$.
VW 5. If $r \neq s$, $xr = xs + t$ has a unique solution x.
VW 6. $x \cdot m \circ b = xm + b$.

If the properties VW 1–5 hold, we may use VW 6 to define a ternary operation, which then determines a plane.

In a finite system, VW 5 is a consequence of the remaining properties, for if $r \neq s$ and $x_1r - x_1s = x_2r - x_2s$, then by VW 4, $(x_1 - x_2)r = (x_1 - x_2)s = w$, say. If $x_1 - x_2 \neq 0$, at least one of r, s is different from zero, and so by VW 2, $w \neq 0$ and we would have $r = s$. Hence, as x ranges over the n values of a finite system, $xr - xs$ also ranges over these n values, and so there is a unique x for which $xr - xs = t$. Of course every division ring is a Veblen-Wedderburn system and, in particular, finite fields satisfy all these properties and more.

A particular kind of VW system (short for Veblen-Wedderburn) is known as a Hall system.

Let F be a field and suppose we have a quadratic polynomial $f(x) = x^2 - rx - s$ irreducible over F. The elements of our VW system S will be all expressions $a + bu$, where $a, b \in F$ and u is an indeterminate. We might, of course, have used ordered pairs (a, b) instead of introducing the indeterminate. Addition in S is defined by

$$\text{A}: (a_1 + b_1u) + (a_2 + b_2u) = (a_1 + a_2) + (b_1 + b_2)u\,.$$

For multiplication we use two rules:

M1. For $a \in F$, put $(e + fu)a = ae + afu$.
M2. For $z = a + bu$, $b \neq 0$, and $w = e + fu$, put

$$\begin{aligned} wz &= (e + fu)(a + bu) \\ &= (ae - b^{-1}a^2f + b^{-1}raf + b^{-1}sf) + (be - af + rf)u\,. \end{aligned}$$

The multiplication rules are more easily understood in the following form:

M1*. For $c \in F$, $c(a + bu) = (a + bu)c = ac + bcu$.
M2*. $(x + y)z = xz + yz$.
M3*. For $z \notin F$, $z^2 = rz + s$.

If F is a finite field GF(q), $q = p^r$, the Hall system defines a finite plane of order q^2. Except for $q^2 = 4$, the plane defined in this way is different from the Desarguesian plane of order q^2 with coordinates from GF(q^2). For a finite F, since VW 5 is a consequence of VW 1–4, the only property at all difficult to check is VW 2.

To prove VW 2, we must show that in $wz = v$, given any two of w, v, z different from zero, the third is unique and different from zero. We write

$$w = e + fu, \qquad z = a + bu, \qquad v = c + du,$$

with $a, b, c, d, e, f \in F$. Given w and z, if $b = 0$, we use rule M1 and if $b \neq 0$, we use rule M2 to determine v. If $b = 0$ and if $w = 0$, $z = a \neq 0$, then clearly $v = 0$, and also if $z = a \neq 0$ and $v = c + du$ are given, we readily determine $w = e + fu$ from $ae = c$, $af = d$, whence w is unique and different from zero if $v \neq 0$. Consider cases with $z = a + bu$ given, $b \neq 0$, when M2* applies. We have

$$\begin{aligned} ae + (-b^{-1}a^2 + b^{-1}ra + b^{-1}s)f &= c, \\ be + (-a + r)f &= d, \end{aligned} \tag{12.4.4}$$

Considering these as linear equations in e and f, the determinant is $-s$, and since $x^2 - rs - s$ is irreducible over F, we certainly have $s \neq 0$. This shows that when $b \neq 0$, and so certainly $z = a + bu \neq 0$, then if $w = e + fu \neq 0$, we have $v = c + du \neq 0$; also if z and v are given, then w exists and is unique. The difficult case consists in having $w = e + fu$ and $v = c + du$ given. First, if $ed - fc = 0$, the equations $ae = c$, $af = d$, are solvable, whence a solution $z = a$ using M1* exists. Furthermore, if a solution using M1* exists, we must have $ed - fc = 0$.

Now suppose $ed - fc \neq 0$. We are now given e, f, c, d and wish to solve (12.4.4) for a and b. If $f = 0$, then $e \neq 0$ and (12.4.4) reduces to $ae = c$, $be = d$, which easily determine a and b. Now suppose $f \neq 0$ and solve the second equation for a; thus

$$a = bef^{-1} - df^{-1} + r. \tag{12.4.5}$$

Substituting this value of a into the first equation gives

$$def^{-1} - b^{-1}d^2f^{-1} + b^{-1}dr + b^{-1}sf = c. \tag{12.4.6}$$

Multiplying by f and transposing the first term give

$$-b^{-1}(d^2 - rdf - sf^2) = cf - de. \tag{12.4.7}$$

But we have $cf - de \neq 0$, and the irreducibility of $x^2 - rx - s$ tells us that $d^2 - rdf - sf^2 \neq 0$ if d and f are not both zero, which is true because $cf - de \neq 0$. Hence, (12.4.7) yields a unique nonzero value for b^{-1} and so for b, and in turn (12.4.5) gives a uniquely. This now completes the proof of the validity of VW 2.

If the multiplication in a VW system is associative, then it forms a group and the VW system is called a "near-field." All finite near-fields have been determined by Zassenhaus [1]. Naturally, these include the finite fields GF(p^r). Apart from seven exceptions, the remaining fields are described in the subsequent discussion.

Let $q = p^h$ be a power of a prime p and let v be an integer, all of whose prime factors divide $q - 1$, where we also require $v \not\equiv 0 \pmod 4$ if $q \equiv 3 \pmod 4$. Then, with $hv = r$, we may construct a near-field K with $n = p^r$ elements from the finite field GF(p^r) in the following way:

1. The elements of K shall be the same as the elements of GF(p^r).
2. Addition in K shall be the same as addition in GF(p^r).

3. A product $w \circ u$ in K can be defined in terms of a product xy in GF(p^r) in the following way: Let z be a fixed primitive root of GF(p^r); then, if $u = z^{kv+j}$, an integer i is uniquely determined modulo v by $q^i \equiv 1 + j(q-1) (\text{mod } v(q-1))$. We define $w \circ u$ by

$$w \circ u = w^{q^i} u. \tag{12.4.8}$$

The seven exceptional cases are all of order $u = p^2$ for an appropriate odd prime p. In all cases the elements of K are of the form $a + bu$, a, b, $\in$ GF(p) and

$$(a_1 + b_1 u) + (a_2 + b_2 u) = (a_1 + a_2) + (b_1 + b_2)u, \tag{12.4.9}$$

and also, for $x, y, z \in K$,

$$(x + y)z = xz + yz, \qquad (xy)z = x(yz), \qquad 1z = z, \tag{12.4.10}$$

together with the particular relation

$$u^2 = -1. \tag{12.4.11}$$

For each of the seven cases, further specific relations are needed:

Case I. $n = 5^2$, $u(1 - 2u) = -1 - 2u$.
Case II. $n = 11^2$, $u(1 + 5u) = -5 - 2u$, $u(4) = 4u$.
Case III. $n = 7^2$, $u(1 + 3u) = -1 - 2u$.
Case IV. $n = 23^2$, $u(1 - 6u) = 12 - 2u$, $u(2) = 2u$.
Case V. $n = 11^2$, $u(2 + 4u) = 1 - 3u$.
Case VI. $n = 29^2$, $u(1 - 7u) = -12 - 2u$, $u(16) = 16u$.
Case VII. $n = 59^2$, $u(9 + 15u) = -10 - 10u$, $u(4) = 4u$.

It is true, though not immediately obvious, that these relations determine the seven exceptions completely.

A VW system in which both distributive laws hold is called a *semifield*, or sometimes a nonassociative division ring. A. A. Albert [1] has given a method for constructing semifields of orders p^r, p an odd prime, r odd, and $r > 1$. In GF(p^r) let us construct a new product (x, y) by the rule

$$(x, y) = \tfrac{1}{2}(xy^p + x^p y). \tag{12.4.12}$$

If there were a product $(x, y) = 0$ with $x \neq 0$, $y \neq 0$, this would lead to

$$y^{p-1} = -x^{p-1}. \tag{12.4.13}$$

But as r and p are both odd, $m = (p^r - 1)/(p - 1)$ is odd. Raising (12.4.13) to the mth power, we get

$$1 = y^{p^r-1} = -x^{p^r-1} = -1, \tag{12.4.14}$$

which is a conflict, since $p \neq 2$. Thus, the product in (12.4.12) has no zero divisors. Hence, for $x \neq 0$, there is a unique $u \neq 0$ such that

$$x = (u, 1) = \tfrac{1}{2}(u + u^p). \tag{12.4.15}$$

We may define a one-to-one mapping α by

$$u = x\alpha \tag{12.4.16}$$

if u and x satisfy (12.4.15). We may now define a semifield K by a product $x \circ y$, where

$$x \circ y = (x\alpha, y\alpha). \tag{12.4.17}$$

The semifield K formed in this way is never isomorphic to GF(p^r) and the corresponding plane is therefore non-Desarguesian.

The two constructions in (12.4.8) and (12.4.17) yield non-Desarguesian planes of all orders $u = p^r$ for p odd and $r \geq 2$, and for $p = 2$, r even, $r \geq 4$. Semifields of orders 2^m, m odd, have been constructed by Donald Knuth. For $m \geq 5$, these are not fields and so yield non-Desarguesian planes. We give his construction.

Let m be odd and in GF(2^m) let x be an element such that $1, x, x^2, \ldots, x^{m-1}$ form a basis over GF(2). We construct a semifield $K(2^m)$ whose elements are the same as those of GF(2^m) and whose addition is the same. The multiplication $y \circ z$ of $K(2^m)$ is determined from the multiplication yz in GF(2^m) by the two distributive laws and the rules for the basis:

$$\begin{aligned} &x^i \circ x^j = x^i x^j, \qquad i, j = 0, 1, \ldots, m-2, \\ &x^i \circ x^{m-1} = x^{m-1} \circ x^i = x^{m-1}x^i + x^{2i} + x^i, \qquad i = 0, \ldots, m-2, \\ &x^{m-1} \circ x^{m-1} = x^{2m-2} + 1. \end{aligned} \tag{12.4.18}$$

From the rules (12.4.18), if y, z are elements in the additive system S with basis $1, x, \ldots, x^{m-2}$, we may derive the following rules:

$$\begin{aligned} y \circ z &= y \cdot z, \\ (x^{m-1} + y) \circ z &= z \circ (x^{m-1} + y) = z(x^{m-1} + z + 1 + y), \\ (x^{m-1} + y) \circ (x^{m-1} + z) &= x^{2m-2} + 1 + z(x^{m-1} + z + 1) \\ &\quad + y(x^{m-1} + y + 1) + yz \\ &= (x^{m-1} + y + 1)^2 \\ &\quad + (x^{m-1} + y + 1)(y + z) + (y + z)^2. \end{aligned} \tag{12.4.19}$$

The distributive laws clearly hold, and clearly 1 is a unit. To show that K is a semifield, we need only prove that there are no divisors of zero. For the three product rules of (12.4.19) in the first $yz \neq 0$, if $y \neq 0$ and $z \neq 0$ from the absence of zero divisors in GF(2^m). For the second rule, if $z(x^{m-1} + z + 1 + y) = 0$ and $z \neq 0$, then $x^{m-1} + z + 1 + y = 0$, which, since y, z, 1 are in S, would make x^{m-1} be in S, contrary to the fact that $1, x, \ldots, x^{m-2}$ are a basis for S. For the third rule, if the product $(x^{m-1} + y) \circ (x^{m-1} + z)$ is zero, we have

$$u^2 + uv + v^2 = 0, \tag{12.4.20}$$

with $u = x^{m-1} + y + 1 \neq 0$, $v = y + z$. Then $w = v/u$ would satisfy

$$w^2 + w + 1 = 0.$$

But in this case, 0, 1, w, $w + 1$ form a subfield of order 2^2 of GF(2^m). This is impossible if m is odd. Hence, K has no zero divisors and is a semifield. If $m \geq 5$,

$$x^m = (x \circ x^{(m-1)/2}) \circ (x^{(m-1)/2}) \neq x \circ (x^{(m-1)/2} \circ x^{(m-1)/2}) = x^m + x^2 + x,$$

and K is not a field.

For $m = 3$ in GF(2^3) the element x may satisfy either $f(x) = x^3 + x + 1 = 0$ of $f(x) = x^3 + x^2 + 1 = 0$. The circle product as determined by one of these choices of $f(x)$ gives the ordinary product as determined by the other $f(x)$. Hence, for $m = 3$, $K(2^3) = \text{GF}(2^3)$. Indeed, it has been established by an exhaustive search on a computer that for order 8 there is no finite plane except the Desarguesian one.

We may use a near-field in quite a different way to construct a non-Desarguesian plane. The first example, of order 9, was constructed by Veblen and Wedderburn [1], and the general construction was given by Hughes [1].

Let F be a field of order $q = p^h$ and K a near-field of order q^2 containing F as its center. Here, K may be any one of the exceptional near-fields or one of the regular ones, with $v = 2$. We begin with the Desarguesian plane π_0 of order q coordinatized by F. We represent the points of π_0 as triples (x, y, z), $x, y, z \in F$ and $(x, y, z) \neq (0, 0, 0)$ together with the scalar equivalence $(x, y, z) = (xu, yu, zu)$, $u \in F$, $u \neq 0$. If a, b, c are not all zero and are elements of F, those points $P = (x, y, z)$ satisfying $ax + by + cz = 0$ are a line, and every line of π_0 is given in this way. By the theorem of Singer (Theorem 11.3.1) π_0 has a collineation α of order $q^2 + q + 1$, which permutes the points of π_0 in a single cycle and also the lines in a single cycle. This collineation α can be represented as a linear transformation with coefficients from F. Thus, for $P = (x, y, z)$ and $P \to P\alpha$

$$\begin{aligned} P\alpha = (x, y, z)\alpha &= (x, y, z)A, \\ A = (a_{ij}), \qquad i, j &= 1, 2, 3, \qquad a_{ij} \in F \end{aligned} \tag{12.4.21}$$

where A is an appropriate 3×3 matrix. Here, $A^m = cI$, $c \in F$ for $m = q^2 + q + 1$, but for no lower power $0 < m < q^2 + q + 1$.

The trick used here is to construct a plane π of order q^2, using coordinates from K and assuming, with no apparent justification, that we may use the matrix A to define a collineation of this larger plane. Fortunately, everything works.

The points of our plane π will be the triples $(x, y, z) \neq (0, 0, 0)$, $x, y, z \in K$ with the equivalence $(x, y, z) = (xu, yu, zu)\, u \neq 0, u \in K$. We consider equations

$$x + ty + z = 0, \tag{12.4.22}$$

where either $t = 1$ or t is one of the $q^2 - q$ elements of K not in F. We define "base lines" L_t by saying $P = (x, y, z)$ is on L_t if and only if x, y, z satisfy (12.4.22). Since we have the right distributive law $(a + b)c = ac + bc$ and associativity of multiplication, it follows from (12.4.22) that

$$\begin{aligned} (x + ty + z)u &= 0, \\ xu + (ty)u + zu &= 0, \\ xu + t(yu) + zu &= 0, \end{aligned} \tag{12.4.23}$$

whence if (x, y, z) satisfies (12.4.22) so also does (xu, yu, zu), $u \neq 0$. Hence, incidence of a point P on L_t is well defined by (12.4.22). We now have $(q^2 - q + 1)$ "base lines" L_t, and it is easy to show that each of these contains $(q^2 + 1)$ points. We define further lines $L_t^{(i)}$, $i = 0, 1, \ldots, q^2 + q$ by

Rule: ■ *If* $(x, y, z) \in L_t$, *then* $(x, y, z)A^i \in L_t^{(i)}$ *and* $L_t^{(i)}$ *contains no further points.*

Note $L_t^{(0)} = L_t$. In this manner we have constructed

$$(q^2 - q + 1)(q^2 + q + 1) = q^4 + q^2 + 1$$

lines, each containing $(q^2 + 1)$ points. The chief difficulty lies in showing that if $L_t^{(i)}$ and $L_s^{(j)}$ are distinct, then they have exactly one point in common.

Let us consider the sets of points $L_t^{(i)}$. Whether or not they form the lines of a plane, they form a system of sets of points carried into each other by the linear transformation A. In particular, $L_t^{(g)}$ and $L_s^{(h)}$ have a point in common if and only if $L_t^{(g-h)}$ and $L_s^{(0)} = L_s$ have a point in common. Suppose $L_t^{(g-h)}$ and L_s have a point (x, y, z) in common. Then (x, y, z) is a point of L_s and $(x, y, z)A^{h-g}$ is a point of $L_t^0 = L_t$. If

$$A^{h-g} = (a_{ij}), \qquad i, j = 1, 2, 3,$$

then

$$(x, y, z)A^{h-g} = (a_{11}x + a_{21}y + a_{31}z, a_{12}x + a_{22}y + a_{32}z, a_{13}x + a_{23}y + a_{33}z). \tag{12.4.24}$$

Since (x, y, z) is on L_s and the point of (12.4.24) is on L_t, we have the two relations

$$(a_{11}x + a_{21}y + a_{31}z) + t(a_{12}x + a_{22}y + a_{32}z) + (a_{13}x + a_{23}y + a_{33}z) = 0, \tag{12.4.25}$$

$$x + sy + z = 0. \tag{12.4.26}$$

Solving (12.4.26) for x and substituting in (12.4.25), we have

$$uy + az + t(vy + bz) = 0, \tag{12.4.27}$$

where

$$u = a_{21} + a_{23} - (a_{11} + a_{13})s, \qquad v = a_{22} - a_{12}s,$$
$$a = a_{31} + a_{33} - (a_{11} + a_{13}), \qquad b = a_{32} - a_{21}. \tag{12.4.28}$$

We note that a, b are in F, the center of K. The proof is subdivided into three cases.

Case 1 ($b \neq 0$): ■ Then (12.4.24) can be rewritten as

$$(u - ab^{-1}v)y + (ab^{-1} + t)(vy + bz) = 0. \tag{12.4.29}$$

If $t = 1$, then (12.4.25) is of the form

$$b_1x + b_2y + b_3z = 0, \qquad b_i \in F \tag{12.4.30}$$

and this has a unique solution point (x, y, z) with (12.4.26) except when $s = 1$ and $b_1 = b_2 = b_3 \neq 0$, whence (12.4.30) and also (12.4.26) are the same as $x + y + z = 0$. But A permutes the lines of the subplane π_0 in a cycle of length $q^2 + q + 1$, and so this exceptional situation can arise only when $s = t = 1$ and $g = h$—when the two given lines are identical. If $t \neq 1$, then $t \notin F$, and so $w = ab^{-1} + t \neq 0$. Multiply (12.4.29) on the left by w^{-1}, and using associativity and the right distributive law, we have

$$[w^{-1}(u - ab^{-1}v) + v]y + bz = 0. \tag{12.4.31}$$

Here, since $b \neq 0$, (12.4.31) and (12.4.26) have a unique point (x, y, z) as a solution to both, and this is the unique point in common to our two lines.

Case 2 ($b = 0, a \neq 0$): ■ Here (12.4.27) takes the form

$$(u + tv)y + az = 0, \tag{12.4.32}$$

and as $a \neq 0$, (12.4.32) with (12.4.26) determines a unique point (x, y, z).

Case 3 ($b = 0, a = 0$): ■ Thus

$$a_{31} + a_{33} = a_{11} + a_{13} \qquad \text{and} \qquad a_{32} = a_{12},$$

whence from (12.4.24) we have

$$(1, 0, -1)A^{h-g} = (a_{11} - a_{31})(1, 0, -1). \tag{12.4.33}$$

Here, $a_{11} - a_{31} \neq 0$, since A is not singular and A^{h-g} fixes the point $(1, 0, -1)$ of the subplane π_0. Hence, $h - g \equiv 0 (\text{mod } q^2 + q + 1)$, and as we chose h, g in the range $0, \ldots, q^2 + q$, we have $h = g$ and our equations (12.4.25) and (12.4.26) are $x + ty + z = 0$ and $x + sy + z = 0$. Since our lines were not identical, we have $t \neq s$ and these equations have the unique point $(1, 0, -1)$ in common. This completes the third case.

We now have $N = q^4 + q^2 + 1$ lines, each containing $(q^2 + 1)$ points. Since two distinct points are on at most one line, a point P_i that is on m_i lines is joined to m_iq^2 other points. As the total number of points is $q^4 + q^2 + 1$, we must have $m_i \leq q^2 + 1$ for every point. The total number of ordered pairs of points on our $N = q^4 + q^2 + 1$ lines is $N(q^2 + 1)q^2$. Hence,

$$\sum_{i=1}^{N} m_i q^2 = N(q^2 + 1)q^2,$$

whence $m_i = q^2 + 1$ in every case and so every point is on $(q^2 + 1)$ lines and is joined to $(q^4 + q^2)$ other points—that is to say, to every other point. Thus, we do have a plane.

The original example of Veblen and Wedderburn is a non-Desarguesian plane of order 9. We have a collineation of order 13 that maps subscripts $i \to i + 1 (\bmod 13)$ of both points and lines. The seven base lines are

$$\begin{aligned}
L_0&: A_0\ A_1\ A_3\ A_9\ B_0\ C_0\ D_0\ E_0\ F_0\ G_0,\\
M_0&: A_0\ B_1\ B_8\ D_3\ D_{11}\ E_2\ E_5\ E_6\ G_7\ G_9,\\
N_0&: A_0\ C_1\ C_8\ E_7\ E_9\ F_3\ F_{11}\ G_2\ G_5\ G_6,\\
P_0&: A_0\ B_7\ B_9\ D_1\ D_8\ F_2\ F_5\ F_6\ G_3\ G_{11},\\
Q_0&: A_0\ B_2\ B_5\ B_6\ C_3\ C_{11}\ E_1\ E_8\ F_7\ F_9,\\
R_0&: A_0\ C_7\ C_9\ D_2\ D_5\ D_6\ E_3\ E_{11}\ F_1\ F_8,\\
S_0&: A_0\ B_3\ B_{11}\ C_2\ C_5\ C_6\ D_7\ D_9\ G_1\ G_8.
\end{aligned} \tag{12.4.34}$$

13

Orthogonal Latin Squares

13.1 Orthogonality and Orthogonal Arrays

It has been observed by Donald Knuth and R. C. Bose that the property of being a Latin square can be expressed as an orthogonality relation. Let us consider two $n \times n$ matrices $A = (a_{ij})$ and $B = (b_{ij})$, whose entries are the numbers $1, \ldots, n$. We say that A is orthogonal to B (and write $A \perp B$) if for every ordered pair (a, b) of numbers, there is at least one set of indices i, j such that $a_{ij} = a, b_{ij} = b$. Since there are n^2 cells and n^2 ordered pairs (a, b), we can replace the words "at least" by "at most" or by "exactly." The two matrices R and C given by

$$R = \begin{bmatrix} 1 & 1 & \cdots & 1 \\ 2 & 2 & \cdots & 2 \\ \cdot & \cdot & \cdots & \cdot \\ \cdot & \cdot & \cdots & \cdot \\ \cdot & \cdot & \cdots & \cdot \\ n & n & \cdots & n \end{bmatrix}, \qquad C = \begin{bmatrix} 1 & 2 & \cdots & n \\ 1 & 2 & \cdots & n \\ \cdot & \cdot & \cdots & \cdot \\ \cdot & \cdot & \cdots & \cdot \\ \cdot & \cdot & \cdots & \cdot \\ 1 & 2 & \cdots & n \end{bmatrix} \tag{13.1.1}$$

are clearly orthogonal. A set of matrices is said to be mutually orthogonal if any two of them are orthogonal. The relation of orthogonality is unchanged by

(1) any substitution P on the numbers $1, \ldots, n$ in one matrix if i is replaced by $P(i)$ throughout, and (2) any permutation of the n^2 cells of all the matrices simultaneously. If $A \perp B$, each of A and B must contain each number $1, \ldots, n$ exactly n times, and so by permuting the cells appropriately we may transform A into R and B into C simultaneously.

From the considerations given above, we see that an $n \times n$ matrix L is a Latin square on $1, \ldots, n$ if and only if $L \perp R$ and $L \perp C$. Conversely, if E, F, G are three mutually orthogonal matrices, we may permute the cells so that two of them become R and C and the third becomes a Latin square L. In this way the property of "latinity" is definable in terms of orthogonality alone. The existence of $(k + 2)$ mutually orthogonal matrices is exactly equivalent to the existence of k mutually orthogonal Latin squares. Furthermore any one of the matrices can be transformed into R and any other into C. This corresponds to the choice of parallel pencils for coordinatization in a finite plane, as was discussed in Chapter 12.

The existence and construction of mutually orthogonal Latin squares has been reduced to the relation of orthogonality. A representation well suited to this approach is called an "orthogonal array." Let us say that two row vectors v_1 and v_2 of length n^2,

$$v_1 = (x_1, \ldots, x_{n^2}) \qquad \text{and} \qquad v_2 = (y_1, \ldots, y_{n^2})$$

are orthogonal if the ordered pairs (x_i, y_i), $i = 1, \ldots, n^2$ include every pair (a, b), a, b from $1, \ldots, n$ exactly once.

DEFINITION. *An orthogonal array* OA(n, s) *of order n and depth s is a matrix with s rows and n^2 columns with entries the numbers* $1, \ldots, n$ *such that every pair of two rows is orthogonal.*

We may associate the columns of OA(n, s) in any order with the n^2 cells of an $n \times n$ matrix and then construct s matrices $A_1, A_2, \ldots, A_s$, using the ith row of OA(n, s) to fill in the cells of A_i in the order given by the association of column with cells. The orthogonality of matrices A_i and A_j is precisely equivalent to the orthogonality of the ith and jth rows of OA(n, s). Conversely, if mutually orthogonal squares $A_1, \ldots, A_s$ of order n are given, we may construct an OA(n, s) from them by reversing this process.

Permuting rows or columns of an OA(n, s) gives an OA(n, s). Also, performing a substitution P on the numbers $1, \ldots, n$ of a row of OA(n, s) gives another OA(n, s). Two orthogonal arrays that may be obtained from each other are called "equivalent." If we drop a row from an OA(n, s), the remaining array is an OA$(n, s - 1)$.

13.2 Main Theorems

From Chapter 12 we know that there is a finite plane of order n if n is a prime power, $n = p^r$. It was shown there that this is equivalent to the existence of

$(n - 1)$ mutually orthogonal Latin squares of order n, or to $(n + 1)$ orthogonal matrices. The corresponding orthogonal array is an OA$(n, n + 1)$. It is easy to see that equivalent orthogonal arrays define the same plane. In general, we may consider an OA(n, s) as defining s pencils of parallel lines, the columns being points and a point (column j) being on the uth line of the ith pencil if $a_{ij} = u$ in OA(n, s). Thus, equivalence of orthogonal arrays consists merely in relabeling points and lines but leaving the incidences unaltered.

THEOREM 13.2.1 (MACNEISH [1]). *If there is an* OA(n_1, s) *and an* OA(n_2, s), *then there is an* OA(n_1n_2, s).

Proof: ∎ Let OA(n_1, s) be the matrix

$$A = (a_{ij}), \qquad i = 1, \ldots, s, \qquad j = 1, \ldots, n_1^2,$$

and OA(n_2, s) be the matrix

$$B = (b_{ij}), \qquad i = 1, \ldots, s, \qquad j = 1, \ldots, n_2^2.$$

Form a new matrix

$$D = (d_{ij}), \qquad i = 1, \ldots, s, \qquad j = 1, \ldots, n_1^2n_2^2$$

by replacing a_{ij} in A by the row vector

$$(b_{i1} + m_{ij}, b_{i2} + m_{ij}, \ldots, b_{in_2^2} + m_{ij}), m_{ij} = (a_{ij} - 1)n_2^2 \qquad \text{for every } i, j.$$

As the numbers a_{ij} run from 1 to n_1 and the numbers b_{ij} from 1 to n_2, the numbers $b_{it} + m_{ij}$ run from 1 to n_1n_2, whence every d_{ij} is one of the numbers $1, \ldots, n_1n_2$. Consider the hth row and the ith row of D and let u, v be any two numbers in the range $1, \ldots, n_1n_2$. Then we may write

$$u = u_1 + (u_2 - 1)n_2, \qquad v = v_1 + (v_2 - 1)n_2,$$

with $1 \leq u_1, v_1 \leq n_2$, $1 \leq u_2, v_2 \leq n_1$ uniquely. In A, let us determine j as that column in which $a_{hj} = u_2$, $a_{ij} = v_2$. In B, let us determine t as that column in which $b_{ht} = u_1$ and $b_{it} = v_1$. Then in D, in column $g = t + n_2(j - 1)$, we have

$$d_{hg} = b_{ht} + (a_{hj} - 1)n_2 = u_1 + (u_2 - 1)n_2 = u$$

and

$$d_{ig} = b_{it} + (a_{ij} - 1)n_2 = v_1 + (v_2 - 1)n_2 = v.$$

This yields the orthogonality of the hth and ith rows of D and so proves that D is an orthogonal array.

An alternate form of the proof of this theorem is as follows: Let $A_1, \ldots, A_s$ be mutually orthogonal of order n_1, where instead of numbers $1, \ldots, n_1$, we place indeterminates x_u, $u = 1, \ldots, n_1$, and similarly $B_1, \ldots, B_s$ mutually orthogonal of order n_2 on indeterminates y_v, $v = 1, \ldots, n_2$. Then the direct products

$$A_1 \times B_1, A_2 \times B_2, \ldots, A_s \times B_s,$$

considered as matrices with the n_1n_2 indeterminates

$$z_{uv} = x_u y_v, \qquad u = 1, \ldots, n_1, \qquad v = 1, \ldots, n_2,$$

are mutually orthogonal. For two given indices h, i in the range $1, \ldots, s$ and $z_{u_1v_1}$ and $z_{u_2v_2}$, there is a cell that contains x_{u_1} in A_h and x_{u_2} in A_i, and a cell that contains y_{v_1} in B_h and y_{v_2} in B_i. The corresponding cell in $A_h \times B_h$ and $A_i \times B_i$ contains $z_{u_1v_1}$ and $z_{u_2v_2}$, respectively.

The next theorem is essentially a corollary of Theorem 13.2.1.

THEOREM 13.2.2. *If $n = p_1^{e_1}p_2^{e_2}\cdots p_r^{e_r}$ is the factorization of the integer n into powers of the distinct primes $p_1, \ldots, p_r$, then there exist at least $N(n)$ mutually orthogonal Latin squares of order n where $N(n) \geq \min(p_i^{e_i} - 1)$, $i = 1, \ldots, r$.*

Proof: ∎ From the existence of a finite plane of order $n_i = p_i^{e_i}$, $i = 1, \ldots, r$, there is an OA(n_i, $n_i + 1$), $i = 1, \ldots, r$. Thus, with s the minimum of all the $n_i + 1$, there is an OA(n,s). By repeated application of Theorem 13.2.1 there is an OA(n, s) and so there are $s - 2 = \min(n_i - 1)$ mutually orthogonal Latin squares of order n. This is the statement of the theorem.

If we let $N(n)$ denote the maximum number of mutually orthogonal squares of order n, then this theorem gives a lower bound for $N(n)$. In particular, if n is of the form $n = 4t + 2$, this theorem gives us only one Latin square. Euler conjectured in 1782 that no pair of orthogonal squares existed for $n = 4t + 2$, saying, "I do not hesitate to conclude that it is impossible to produce any complete square of 36 entries, and the same impossibility extends to the cases of $n = 10$, $n = 14$, and in general to all unevenly even numbers." MacNeish in 1922 went so far as to conjecture that

$$N(n) = \min(p_i^{e_i} - 1), \qquad i = 1, \ldots, r \qquad \text{for} \qquad n = p_1^{e_1} \cdots p_r^{e_r},$$

and published an erroneous proof that $N(4t + 2) = 1$, using a topological method. Both the Euler and MacNeish conjectures were finally disproved in 1959, as we shall see. Indeed, Euler was correct only for the cases $n = 2$ and $n = 6$. The case $n = 6$ was settled by Tarry [1] in 1900, who enumerated all possible cases.

Not every Latin square has another Latin square orthogonal to it, or, as we shall say, an orthogonal mate. A condition for the existence of an orthogonal mate is due to Mann [1].

Let the n numbers be divided into two categories A and B, where A contains k letters and B contains $(n - k)$ letters. We shall further require that k be odd. The columns of the OA(n, 3) corresponding to a Latin square of order n can be divided into eight types according to the category of the number appearing in each row. We write

$$\begin{array}{cccccccc} x_1 & x_2 & x_3 & x_4 & x_5 & x_6 & x_7 & x_8 \\ A & A & A & A & B & B & B & B \\ A & A & B & B & A & A & B & B \\ A & B & A & B & A & B & A & B \end{array} \tag{13.2.1}$$

for which x_i denotes the number of columns of the type. The orthogonality of the rows gives a number of relations on the x's, since between any two rows there are k^2 combinations AA, $k(n-k)$ combinations BA or AB, and $(n-k)^2$ combinations BB. This leads to the following relations, among others:

$$\begin{aligned} &x_1 + x_2 = k^2, \qquad x_2 + x_4 = k(n-k), \qquad x_7 + x_8 = (n-k)^2, \\ &x_1 + x_3 = k^2, \qquad x_5 + x_6 = k(n-k), \\ &x_1 + x_5 = k^2, \qquad x_5 + x_7 = k(n-k). \end{aligned} \tag{13.2.2}$$

These yield

$$\begin{aligned} x_2 &= k^2 - x_1, & x_6 &= nk - 2k^2 + x_1, \\ x_3 &= k^2 - x_1, & x_7 &= nk - 2k^2 + x_1, \\ x_4 &= nk - 2k^2 + x_1, & x_8 &= n^2 - 3nk + 3k^2 - x_1. \\ x_5 &= k^2 - x_1, \end{aligned} \tag{13.2.3}$$

The other relations are redundant. Let us call odd columns those containing an odd number of A's, namely, those numbered x_1, x_4, x_6, and x_7. The total number of odd columns is $4x_1 + 3nk - 6k^2$. Let us suppose that this number is less than n:

$$4x_1 + 3nk - 6k < n. \tag{13.2.4}$$

Now, if a fourth row can be added to make an OA(n, 4), since there are less than n odd columns there must be a number (say, u) in the fourth row that appears only in the even columns:

$$\begin{matrix} y_2 & y_3 & y_5 & y_8 \\ A & A & B & B \\ A & B & A & B \\ B & A & A & B \\ u & u & u & u \end{matrix} \tag{13.2.5}$$

Here the y's represent the number of columns of each kind. Since this accounts for all occurrences of u, we have

$$\begin{gathered} y_2 + y_3 + y_5 + y_8 = n, \\ y_2 + y_3 = k, \qquad y_3 + y_5 = k, \\ y_2 + y_5 = k, \end{gathered} \tag{13.2.6}$$

where the first counts the total number of occurrences of u, and the others count the pairing of u with A's in the first, second, and third rows, respectively. But these conditions yield

$$2(y_2 + y_3 + y_5) = 3k, \tag{13.2.7}$$

which conflicts with our assumption that k is odd. Hence, with k odd, the condition (13.2.4) leads to a contradiction. Appropriate choices of k, n, and x give

THEOREM 13.2.3 (MANN). *Let L be a Latin square of side* $4t + 2$ *(alternately* $4t + 1$*) with a subsquare of side* $2t + 1(2t)$ *in which all entries are from a set of* $2t + 1(2t)$ *numbers except for* $t([t - 1/2])$ *or less of the cells. Then there is no Latin square orthogonal to L.*

Proof: ■ In the first case take $k = 2t + 1$. Condition (13.2.4) reduces to $4x_1 < 4t + 2$ or $x_1 \leq t$. This means that in the subsquare of side $2t + 1$ corresponding to the columns headed x_1 and x_2, all entries are B's except at most t A's. In the second case, with $n = 4t + 1$, $k = 2t + 1$, condition (13.2.4) reduces to $4x_1 < 10t + 4$, whence the subsquare of side $2t$ of the two columns headed by x_7 and x_8 contains at most $x_7 = x_1 - 2t - 1$ A's. But our condition is equivalent to

$$x_7 < \frac{t}{2} \qquad \text{or} \qquad x_7 \leq \left[\frac{t-1}{2}\right],$$

the statement of the theorem. For n of the form $4t$ or $4t + 3$ there is no comparable result.

For orthogonal Latin squares of order 10, Ostrowski and Van Duren have found, using a computer, a pair in which one of the squares has a subsquare of order 5 in which all but three of the entries are the numbers 0 through 4, showing that for $n = 10$. Mann's inequality is the best possible of its kind. The squares are as follows:

01234	56789	01923	84657
34012	79865	67895	23104
43120	97658	93746	58210
12407	85396	38254	79061
20375	68941	14507	36982
57698	34120	25619	40873
89756	12034	40138	62795
65981	43207	56480	17329
98563	01472	82071	95436
76849	20513	79362	01548

(13.2.8)

13.3 Constructions of Orthogonal Squares

We begin with a construction that will prove $N(12t + 10) \geq 2$. Let m be a number such that a pair of orthogonal squares of order m exists. Define the following vectors of length m of residues modulo $2m + 1$ for $i = 0, 1, \ldots, 2m$:

$$\begin{aligned} a_i &= (i, i, i, \ldots), \\ b_i &= (i + 1, i + 2, \ldots, i + m), \\ c_i &= (i - 1, i - 2, \ldots, i - m). \end{aligned} \tag{13.3.1}$$

The differences of these vectors are

$$\begin{aligned} d_1 &= a_i - b_i = (2m, 2m - 1, \ldots, m + 1), \\ d_1' &= b_i - a_i = (1, 2, \ldots, m), \\ d_2 &= a_i - c_i = (1, 2, \ldots, m), \\ d_2' &= c_i - a_i = (2m, 2m - 1, \ldots, m + 1), \\ d_3 &= b_i - c_i = (2, 4, \ldots, 2m), \\ d_3' &= c_i - b_i = (2m - 1, 2m - 3, \ldots, 1). \end{aligned} \tag{13.3.2}$$

Here, d_j and d_j' for $j = 1, 2, 3$ together contain all nonzero residues modulo $2m + 1$. Now construct the vectors of length $m(2m + 1)$:

$$\begin{aligned} A &= (a_0, a_1, a_2, \ldots, a_{2m}), \\ B &= (b_0, b_1, b_2, \ldots, b_{2m}), \\ C &= (c_0, c_1, c_2, \ldots, c_{2m}). \end{aligned} \tag{13.3.3}$$

We take m new letters $x_1, \ldots, x_m$ and form a vector X of length $m(2m + 1)$:

$$X = (x_1, \ldots, x_m, x_1, \ldots, x_m, \ldots, x_1, \ldots, x_m). \tag{13.3.4}$$

Now we form the $4 \times 4m(2m + 1)$ matrix D:

$$D = \begin{bmatrix} A & B & C & X \\ B & A & X & C \\ C & X & A & B \\ X & C & B & A \end{bmatrix} \tag{13.3.5}$$

Consider any two rows of this array. They contain one of $\begin{smallmatrix} AB \\ BA \end{smallmatrix}$, $\begin{smallmatrix} AC \\ CA \end{smallmatrix}$, or $\begin{smallmatrix} BC \\ CB \end{smallmatrix}$. Thus, if u, v are distinct residues modulo $2m + 1$, there is a residue

$$e \equiv u - v \pmod{2m + 1},$$

such that e occurs in the appropriate d_j or d_j'. Thus,

$$e \equiv (i + g) - (i + h) \equiv g - h \equiv u - v \pmod{2m + 1},$$

where $(i + g) - (i + h)$ is the difference in d_j or d_j'. Determine i by $i + h \equiv v \pmod{2m + 1}$ and then,

$$i + g \equiv (i + h) + u - v \equiv u \pmod{2m + 1},$$

whence the pair $\binom{u}{v}$ occurs as a column in these two rows. Similarly, one of

$\frac{AX}{XA}$, $\frac{BX}{XB}$, or $\frac{CX}{XC}$ occurs, and from these we get all columns $\binom{x_j}{u}$ and $\binom{u}{x_j}$, with u any residue modulo $2m + 1$ and x_j any one of $x_1, \ldots, x_m$. If we suppose that there exists a pair of orthogonal squares of order m, there exists an OA$(m, 4) = E$ with entries $x_1, \ldots, x_m$, a matrix of size $4 \times m^2$ whose rows are orthogonal in the x's. We now form the matrix F:

$$F = \begin{bmatrix} 0 & 1 & 2 & \cdots & 2m \\ 0 & 1 & 2 & \cdots & 2m \\ 0 & 1 & 2 & \cdots & 2m \\ 0 & 1 & 2 & \cdots & 2m \end{bmatrix} \quad DE. \tag{13.3.6}$$

Here F has 4 rows and $2m + 1 + 4m(2m + 1) + m^2 = (3m + 1)^2$ columns. Between any two rows of F we get columns $\binom{u}{u}$, u a residue modulo $2m + 1$ from the first $2m + 1$ columns, columns $\binom{x_i}{x_j}$ from E and the rest from D. Hence, F is an OA$(3m + 1, 4)$. We state our result as

THEOREM 13.3.1. *If $N(m) \geq 2$, then $N(3m + 1) \geq 2$. In particular, since $N(4t + 3) \geq 2$, it follows that $N(12t + 10) \geq 2$.*

The main theorem on construction of orthogonal squares is a recursive construction depending on the existence of a pairwise balanced design. A *pairwise balanced design* BIB$(v, k_1, \ldots, k_m, \lambda)$ is an arrangement of v objects into b blocks such that (1) each block contains $(k_i < v)$ distinct objects for some $i = 1, \ldots, m$; and (2) each pair of objects occurs together in exactly λ blocks. Let b_i be the number of blocks with k_i elements. Then

$$\sum b_i = b, \qquad \lambda v(v - 1) = \sum b_i k_i(k_i - 1). \tag{13.3.7}$$

The b_i blocks with k_i elements will be called the ith *equiblock component.* A *clear set* is a set of equiblock components in which no two blocks have an element in common. We write

$$\text{BIB}(v, k_1, \ldots, k_{\bar{r}}, \ldots, k_m, \lambda),$$

if the first r equiblock components form a clear set. For our constructions we use only pairwise balanced designs with $\lambda = 1$.

The following theorem, in a more restricted form, was first given by Parker [2], later generalized by Bose and Shrikhande [1], and still further generalized by all three authors together [1].

THEOREM 13.3.2. *If there is a pairwise balanced design*

$$\text{BIB}(v, k_1, \ldots, k_{\bar{r}}, \ldots, k_m, 1),$$

then

$$N(v) \geq \min(N(k_1), \ldots, N(k_r), \quad N(k_{r+1}) - 1, \ldots, N(k_m) - 1).$$

Proof: ■ The clear set is treated differently from the rest of the design and the conclusion of the theorem is correspondingly different. The clear set is roughly analogous to parallel lines in an affine plane, the nonclear set to lines in a projective plane.

Let us put

$$c = \min(N(k_1) + 2, \ldots, N(k_r) + 2, N(k_{r+1}) + 1, \ldots, N(k_m) + 1).$$

For $i = 1, \ldots, r$, put $A_i = \mathrm{OA}(k_i, c)$. For $i = r + 1, \ldots, n$, an

$$\mathrm{OA}(k_i, c + 1) = D_i$$

exists. In D_i, a matrix on numbers $1, \ldots, k_i$, let us move the k_i columns with 1 in the first row to the beginning and renumber the remaining rows so that the first k_i columns of D_i are of the form

$$\begin{bmatrix} 1 & 1 & \cdots & 1 \\ 1 & 2 & \cdots & k_i \\ 1 & 2 & \cdots & k_i \\ \cdot & \cdot & \cdots & \cdot \\ \cdot & \cdot & \cdots & \cdot \\ \cdot & \cdot & \cdots & \cdot \\ 1 & 2 & \cdots & k_i \end{bmatrix}. \tag{13.3.8}$$

From D_i, $i = r + 1, \ldots, m$, we form a matrix A_i by deleting the first row and first k_i columns of D_i. A_i is a $c \times (k_i^2 - k_i)$ matrix, which has in every two rows all columns $\binom{u}{w}$ with $u \neq w$, $u, w = 1, \ldots, k_i$, but no columns $\binom{u}{u}$. Let the b blocks of the given pairwise balanced design be $S_1, S_2, \ldots, S_b$. If S_j has k_i elements, form an array B_j from A_i, replacing the numbers $1, \ldots, k_i$ of A_i by the numbers of the objects in S_j. Finally, form the array

$$C = (B_1, B_2, \ldots, B_b, E), \tag{13.3.9}$$

where E is an extra set of columns, each column of E consisting of the same number repeated c times and E has one column for each number that did not appear in the clear set.

We assert that the array C of (13.3.9) is an $\mathrm{OA}(v, c)$. Choose two rows of C. If $u \neq w$ are two distinct elements of the v elements of our pairwise balanced design, there is exactly one block S_j containing both elements, and in the corresponding B_j there will be a column $\binom{u}{w}$ in the two rows. Also, u and w do not both occur in any other B, nor can both of them occur in a column of E. If u is an element of the clear set in the block S_j the matrix B_j contains a column $\binom{u}{u}$ in the two chosen rows. If u is not in the clear set, there is a column $\binom{u}{u}$ in E. Thus, the two rows are orthogonal and C is an $\mathrm{OA}(v, c)$. Hence, there are at least $(c - 2)$ mutually orthogonal squares of order v, and $N(v) \geq c - 2$ is the statement of our theorem, which was to be proved.

The finite plane of order 4 is a BIB(21, 5, 1), and applying Theorem 13.3.2 to it yields $N(21) \geq N(5) - 1 = 3$. This is an improvement of the MacNeish value $n(21) = 2$, and was the first case in which MacNeish's conjecture was disproved, appearing in the paper by Parker.

The next two theorems are almost immediate applications of the main theorem.

THEOREM 13.3.3. *If there is a* BIB$(v, k, 1)$, *then* (1)

$$N(v - 1) \geq min(N(k - 1), N(k) - 1)$$

and (2)

$$N(v - x) \geq min(N(k - x), N(k - 1) - 1, N(k) - 1),$$

for any x such that $1 \leq x \leq k$.

In this theorem we use the convention $N(0) = N(1) = \infty$.

Proof: ■ Delete x objects of a single block from this block and all others. If $x = 1$, the blocks from which x has been deleted form a clear set of blocks with $(k - 1)$ objects, and we have a BIB$(v - 1, k - 1, k, 1)$. Application of Theorem 13.3.2 gives part (1). If $x > 1$, the block with $(k - x)$ objects remaining is a clear set and all other blocks have $(k - 1)$ or k objects. Application of Theorem 13.3.2 gives part (2).

THEOREM 13.3.4. *If there is a* BIB$(v, k, 1)$, *then*

$$N(v - 3) \geq min(N(k - 2), N(k - 1) - 1, N(k) - 1).$$

Proof: ■ Delete from BIB$(v, k, 1)$ three objects that are not in the same block. This gives a BIB$(v - 3, k - 2, k - 1, k, 1)$, since the three blocks from which two objects have been deleted form a clear set and all remaining blocks have $(k - 1)$ or k objects. Applying the main theorem gives us our result. In particular, the BIB(21, 5, 1) gives

$$N(18) \geq \min(N(3), N(4) - 1, N(5) - 1) = \min(2, 2, 3) = 2,$$

proving the existence of a pair of orthogonal squares of order 18.

A design is called "resolvable" if the blocks can be separated into r sets (replications) such that each set contains each object exactly once. For example, the lines of an affine plane of order n divided into $(n + 1)$ sets of parallel lines form a resolvable design.

THEOREM 13.3.5. *If there is a resolvable* BIB$(v, k, 1)$ *with r replications, then*

1. $N(v + x) \geq \min(N(x), N(k) - 1, N(k + 1) - 1)$ for $1 \leq x < r - 1$.
2. $N(v + r - 1) \geq \min(N(r - 1), N(k), N(k + 1) - 1)$.
3. $N(v + r) \geq \min(N(r), N(k + 1) - 1)$.

Proof: ■ Let $1 \leq x \leq r$. To each block of the ith replication add a new object Y_i, $1 \leq i \leq x$. Then add the block $Y_1, \ldots, Y_x$. If $1 \leq x < r - 1$, we have a BIB($v + x, x, k, k + 1, 1$). If $x = r - 1$, we have a BIB($v + r - 1, r - 1, k, k + 1, 1$). If $x = r$, we have a BIB($v + r, r, k + 1, 1$). The results now come from the main theorem. As an application of this, the affine plane of order 7 is a BIB(49, 7, 1). With $x = 1$, we find $N(50) \geq 5$ and with $x = 5$, we find $N(54) \geq 4$.

Another kind of design intimately related to orthogonal arrays is a *group divisible design* GD($v; k, m; \lambda_1, \lambda_2$). This is a design in which the objects are divided into m groups, each containing m elements, such that (1) each block contains k objects, and (2) any two objects from the same group occur together in λ_1 blocks while any two objects from different groups occur together in λ_2 blocks. The only case that will be needed here is that in which $\lambda_1 = 0$, $\lambda_2 = 1$.

THEOREM 13.3.6. *If $k \leq N(m) + 1$, there exists a resolvable* GD($km; k, m; 0, 1$).

Proof: ■ We assume that an orthogonal array OA($m, k + 1$) exists. Arrange the columns so that the last row is of the form

$$1, \ldots, 1, 2, \ldots, 2, 3, \ldots, 3, \ldots, m, \ldots, m. \tag{13.3.10}$$

This divides the columns into m sets of m each, the ith set being that for which the entries in the last row are i. This divides the first k rows into m sections. Now drop the last row and replace the number i in the jth row by the ordered pair (i, j), these being km objects. Let the columns be the blocks of a new design. Objects with the same second coordinate never occur in the same block and objects whose second coordinates differ occur together exactly once. Furthermore, the design is resolvable, since the orthogonality of the (k + 1st) row to the others in our original array assures us that in each of the m sections, we have each of the km pairs (i, j) exactly once.

From Theorems 13.3.5 and 13.3.6 we have relations between orthogonal arrays and resolvable designs so that from each we may construct the other.

THEOREM 13.3.7. *If there is a resolvable* GD($v; k, m; 0, 1$) *with r replications, then*

1. $N(v + x) \geq \min(N(m), N(x), N(k) - 1, N(k + 1) - 1)$ for $1 \leq x < r$.
2. $N(v + r) \geq \min(N(m), N(r), N(k + 1) - 1)$.
3. $N(v + r) \geq \min(N(k), N(r), N(k + 1) - 1, N(m + 1) - 1)$.
4. $N(v + r + 1) \geq \min(N(r + 1), N(k + 1) - 1, N(m + 1) - 1)$.

Proof: ■ To the GD($v; k, m; 0, 1$) let us first adjoin k blocks, each consisting of the m objects in a group. This is a BIB($v, m, k, 1$). Let us also adjoin a new object Y_i to each block of the ith group, $i = 1, \ldots, x$, but not to the block consisting of the entire group, and also adjoin a new block $Y_1, \ldots, Y_x$. For $x < r$, this gives us a

$$\text{BIB}(v + x, x, m, k, k + 1, 1).$$

For $x = r$, it gives a

$$\text{BIB}(v + r, r, m, k + 1, 1).$$

The main theorem now gives us the first two parts of our theorem. For the third part, take $x = r - 1$, leaving the last replication clear, adjoin Y_0 to each of the group blocks and take the single block $Y_0, Y_1, \ldots, Y_{r-1}$, giving a

$$\text{BIB}(v + r, k, r, k + 1, m + 1, 1).$$

For the fourth part, add Y_i, $i = 1, \ldots, r$ to the blocks of the ith group, Y_0 to the group blocks, and take the single block $Y_0, Y_1, \ldots, Y_r$. This gives us a

$$\text{BIB}(v + r + 1, r + 1, k + 1, m + 1, 1),$$

and so the fourth part of the theorem follows.

Theorems 13.3.6 and 13.3.7 may be combined to yield the important result

THEOREM 13.3.8. *If* $k \leq N(m) + 1$, *then for* $1 \leq x < m$,

$$N(km + x) \geq \min(N(m), N(x), N(k) - 1, N(k + 1) - 1).$$

Proof: ■ This uses Theorem 13.3.6, and we take the first part of Theorem 13.3.7 with $v = km$.

13.4 The End of the Euler Conjecture

Euler's conjecture, that there does not exist a pair of orthogonal squares of order n when n is of the form $4t + 2$, turns out to be true for $n = 2$ and $n = 6$, but in no other case. This section is devoted to proving this, and so completely disposing of the conjecture. This we state as

THEOREM 13.4.1. *For every order* $n > 6$, *there exists a pair of orthogonal Latin squares.*

Proof: ■ We must prove $N(n) \geq 2$ for $n > 6$. From Theorem 13.2.2 it is sufficient to prove this for n of the form $n = 4t + 2$. Our first lemma reduces the proof to a finite number of values of n.

LEMMA 13.4.1. *If* $N(4t + 2) \geq 2$ *for* $10 \leq 4t + 2 \leq 726$, *then* $N(v) \geq 2$ *for all* $v > 6$.

Proof: ■ We must prove this for $v = 4t + 2$, $v \geq 730$. Here, $v - 10 \geq 720$ and so we may write

$$v - 10 = 144g + 4u, \qquad g \geq 5, \qquad 0 \leq u \leq 35. \tag{13.4.1}$$

This gives

$$v = 4(36g) + 4u + 10, \qquad g \geq 5, \qquad 0 \leq u \leq 35. \qquad (13.4.2)$$

Now consider Theorem 13.3.8 with $k = 4$, $m = 36g$, and $x = 4u + 10$. Here, $N(4) = 3$, $N(5) = 4$. With $m = 36g$, the smallest prime power dividing m is at least 4, and so $N(m) \geq 3$ and the condition $k \leq N(m) + 1$ is satisfied for $k = 4$. As $0 \leq u \leq 35$, we have $10 \leq x \leq 150$; and as $g \geq 5$, $m \geq 180$ and the condition $1 \leq x < m$ is satisfied. Thus, application of Theorem 13.3.8 gives

$$N(v) \geq \min(N(m), N(x), 2, 3) = \min(N(x), 2)$$

and so if $N(x) \geq 2$, then $N(v) \geq 2$. This shows that if $N(x) \geq 2$ for $10 \leq x \leq 726$, then

$$N(v) = N(4t + 2) \geq 2 \qquad \text{for all} \qquad v = 4t + 2 \geq 730,$$

as stated in our lemma.

We begin with three special examples.

Example 1: ■ For $n = 14$, consider the matrix

$$P_0 = \begin{bmatrix} 0 & x_1 & x_2 & x_3 \\ 1 & 0 & 0 & 0 \\ 4 & 4 & 6 & 9 \\ 6 & 1 & 2 & 8 \end{bmatrix},$$

whose elements are residues modulo 11 and the three indeterminates x_1, x_2, x_3. Let P_1, P_2, P_3 be obtained from P_0 by cyclic permutation of the rows. Now put $A_0 = (P_0, P_1, P_2, P_3)$ and let A_i be obtained from A_0 by adding i to every residue modulo 11 in A_0. If A^* is the OA(3, 4) on x_1, x_2, x_3 and E is the 4×11 matrix whose ith column contains i in every place, then the matrix

$$D = (E, A, A^*)$$

is a matrix of order 4×196 and may be verified to be an OA(14, 4), whence $N(14) \geq 2$,

Example 2: ■ Starting with the matrix

$$P_0 = \begin{bmatrix} 0 & 0 & 0 & 0 & x_1 & x_2 & x_3 \\ 3 & 6 & 2 & 1 & 0 & 0 & 0 \\ 8 & 20 & 12 & 16 & 20 & 17 & 8 \\ 12 & 16 & 7 & 2 & 19 & 6 & 21 \end{bmatrix}$$

with residues modulo 23, as in Example 1 we may construct a matrix and show it to be an OA(26, 4) whence $N(26) \geq 2$.

Example 3: ■ If we take the residues modulo 41 as objects, we can construct a design with $v = 41$, $b = 82$, $r = 10$, $k = 5$, $\lambda = 1$, and our blocks are

$$\begin{aligned} A_i &= (i, i + 1, i + 4, i + 11, i + 29), \\ B_i &= (i + 1, i + 10, i + 16, i + 18, 1 + 37), \end{aligned} \qquad i = 0, \ldots, 40 (\text{mod } 41).$$

We may verify this by observing that the two sets of residues modulo 41 (0, 1, 4, 11, 29) and (1, 10, 16, 18, 37) together have the property that every nonzero residue d modulo 41 is expressible in exactly one way as

$$d \equiv x_i - x_j (\text{mod } 41),$$

where both x_i and x_j are either in the first set or in the second set. Then, if u and v are two distinct residues modulo 41, let x_i and x_j be determined by $u - v \equiv x_i - x_j (\text{mod } 41)$. Putting $t \equiv u - x_i (\text{mod } 41)$, then

$$u \equiv x_i + t, \qquad v \equiv x_j + t (\text{mod } 41),$$

and so both u and v are in A_t or in B_t. Thus, every pair of distinct objects occurs together in one block, and as there are 82 blocks, each containing five objects, this is enough to establish that we do have a design with

$$b = 82, \qquad v = 41, \qquad k = 5, \qquad r = 10, \qquad \lambda = 1.$$

This design is in turn a BIB(41, 5, 1). Applying Theorem 13.3.4 gives

$$N(38) \geq \min(N(3), N(4) - 1, N(5) - 1) = 2,$$

and so yields a pair of orthogonal squares of order 38.

We are now in a position to show that $N(n) \geq 2$ for all $n = 4t + 2$ in the range $10 \leq 4t + 2 \leq 726$. The three examples above, Theorems 13.2.2, 13.3.1, and 13.3.8, and one further application of Theorem 13.3.4 cover all the cases. In the following list, the form given for the number indicates which of the theorems is relevant. Thus, $50 = 5.10$ means that from Theorem 13.2.2, $N(50) \geq \min(N(5), N(10)) \geq 2$. Also, $34 \equiv 10 (\text{mod } 12)$ means that Theorem 13.3.1 is applicable and $N(34) \geq 2$. A representation $n = 4m + x$ means that Theorem 13.3.8 is being applied with $k = 4$. This requires that $N(m) \geq 3$ and $1 \leq x < m$, as is easily checked in every case.

$10 \equiv 10 (\text{mod } 12)$.
14 Example 1.
$18 = 21 - 3$. (Theorem 13.3.4 applied to BIB(21, 5, 1).)
$22 \equiv 10 (\text{mod } 12)$.
26 Example 2.
$30 = 3.10$.
$34 \equiv 10 (\text{mod } 12)$.
38 Example 3.
$42 = 3.14$.
$46 \equiv 10 (\text{mod } 12)$.
$50 = 5.10$.
$54 = 3.18$.
$58 \equiv 10 (\text{mod } 12)$.
$62 = 4.13 + 10$.
$66 = 3.22$.
$70 \equiv 10 (\text{mod } 12)$.
$74 = 4.16 + 10$.
$78 = 3.26$.
$82 \equiv 10 (\text{mod } 12)$.
$86 = 4.19 + 10$.
$90 = 4.19 + 14$.
$94 = 4.19 + 18$.
$98 = 7.14$.

For values of n greater than 100, application of Theorem 13.3.8 with $k = 4$ and $n = 4m + x$ is sufficient to prove $N(n) \geq 2$ in every case. Each value of

m can be used with a range of x's of the form $4t + 2$, and there will be a corresponding range for $n = 4m + x$.

m	Range of x	Range of $n = 4m + x$
23	10–22	102–114
27	10–26	118–134
31	14–30	138–154
37	10–34	158–182
44	10–42	186–218
53	10–50	222–262
64	10–62	266–318
77	14–74	322–382
92	18–90	386–458
113	10–110	462–562
139	10–138	556–694
172	10–38	698–726

There is no reason to believe that the results found so far are even close to being the best possible. $N(12) \geq 2$ is all that has been proved. But, by direct construction, Johnson, Dulmage, and Mendelsohn [1] have found five mutually orthogonal Latin squares of order 12. These are based on the Abelian group of order 12 which is the direct product of cyclic groups of orders 6 and 2. Let the generators be a and b with $a^6 = 1$, $b^2 = 1$, and let us write

$$a_i = a^i, \qquad i = 0, \ldots, 5, \qquad b_i = ba^i, \qquad i = 0, \ldots, 5.$$

Construct the five rows:

$$\begin{array}{cccccccccccc}
a_0 & a_1 & a_2 & a_3 & a_4 & a_5 & b_0 & b_1 & b_2 & b_3 & b_4 & b_5, \\
a_0 & b_0 & b_2 & a_2 & b_1 & a_1 & b_3 & b_5 & a_4 & b_4 & a_5 & a_3, \\
a_0 & a_3 & b_0 & a_1 & b_3 & b_5 & a_2 & b_2 & a_5 & a_4 & b_1 & b_4, \\
a_0 & b_2 & a_1 & b_5 & a_5 & b_3 & a_3 & b_4 & a_2 & b_1 & b_0 & a_4, \\
a_0 & a_4 & b_5 & b_4 & a_2 & b_1 & b_2 & b_0 & b_3 & a_1 & a_3 & a_5.
\end{array}$$

Take these rows as the first rows of five squares. The second and third to twelfth rows are formed by multiplying the first rows throughout by $a_1, \ldots, a_4$, $b_0, \ldots, b_5$, respectively. No other square of order 12 is orthogonal to all these, and no set of six mutually orthogonal squares of order 12 has been found.

The following theorem has been proved by Chowla, Erdös, and Straus [1].

THEOREM 13.4.2. *There exists a number v_0 such that $N(v) \geq \frac{1}{3}v^{(1/91)}$ whenever $v > v_0$.*

The proof rests on Theorem 13.3.8 and a variety of complicated arithmetical considerations.

14

Hadamard Matrices

14.1 Paley's Constructions

A Hadamard matrix of order m is an $m \times m$ matrix of $+1$'s and -1's such that

$$HH^T = mI. \tag{14.1.1}$$

This equation is equivalent to the assertion that any two rows of H are orthogonal. Clearly, permuting rows or columns of H or multiplying rows or columns of H by -1 leaves this property unchanged, and we consider such matrices equivalent. If H_1 and H_2 are equivalent Hadamard matrices, then

$$H_2 = PH_1Q, \tag{14.1.2}$$

where P and Q are monomial permutation matrices of $+1$'s and -1's. By this we mean that P and Q have exactly one nonzero entry in every row and in every column, and this nonzero entry is $+1$ or -1. P gives the permutation and change of sign of rows; Q of columns. Given a Hadamard matrix, we can always find one equivalent to it whose first row and first column consist entirely of $+1$'s. Such a Hadamard matrix is called "normalized." Permuting rows except the first, or columns except the first, leaves a normalized matrix normalized, but in general there may be equivalent normalized matrices that are not equivalent by merely permuting rows and columns.

We shall say "H matrix" for Hadamard matrix as an abbreviation. For orders 1 and 2 there are H matrices

$$H_1 = [1], \qquad H_2 = \begin{bmatrix} 1 & 1 \\ 1 & -1 \end{bmatrix}. \tag{14.1.3}$$

Let H be an H matrix of order $m > 2$, which we shall suppose normalized so that its first row consists entirely of 1's. Hence, three rows of H may be put into the form

$$\begin{array}{ccc|ccc|ccc|ccc} 1 & \cdots & 1 & 1 & \cdots & 1 & 1 & \cdots & 1 & 1 & \cdots & 1 \\ 1 & \cdots & 1 & 1 & \cdots & 1 & -1 & \cdots & -1 & -1 & \cdots & -1 \\ 1 & \cdots & 1 & -1 & \cdots & -1 & 1 & \cdots & 1 & -1 & \cdots & -1 \end{array}. \tag{14.1.4}$$

Let us suppose the number of columns of each type is x, y, z, w, respectively. Then

$$\begin{aligned} x + y + z + w &= m, \\ x + y - z - w &= 0, \\ x - y + z - w &= 0, \\ x - y - z + w &= 0, \end{aligned} \tag{14.1.5}$$

The first of these equations counts the total number of columns. The other three express in turn the orthogonality of the first two rows, of the first and third, and of the second and third. We may easily solve equations (14.1.5) to find

$$x = y = z = w = \frac{m}{4}. \tag{14.1.6}$$

We conclude that if H is a Hadamard matrix of order $m > 2$, then m is necessarily a multiple of 4.

From a normalized H matrix of order $m = 4t$, we may construct a symmetric block design D with parameters $v = 4t - 1$, $k = 2t - 1$, $\lambda = t - 1$, and conversely, from such a design D we may construct a normalized H. First suppose we are given a normalized H matrix of order $4t$. Number the rows and columns of H by $0, 1, \ldots, 4t - 1$ so that the row and column numbered zero consist entirely of $+1$'s. With the remaining rows, associate objects a_i, $i = 1, \ldots, 4t - 1$, and with the remaining columns, associate blocks $B_j j = 1, \ldots, 4t - 1$. We now say that $a_i \in B_j$ if the entry $b_{ij} = +1$ in H and $a_i \notin B_j$ if $b_{ij} = -1$. Thus, we have constructed an incidence system of $(v = 4t - 1)$ objects and v blocks with an incidence matrix

$$A = (a_{ij}), \qquad i, j = 1, \ldots, v$$

$$\text{where } a_{ij} = +1, \quad \text{if } b_{ij} = +1 \quad \text{and} \quad a_{ij} = 0, \quad \text{if } b_{ij} = -1.$$

In (14.1.4), (14.1.5), and (14.1.6), it has been shown that in a normalized Hadamard matrix, every row except row zero contains $2t$ $+1$'s, and two such rows both have $+1$'s in t columns. Deleting column zero, these values become $2t - 1$ and $t - 1$, respectively. Hence, A satisfies

$$AA^T = tI + (t - 1)J, \qquad AJ = (2t - 1)J. \tag{14.1.7}$$

From Theorem 10.2.3 it follows that A also satisfies

$$A^TA = tI + (t - 1)J, \qquad JA = (2t - 1)J, \tag{14.1.8}$$

and as an incidence matrix it follows that A is the incidence matrix of a symmetric design D with $v = 4t - 1$, $k = 2t - 1$, $\lambda = t - 1$. Conversely, if we are given a symmetric design D with $v = 4t - 1$, $k = 2t - 1$, $\lambda = t - 1$, its incidence matrix A satisfies (14.1.7) and (14.1.8). We now construct a matrix $H = (b_{ij})$, $i, j = 0, \ldots, 4t - 1$ by putting $b_{0j} = b_{i0} = +1$ and for $i, j = 1, \ldots, 4t - 1$,

$$b_{ij} = +1, \quad \text{if } a_{ij} = +1 \quad \text{and} \quad b_{ij} = -1, \quad \text{if } a_{ij} = 0,$$

From the fact that A satisfies (14.1.7) it now follows that H satisfies (14.1.1). and is a Hadamard matrix.

Thus, the construction of normalized Hadamard matrices is equivalent to the construction of symmetric designs with $v = 4t - 1$, $k = 2t - 1$, $\lambda = t - 1$. But it is to be noted that inequivalent designs may yield equivalent matrices, since a Hadamard matrix may be normalized in many ways. The Bruck-Ryser-Chowla condition (Theorem 10.3.1) for these designs is that there be solutions in integers x, y, z not all zero of

$$z^2 = tx^2 - (t - 1)y^2. \tag{14.1.9}$$

Clearly, $x = y = z = 1$ is a solution, and so it is conceivable that a Hadamard matrix of order m exists for every m that is a multiple of 4. Whether or not this is the case remains a challenging problem.

It is possible to construct Hadamard matrices from designs in another way. Let us suppose that H_m is not normalized and that the $+1$'s in *all* columns determine the incidence of objects in blocks. Then $v = m = 4t$ and we have $k + 1$'s and $(v - k) - 1$'s in every column and so also in every row. The condition

$$H_mH_m^T = H_m^TH_m = mI \tag{14.1.10}$$

requires for the inner product of two different rows

$$\lambda - (k - \lambda) - (k - \lambda) + (v - 2k + \lambda) = 0, \tag{14.1.11}$$

counting $+1$'s over $+1$'s, plus over minus, minus over plus, and minus over minus. As v is even, $k - \lambda$ must be a square (Theorem 10.3.1). Thus, we have

$$k - \lambda = u^2, \tag{14.1.12}$$

and (14.1.11) as well as the basic

$$k(k - 1) = \lambda(v - 1). \tag{14.1.13}$$

Together these relations give, taking $-H_m$ instead of H_m if necessary to make k less than half of v,

$$v = 4u^2, \qquad k = 2u^2 - u, \qquad \lambda = u^2 - u. \tag{14.1.14}$$

Such designs exist; thus, with $v = 16, k = 6, \lambda = 2$, an example was constructed at the beginning of Section 11.4. We take the elementary Abelian group A of order 16 with generators a, b, c, d, all of order 2, and the group difference set a, b, c, d, ab, cd. If we take the elements of A in the order $1, a, b, c, d, ab, ac, ad, bc, bd, cd, abc, abd, acd, bcd, abcd$, numbering them from 1 to 16, the block design corresponding to the $+1$'s is

$$\begin{array}{llll}
B_1: & 2, 3, 4, 5, 6, 11, & B_9: & 3, 4, 7, 10, 12, 15, \\
B_2: & 1, 3, 6, 7, 8, 14, & B_{10}: & 3, 5, 8, 9, 13, 15, \\
B_3: & 1, 2, 6, 9, 10, 15, & B_{11}: & 1, 4, 5, 14, 15, 16, \\
B_4: & 1, 5, 7, 9, 11, 12, & B_{12}: & 4, 6, 7, 9, 13, 16, \\
B_5: & 1, 4, 8, 10, 11, 13, & B_{13}: & 5, 6, 8, 10, 12, 16, \\
B_6: & 1, 2, 3, 12, 13, 16, & B_{14}: & 2, 7, 8, 11, 15, 16, \\
B_7: & 2, 4, 8, 9, 12, 14, & B_{15}: & 3, 9, 10, 11, 14, 16, \\
B_8: & 2, 5, 7, 10, 13, 14, & B_{16}: & 6, 11, 12, 13, 14, 15.
\end{array} \tag{14.1.15}$$

The corresponding H matrix is symmetric, since $i \in B_j$ if and only if $j \in B_i$.

There are a number of methods for constructing Hadamard matrices. It is possible to construct Hadamard matrices of the following orders N (in this list, p is an odd prime):

I. $N = 2^r$.
II. $N = p^r + 1 \equiv 0 (\text{mod } 4)$.
III. $N = h(p^r + 1), h \geq 2$, order of a Hadamard matrix.
IV. $N = h(h - 1)$, h a product of numbers of forms I and II.
V. $N = 92, 116, 172$.
VI. $N = h(h + 3)$, h and $h + 4$, both products from I and II.
VII. $N = h_1 h_2 (p^r + 1) p^r$, $h_1 > 1$ and $h_2 > 1$, orders of Hadamard matrices.
VIII. $N = h_1 h_2 s(s + 3)$, $h_1 > 1$, $h_2 > 1$ orders of Hadamard matrices and s, $s + 4$, both of form $p^r + 1$.
IX. $N = q(q + 2) + 1$, q, $q + 2$, both of form p^r.
X. N a product of numbers in I to IX.

In Section 11.6 the difference sets Q give type II, those of type H_6 are a subset of the numbers of type II, and T is type IX. As presented here, the rest of the cases are due to Williamson [1], [2], who generalized earlier methods, chiefly those of Paley [1].

These constructions lean very heavily on the direct product (or Kronecker product) of two matrices. If $A = (a_{ij})$ is an $m \times m$ matrix and $B = (b_{rs})$ is an

$n \times n$ matrix, then the direct product $A \times B$ is the $mn \times mn$ matrix given by

$$A \times B = \begin{bmatrix} a_{11}B & a_{12}B & \cdots & a_{1m}B \\ a_{21}B & a_{22}B & \cdots & a_{2m}B \\ \cdots & \cdots & \cdots & \cdots \\ a_{i1}B & a_{i2}B & \cdots & a_{im}B \\ \cdots & \cdots & \cdots & \cdots \\ a_{m1}B & a_{m2}B & \cdots & a_{mm}B \end{bmatrix}. \tag{14.1.16}$$

It is easy to see that

$$A \times B = (A \times I_n)(I_m \times B) = (I_m \times B)(A \times I_n). \tag{14.1.17}$$

Here, $(A \times B) \times C = A \times (B \times C)$ in all cases. The matrix $B \times A$ is equivalent to $A \times B$ by an appropriate permutation of rows and columns. The direct product has a number of useful properties:

$$\begin{aligned} a(A \times B) &= (aA) \times B = A \times (aB), \qquad \text{a scalar}, \\ (A_1 + A_2) \times B &= A_1 \times B + A_2 \times B, \\ A \times (B_1 + B_2) &= A \times B_1 + A \times B_2, \\ (A \times B)(C \times D) &= AC \times BD, \\ (A \times B)^T &= A^T \times B^T. \end{aligned} \tag{14.1.18}$$

Here, A, A_1, A_2, C are $m \times m$ matrices and B, B_1, B_2, D are $n \times n$ matrices.

THEOREM 14.1.1. *If there exist H matrices of orders m and n, their direct product is an H matrix of order mn.*

Proof: ■ Let H_m and H_n be H matrices of orders m and n, respectively. Then

$$\begin{aligned} (H_m \times H_n)(H_m \times H_n)^T &= (H_m \times H_n)(H_m^T \times H_n^T) \\ &= H_m H_m^T \times H_n H_n^T \\ &= mI_m \times nI_n = mnI_{mn}. \end{aligned} \tag{14.1.19}$$

For some of the methods we need H matrices with additional special properties. We begin with some matrices based on finite fields, where p is an odd prime. Let $\chi(x)$ be the character defined on $GF(p^r)$, where $\chi(0) = 0$, $\chi(x) = +1$ if x is a square and $\chi(x) = -1$ if x is not a square. Then, from the properties of finite fields $\chi(xy) = \chi(x)\chi(y)$ in all cases.

LEMMA 14.1.1. $\sum_b \chi(b)\chi(b + c) = -1$, *if* $c \neq 0$.

Proof: ■ $\chi(0)\chi(0 + c) = 0$. For $b \neq 0$, there is a unique $z \neq 0$ such that $b + c = bz$. As b ranges over the nonzero elements of F, z ranges over all elements of F except 1. (For $b = -c$, $z = 0$.) Hence,

$$\begin{aligned}\sum_b \chi(b)\chi(b+c) &= \sum_{b\neq 0} \chi(b)\chi(b+c)\\ &= \sum_{b\neq 0} \chi(b)^2\chi(z) = \sum_{z\neq 1}\chi(z)\\ &= \sum_z \chi(z) - \chi(1) = 0 - 1 = -1. \qquad (14.1.20)\end{aligned}$$

With $q = p^r$, let $a_0 = 0, a_1, \ldots, a_{q-1}$ be the elements of F numbered so that $a_0 = 0$ and $a_{q-i} = -a_i$, $i = 1, \ldots, q-1$. Now put

$$Q = (q_{ij}), \qquad q_{ij} = \chi(a_i - a_j). \qquad (14.1.21)$$

Here,

$$q_{ji} = \chi(a_j - a_i) = \chi(-1)\chi(a_i - a_j),$$

and since -1 is a square if $q \equiv 1 (\mathrm{mod}\ 4)$ and a nonsquare if $q \equiv 3 (\mathrm{mod}\ 4)$, it follows that Q is symmetric if $q \equiv 1 (\mathrm{mod}\ 4)$ and is skew-symmetric if $q \equiv 3 (\mathrm{mod}\ 4)$.

LEMMA 14.1.2. $QQ^T = qI_q - J$, $QJ = JQ = 0$. *Here, J is (as usual) the matrix consisting entirely of* 1's. If $QQ^T = B = (b_{ij})$, *then*

$$\begin{aligned}b_{ij} &= \sum_t \chi(a_i - a_t)\chi(a_j - a_t),\\ &= q - 1, \qquad \text{if } i = j,\\ &= -1, \qquad \text{if } i \neq j. \qquad (14.1.22)\end{aligned}$$

The last relation comes from Lemma 14.1.1 with $b = a_i - a_t$, $c = a_j - a_i$. These relations prove the first part of Lemma 14.1.2. The relations $QJ = JQ = 0$ are consequences of $\sum_z \chi(z) = 0$.

Let $e = e_q = (1, \ldots, 1)$ be the vector of q 1's. Then, if $q = p^r \equiv -1 (\mathrm{mod}\ 4)$, the matrix

$$S = \begin{bmatrix} 0 & e \\ -e^T & Q \end{bmatrix} \qquad (14.1.23)$$

has the properties

$$S^T = -S, \qquad SS^T = qI_{q+1}. \qquad (14.1.24)$$

Here,

$$H_{q+1} = (I_{q+1} + S) \qquad (14.1.25)$$

is an H matrix of order $q + 1$, since every element in H is ± 1, and also

$$\begin{aligned}HH^T &= (I + S)(I + S^T) = I + S + S^T + SS^T\\ &= I + qI = (q+1)I. \qquad (14.1.26)\end{aligned}$$

Let us call an H matrix of "skew type" if it is of the form

$$H_m = I_m + S_m, \qquad S_m^T = -S_m. \qquad (14.1.27)$$

Williamson [1] calls such an H matrix of "type I." He gives the following useful lemma.

LEMMA 14.1.3. *Let S be a matrix of order n such that $S^T = \epsilon S$, $\epsilon = \pm 1$, $SS^T = (n-1)I_n$. Further, let A and B be matrices of order m satisfying*

$$AA^T = BB^T = mI_m \qquad \text{and} \qquad AB^T = -\epsilon BA^T.$$

Then the matrix $K = A \times I_n + B \times S$ satisfies $KK^T = mnI_{mn}$.

Proof: ■ We calculate

$$\begin{aligned} KK^T &= (A \times I_n + B \times S)(A^T \times I_n + B^T \times S^T) \\ &= AA^T \times I_n + AB^T \times S^T + BA^T \times S + BB^T \times SS^T \\ &= mI_m \times I_n + (-\epsilon BA^T) \times (\epsilon S) + BA^T \times S + mI_m \times (n-1)I_n \\ &= mI_{mn} + m(n-1)I_{mn} = mnI_{mn}. \end{aligned} \tag{14.1.28}$$

If $q \equiv p^r \equiv 1 \pmod 4$, then with $n = q + 1$ the matrix S_n given by

$$S_n = \begin{bmatrix} 0 & e \\ e^T & Q \end{bmatrix}, \qquad e = e_{n-1}, \tag{14.1.29}$$

where Q is given by (14.1.21) and is symmetric. Because of Lemma 14.1.2, S_n satisfies

$$S_n^T = S_n, \qquad S_nS_n^T = (n-1)I_n. \tag{14.1.30}$$

Now let A be any H matrix of order $m > 1$. Here, m is even, and so we can construct a matrix U_m that has $m/2$ matrices $\left[\begin{smallmatrix} 0 & 1 \\ -1 & 0 \end{smallmatrix}\right]$ down the main diagonal, or

$$U_m = I_{m/2} \times \begin{bmatrix} 0 & 1 \\ -1 & 0 \end{bmatrix}. \tag{14.1.31}$$

Now define B by

$$B = U_mA. \tag{14.1.32}$$

Then we find

$$\begin{aligned} BB^T &= U_mAA^TU_m^T = U_mmI_mU_m^T = mI_m, \\ AB^T &= AA^TU_m^T = mI_mU_m^T = -mU_m, \\ BA^T &= U_mAA^T = U_m(mI_m) = mU_m. \end{aligned} \tag{14.1.33}$$

Thus, S_n in (14.1.29), A and B satisfy the hypotheses of Lemma 14.1.3. Let us note that S_n is 0 on the main diagonal and ± 1 elsewhere. Hence, the matrix K of the lemma has all its entries ± 1 and so is an H matrix of order mn. We state this as

THEOREM 14.1.2. *If $p^r \equiv 1 \pmod 4$, p prime and $h > 1$ is the order of an H matrix, then there is an H matrix of order $h(p^r + 1)$.*

This covers case III of the list of the orders N, since if $p^r \equiv -1 \pmod 4$, there is an H matrix of order $p^r + 1$, and the result follows from Theorem 14.1.1. This result of Williamson generalizes a result of Paley [1], who proves that there is an H matrix of order $2(p^r + 1)$ when $p^r \equiv 1 \pmod 4$. Paley's construction is equivalent to taking the S_n of (14.1.29) and forming

$$H_{2n} = S_n \times \begin{bmatrix} 1 & 1 \\ 1 & -1 \end{bmatrix} + I_n \times \begin{bmatrix} 1 & -1 \\ -1 & -1 \end{bmatrix}. \tag{14.1.34}$$

From (14.1.30) it readily follows that H_{2n} is an H matrix and is symmetric.

LEMMA 14.1.4. *If n is of the form $n = 2^t k_1 \cdots k_s$, where either $k_i = p_i^{r_i} + 1 \equiv 0 \pmod 4$ or $k_i = 2(p_i^{u_i} + 1)$, $p_i^{u_i} \equiv 1 \pmod 4$, there is a symmetric H matrix of order n.*

Proof: ■ The direct product of symmetric H matrices is a symmetric H matrix. There is a symmetric H matrix of order 2, and in (14.1.34) we have a symmetric H matrix of order $2(p^r + 1)$ where $p^r \equiv 1 \pmod 4$. It remains to show that if $n = p^r + 1 \equiv 0 \pmod 4$, there is a symmetric matrix of order n. There is an H matrix A of order n, of skew type, as given in (14.1.25):

$$A = I_n + S = I_n + \begin{bmatrix} 0 & e \\ -e^T & Q \end{bmatrix}. \tag{14.1.35}$$

Let U be the matrix of order $n - 1 = p^r = q$, where

$$\begin{aligned} &U = (u_{ij}), \qquad i, j = 0, \ldots, q-1, \qquad u_{00} = 1, \\ &u_{i,q-i} = 1, \qquad i = 1, \ldots, q-1, \qquad u_{ij} = 0, \qquad \text{otherwise}. \end{aligned} \tag{14.1.36}$$

Now define a matrix B by the equation

$$B = \begin{bmatrix} -1 & 0 \\ 0 & U \end{bmatrix} A = \begin{bmatrix} -1 & 0 \\ 0 & U \end{bmatrix} + \begin{bmatrix} 0 & -e \\ -e^T & UQ \end{bmatrix}. \tag{14.1.37}$$

Since

$$Q = (b_{ij}),\ i, j = 0, \ldots, q-1 \qquad \text{and} \qquad b_{ij} = \chi(a_i - a_j),$$

with $UQ = (c_{ij})$, then $c_{0j} = \chi(0 - a_j)$ and

$$c_{ij} = \chi(a_{q-i} - a_j) = \chi(-a_i - a_j), \qquad i = 1, \ldots, q-1,$$

whence in all cases $c_{ij} = \chi(-a_i - a_j)$ and so UQ is symmetric, whence also B is symmetric:

$$B^T = B. \tag{14.1.38}$$

In addition, noting that U is symmetric and $U^2 = 1$, we find

$$BB^T = \begin{bmatrix} -1 & 0 \\ 0 & U \end{bmatrix} AA^T \begin{bmatrix} -1 & 0 \\ 0 & U \end{bmatrix} = nI_n. \tag{14.1.39}$$

The matrix B is a symmetric H matrix of order $n = p^r + 1$ and its existence is sufficient to prove our lemma.

LEMMA 14.1.5. *If there is a skew-type H matrix of order n, and if $p^r + 1 \equiv 0 \pmod 4$ p prime, then there is a skew-type matrix of order $n(p^r + 1)$.*

Proof: ■ By hypothesis $H_n = I_n + S$, where $S^T = -S$. With A as in (14.1.35) and B as in (14.1.37), we have

$$H_n H_n^T = (I_n + S)(I_n - S) = nI_n. \tag{14.1.40}$$

whence

$$SS^T = -S^2 = (n-1)I_n \tag{14.1.41}$$

and also, with $m = p^r + 1$,

$$AA^T = BB^T = mI_m,$$
$$AB^T = mI_m\begin{bmatrix} -1 & 0 \\ 0 & U \end{bmatrix} = BA^T. \tag{14.1.42}$$

Thus, A, B, S satisfy the requirements of Lemma 14.1.3 with $\epsilon = +1$. Hence,

$$K = A \times I_n + B \times S \tag{14.1.43}$$

is an H matrix of order mn, and as $A = I_m + C$, $C^T = -C$, we have

$$(K - I_{mn})^T = (C \times I_n + B \times S)^T$$
$$= (-C) \times I_n + B \times (-S) = -(K - I_{mn}). \tag{14.1.44}$$

Thus, K is of skew type, and our lemma is proved.

LEMMA 14.1.6. *There is an H matrix of skew type of order n if $n = 2^t k_1 \cdots k_s$, each k_i of the form $p^r + 1 \equiv 0 \pmod 4$.*

For order $p^r + 1 \equiv 0 \pmod 4$, the matrix H_{q+1} of (14.1.25) satisfies our requirement. Furthermore, if H is of skew type of order n (including the matrix (1) of order 1) then

$$\begin{bmatrix} 1 & 1 \\ -1 & 1 \end{bmatrix} \times H$$

is also an H matrix of skew type of order $2n$. Together with Lemma 14.1.5 this covers all values of n in this lemma.

THEOREM 14.1.3. *If there is an H matrix of skew type of order n, there is an H matrix of order $n(n-1)$.*

Proof: ■ Let $H = (b_{ij})$, $i, j = 1, \ldots, n$. Since H is of skew type, $b_{ii} = +1$, $i = 1, \ldots, n$. If we multiply the rth row and rth column by -1, the property of being of skew type is not altered. Hence, we may suppose also $b_{1j} = +1$, $j = 2, \ldots, n$ without loss of generality. Thus,

$$H = I_n + S, \qquad S^T = -S$$
$$H = \begin{bmatrix} 1 & e \\ -e^T & D \end{bmatrix}. \tag{14.1.45}$$

Since $HH^T = nI_n$, we have

$$SS^T = (n-1)I_n \tag{14.1.46}$$

and also

$$\begin{gathered} DD^T = nI_{n-1} - J_{n-1}, \qquad (D - I_{n-1})^T = -(D - I_{n-1}), \\ eD^T = e, \qquad De^T = e^T, \qquad e^Te = J_{n-1}, \\ J_{n-1}D^T = e^T(eD^T) = e^Te = J_{n-1}. \end{gathered} \tag{14.1.47}$$

If we put

$$K = J_{n-1} \times I_n + D \times S, \tag{14.1.48}$$

we find

$$\begin{aligned} KK^T &= (J_{n-1} \times I_n + D \times S)(J_{n-1}^T \times I_n + D^T \times S^T) \\ &= J_{n-1}^2 \times I_n + J_{n-1}D^T \times (-S) + J_{n-1}D \times S \\ &\quad + (nI_{n-1} - J_{n-1}) \times ((n-1)I_n) \\ &= (n-1)J_{n-1} \times I_n + J_{n-1} \times (-S) + J_{n-1}(2I_{n-1} - D^T) \\ &\quad \times S + n(n-1)I_{n-1} \times I_n - (n-1)J_{n-1} \times I_n \\ &= -J_{n-1} \times S + (2J_{n-1} - J_{n-1}) \times S + n(n-1)I_{n-1} \times I_n \\ &= n(n-1)I_{n(n-1)}. \end{aligned} \tag{14.1.49}$$

and so K is an H matrix of order $n(n-1)$ and our theorem is proved. This theorem of Williamson generalizes a result of Scarpis [1], who showed for $p \equiv 3 \pmod 4$ the existence of an H matrix of order $p(p+1)$. Together with Lemma 14.1.6, this establishes case IV in our list.

THEOREM 14.1.4. *If there is a skew-type H matrix of order n and a symmetric H matrix of order $m = n + 4$, there is an H matrix of order $n(n+3)$.*

Proof: ■ Let H_m be symmetric and H_n be of skew type. Then, changing signs of rows and columns simultaneously if necessary, and replacing H_m by $-H_m$ if necessary, we have

$$H_m = \begin{bmatrix} 1 & e \\ e^T & D \end{bmatrix}, \qquad e = e_{m-1}, \qquad H_n = I_n + S_n. \tag{14.1.50}$$

Here, since $mI_m = H_mH_m^T = H_m^2$,

$$\begin{bmatrix} m & e + eD \\ e^T + De^T & e^Te + D^2 \end{bmatrix} = \begin{bmatrix} m & 0 \\ 0 & mI_{m-1} \end{bmatrix} \tag{14.1.51}$$

and so

$$D^T = D, \quad eD = -e, \quad De^T = -e^T, \quad D^2 = mI_{m-1} - J_{m-1}. \tag{14.1.52}$$

Since $e^Te = J_{m-1}$, we find

$$J_{m-1}D = -J_{m-1} = DJ_{m-1}. \tag{14.1.53}$$

Now put $F_{m-1} = 2I_{m-1} - J_{m-1}$. Then

$$F_{m-1}D = DF_{m-1}, \qquad F_{m-1}^2 = 4I_{m-1} + (m-5)J_{m-1}. \tag{14.1.54}$$

Also, since H_n is of skew type,

$$S_n^T = -S_n, \qquad S_nS_n^T = (n-1)I_n. \tag{14.1.55}$$

Now define W by

$$W = F_{m-1} \times I_n + D \times S_n, \tag{14.1.56}$$

and note that every element of W is ± 1. Then

$$\begin{aligned} WW^T &= (F_{m-1} \times I_n + D \times S_n)(F_{m-1} \times I_n - D \times S_n) \\ &= F_{m-1}^2 \times I_n - D^2 \times S_n^2 \\ &= (4I_{m-1} + (m-5)J_{m-1}) \times I_n + (mI_{m-1} - J_{m-1}) \\ &\quad \times ((m-1)I_n) \\ &= (mn - m + 4)I_{m-1} \times I_n + (m - n - 4)J_{m-1} \times I_n. \end{aligned} \tag{14.1.57}$$

Hence, if $m = n + 4$, W is an H matrix of order $n(n+3)$ and our theorem is proved. This theorem, with Lemmas 14.1.4 and 14.1.6, proves case VI of our list and somewhat more.

LEMMA 14.1.7. *If there exists an H matrix A of order $n > 1$, there exist two H matrices B and C of order n such that $AB^T = -BA^T$, $AC^T = CA^T$, $BC^T = CB^T$.*

Proof: ■ Here n is even; $n = 2m$. Putting

$$X = I_m \times \begin{bmatrix} 0 & 1 \\ -1 & 0 \end{bmatrix}, \qquad Y = I_m \times \begin{bmatrix} 0 & 1 \\ 1 & 0 \end{bmatrix}, \tag{14.1.58}$$

we verify directly that $B = XA$, $C = YA$ satisfy the conditions of the lemma.

Now suppose $p^r \equiv 1 \pmod 4$, p prime, and write $m = p^r + 1$. Then the matrix Q of (14.1.21) is symmetric, and from Lemma (14.1.2),

$$QQ^T = Q^2 = (m-1)I_{m-1} - J_{m-1}, \qquad QJ_{m-1} = J_{m-1}Q = 0. \tag{14.1.59}$$

Let A_1 and B_1 be H matrices of order n, as in Lemma 14.1.7, such that

$$A_1B_1^T = -B_1A_1^T. \tag{14.1.60}$$

Define the matrix K of order $(m-1)n_1$ by

$$K = A_1 \times I_{m-1} + B_1 \times Q. \tag{14.1.61}$$

Then every element of K has the value ± 1 and

$$\begin{aligned} KK^T &= A_1A_1^T \times I_{m-1} + B_1B_1^T \times Q^2 \\ &= n_1I_{n_1} \times (mI_{m-1} - J_{m-1}). \end{aligned} \tag{14.1.62}$$

If $M = A_1 \times J_{m-1}$,

$$MM^T = n_1I_{n_1} \times (m-1)J_{m-1}. \tag{14.1.63}$$

Now let A_2 and B_2 be H matrices of order n_2 satisfying

$$A_2B_2^T = -B_2A_2^T. \tag{14.1.64}$$

Now define W by

$$W = (A_2 \times M) \times I_m + (B_2 \times K) \times S, \tag{14.1.65}$$

where

$$S = \begin{bmatrix} 0 & e \\ e^T & Q \end{bmatrix}, \qquad S^2 = (m-1)I_m. \tag{14.1.66}$$

Then

$$\begin{aligned} WW^T &= (A_2A_2^T \times MM^T) \times I_m + (B_2B_2^T \times KK^T) \times S^2 \\ &= [n_2I_{n_2} \times (n_1I_{n_1} \times (m-1)J_{m-1}) \\ &\quad + (m-1)n_2I_{n_2} \times (n_1I_{n_1} \times (mI_{m-1} - J_{m-1}))] \times I_m \\ &= [n_2I_{n_2} \times n_1I_{n_1} \times ((m-1)J_{m-1} + (m-1)mI_{m-1} \\ &\quad - (m-1)J_{m-1})] \times I_m \\ &= NI_N, \qquad N = n_1n_2m(m-1). \end{aligned} \tag{14.1.67}$$

We have proved the following theorem.

THEOREM 14.1.5. *If H matrices of orders n_1 and n_2 exist, $n_1 > 1$, $n_2 > 1$, and p is a prime such that $p^r \equiv 1 (\mathrm{mod}\ 4)$, there exists an H matrix of order $n_1n_2(p^r+1)p^r$.*

By case IV of our list, if $p^r \equiv 3 (\mathrm{mod}\ 4)$, there is an H matrix of the order $(p^r+1)p^r$. This observation together with our theorem establishes case VII of the list.

The proof of case VIII is much like that of case VII. Let A_1, B_1, C_1, be H matrices of order n_1 satisfying the relations of Lemma 14.1.7 and similarly, A_2, B_2, C_2 of order n_2. (We shall not need C_2.) With $Q = Q_1$ from (14.1.59) of order $m - 1 = p_1^r$, p_1 prime, $p_1^r \equiv 1 (\mathrm{mod}\ 4)$, put

$$\begin{aligned} U &= C_1 \times (2I_{m-1} - J_{m-1}) \\ V &= A_1 \times I_{m-1} + B_1 \times Q_1. \end{aligned} \tag{14.1.68}$$

Similarly, let S_n be, from (14.1.29), of order $n = p_2^s + 1$ with p_2 prime, $p_2^s \equiv 1 (\text{mod } 4)$. Finally, put

$$W = (A_2 \times U) \times I_n + (B_2 \times V) \times S_n. \tag{14.1.69}$$

Then we find

$$\begin{aligned} UU^T &= n_1 I_{n_1} \times (4I_{m-1} + (m-5)J_{m-1}), \\ VV^T &= n_1 I_{n_1} \times (mI_{m-1} - J_{m-1}), \\ UV^T &= C_1 A_1^T \times (2I_{m-1} - J_{m-1}) + C_1 B_1^T \times 2Q_1. \end{aligned} \tag{14.1.70}$$

Since A_1, B_1, C_1 satisfy Lemma 14.1.7, we see that UV^T is symmetric, or

$$UV^T = VU^T. \tag{14.1.71}$$

Hence, we find

$$\begin{aligned} WW^T &= n_2 I_{n_2} \times (UU^T \times I_n + VV^T \times S_n^2) \\ &= n_1 n_2 I_{n_1 n_2} \times [(4I_{m-1} + (m-5)J_{m-1} \times I_n) \\ &\quad + (mI_{m-1} - J_{m-1}) \times (n-1)I_n] \\ &= n_1 n_2 I_{n_1 n_2} \times [(4 + mn - m)I_{m-1} \\ &\quad + (m - 4 - n)J_{n-1}] \times I_n. \end{aligned} \tag{14.1.72}$$

Hence, if $m = n + 4$, we have

$$WW^T = NI_N, \qquad N = n_1 n_2 n(n+3), \tag{14.1.73}$$

and W is an H matrix of order N.

We have now proved

THEOREM 14.1.6. *If both n and $n + 4$ are of the form $p^r + 1$, p prime, and if $n_1 > 1$ and $n_2 > 1$ are orders of H matrices, then there is an H matrix of order $n_1 n_2 n(n + 3)$.*

Note that if $n \equiv 0 (\text{mod } 4)$, this is a consequence of case VI. The preceding proof covers the case $n \equiv 2 (\text{mod } 4)$ and proves case VIII of the list.

14.2 Williamson's Method

Another method for constructing H matrices is also due to Williamson. This method has been successful in a number of special cases, in particular giving the numbers 92, 116, and 172 of case V.

Consider the matrix H given by

$$H = \begin{bmatrix} A & B & C & D \\ -B & A & -D & C \\ -C & D & A & -B \\ -D & -C & B & A \end{bmatrix}. \tag{14.2.1}$$

Here, if A, B, C, D are numbers,

$$HH^T = (A^2 + B^2 + C^2 + D^2) \times I_4. \tag{14.2.2}$$

Under certain circumstances, (14.2.2) is valid if A, B, C, D are matrices of order n. A sufficient condition for this is that A, B, C, D be symmetric and commute with each other. We shall assume this to be the case.

Let U be the matrix of order n corresponding to a cyclic permutation of order n. Here,

$$U = \begin{bmatrix} 0 & 1 & 0 & \cdots & 0 \\ 0 & 0 & 1 & \cdots & 0 \\ \cdots & \cdots & \cdots & \cdots & \cdots \\ 0 & 0 & 0 & \cdots & 1 \\ 1 & 0 & 0 & \cdots & 0 \end{bmatrix}, \qquad U^n = I. \tag{14.2.3}$$

If A, B, C, D are polynomials in U, they commute with each other. Let us take

$$\begin{aligned} A &= a_0I + a_1U + \cdots + a_{n-1}U^{n-1}, \\ B &= b_0I + b_1U + \cdots + b_{n-1}U^{n-1}, \\ C &= c_0I + c_1U + \cdots + c_{n-1}U^{n-1}, \\ D &= d_0I + d_1U + \cdots + d_{n-1}U^{n-1}. \end{aligned} \tag{14.2.4}$$

Since $U^T = U^{-1}$, the matrices A, B, C, D will be symmetric if

$$a_{n-i} = a_i, \quad b_{n-i} = b_i, \quad c_{n-i} = c_i, \quad d_{n-i} = d_i, \qquad i = 1, \ldots, n-1. \tag{14.2.5}$$

If every coefficient a, b, c, d is ± 1, then H will have every entry ± 1 and we shall have

$$HH^T = 4nI_{4n}, \tag{14.2.6}$$

providing

$$A^2 + B^2 + C^2 + D^2 = 4nI_n. \tag{14.2.7}$$

Thus, H of (14.2.1) will be an H matrix of order $4n$ if A, B, C, D are given by (14.2.4) with the coefficients ± 1 subject to conditions (14.2.5), providing (14.2.6) holds. For example, if $n = 3$, we may take

$$A = \begin{bmatrix} 1 & 1 & 1 \\ 1 & 1 & 1 \\ 1 & 1 & 1 \end{bmatrix}, \qquad B = C = D = \begin{bmatrix} 1 & -1 & -1 \\ -1 & 1 & -1 \\ -1 & -1 & 1 \end{bmatrix}. \tag{14.2.8}$$

Since from (14.2.8) we find

$$A^2 + B^2 + C^2 + D^2 = 12I_3, \tag{14.2.9}$$

the matrix H of (14.2.1) is an H matrix of order 12.

From here on we shall assume that n is odd, and also change the signs of A, B, C, D if necessary so that

$$a_0 = b_0 = c_0 = d_0 = +1. \tag{14.2.10}$$

The relations (14.2.4) and (14.2.7) may be considered as in the group ring $R(G)$ over the integers, where G is the cyclic group $1, u, \ldots, u^{n-1}$, with $u^n = 1$, u corresponding to U. Thus, we write

$$A = a_0 + a_1 u + \cdots + a_{n-1}u^{n-1} \tag{14.2.11}$$

and similarly express B, C, and D. Let us further write

$$A = P_1 - N_1 \tag{14.2.12}$$

with P_1 the sum of the positive terms in A, $-N_1$ the sum of the negative terms in A, whence

$$P_1 = \sum_j a_j u^j, \qquad a_j = +1, \qquad -N_1 = \sum_j a_j u^j, \qquad a_j = -1. \tag{14.2.13}$$

In the same way we write

$$B = P_2 - N_2, \qquad C = P_3 - N_3, \qquad D = P_4 - N_4. \tag{14.2.14}$$

Since $a_0 = +1$ and, by (14.2.5), $a_{n-i} = a_i$, $i = 1, \ldots, n-1$, the positive terms except for a_0 occur in pairs, whence p_1, the number of terms in P_1, is an odd number. Similarly, p_2, p_3, p_4 are odd numbers. Let us write

$$T = 1 + u + u^2 + \cdots + u^{n-1}. \tag{14.2.15}$$

Then

$$P_i + N_i = T, \qquad i = 1, 2, 3, 4. \tag{14.2.16}$$

Then the relation corresponding to (14.2.7) is

$$A^2 + B^2 + C^2 + D^2 = 4n, \tag{14.2.17}$$

and by (14.2.12), (14.2.13), and (14.2.14), this becomes

$$(2P_1 - T)^2 + (2P_2 - T)^2 + (2P_3 - T)^2 + (2P_4 - T)^2 = 4n. \tag{14.2.18}$$

Since $u^j T = T$ for all j, we have $P_i T = p_i T$ and $T^2 = nT$, and so (14.2.18) takes the form

$$4(P_1^2 + P_2^2 + P_3^2 + P_4^2) - 4(p_1 + p_2 + p_3 + p_4)T + 4nT = 4n. \tag{14.2.19}$$

Dividing by 4 this becomes

$$P_1^2 + P_2^2 + P_3^2 + P_4^2 = (p_1 + p_2 + p_3 + p_4 - n)T + n. \tag{14.2.20}$$

Since each p_i is odd and n is odd, we see that every term $u^j, j \neq 0$, must occur an odd number of times (namely, $p_1 + p_2 + p_3 + p_4 - n$) on the left of (14.2.20). For each $t = 1, \ldots, n-1$, there is a unique s such that $(u^s)^2 = u^t$. In $P_i^2 = (\sum u^k)^2$, k in a subset of $0, 1, \ldots, n-1$, we have $P_i^2 \equiv \sum u^{2k} \pmod 2$. Hence, every $u^t = (u^t)^2$, to appear with an odd coefficient on the left of (14.2.20), must occur in exactly one or three of P_1, P_2, P_3, P_4. The corresponding u^s appears in exactly one or three of P_1, P_2, P_3, P_4. This leads to

LEMMA 14.2.1. *If n is odd, and if the matrices A, B, C, D of* (14.2.4) *are chosen with signs so that $a_0 = b_0 = c_0 = d_0 = 1$ and satisfy $A^2 + B^2 + C^2 + D^2 = 4_n I_n$, then for each $i = 1, \ldots, n-1$ of a_i, b_i, c_i, d_i, exactly three are of the same sign.*

Proof: ■ Here $A = P_1 - N_1$, $B = P_2 - N_2$, $C = P_3 - N_3$, $D = P_4 - N_4$. The term u^i occurs in one or three of P_1, P_2, P_3, P_4, as we have just shown, and so either u^i occurs in exactly three of A, B, C, D, or $-u^i$ in exactly three of them. Let us now define W_1, W_2, W_3, W_4 by

$$\begin{aligned} 2W_1 &= A + B + C - D, & 2W_2 &= A + B - C + D, \\ 2W_3 &= A - B + C + D, & 2W_4 &= -A + B + C + D. \end{aligned} \tag{14.2.21}$$

Then, substituting in (14.2.17), we find

$$W_1^2 + W_2^2 + W_3^2 + W_4^2 = 4n, \tag{14.2.22}$$

and from Lemma 14.2.1, each W_i has integral coefficients and

$$W_i = 1 \pm 2u^j \pm \cdots \pm 2u^k, \tag{14.2.23}$$

where for each $j = 1, \ldots, n-1$, we have a term $\pm 2u^j$ in exactly one of the W's. For example, if u^j occurs with the coefficient -1 in A, B, and C and with the coefficient $+1$ in D, then we have $-2u^j$ in W_1 and no term in u^j in W_2, W_3, or W_4. And, as u^j and u^{n-j} have the same coefficient in each of A, B, C, D, the same holds true in the W's. Conversely, if the W's are of the form (14.2.23) with u^j and u^{n-j} with the same coefficient and also satisfy (14.2.22), then A, B, C, D satisfy Lemma 14.2.1 and yield an H matrix in (14.2.1). Here, A, B, C, D are given by

$$\begin{aligned} 2A &= W_1 + W_2 + W_3 - W_4, & 2C &= W_1 - W_2 + W_3 + W_4, \\ 2B &= W_1 + W_2 - W_3 + W_4, & 2D &= -W_1 + W_2 + W_3 + W_4. \end{aligned} \tag{14.2.24}$$

Since (14.2.22) must hold in the group ring $R(G)$ of the cyclic group $\{u\}$ with $u^n = 1$, it must be valid when u is replaced by any nth root of unity—in particular when we put $u = 1$. In this case, each of the W_i is an odd integer. By the theorem of Lagrange (Hardy and Wright [1]), every integer can be written as the sum of four squares of integers. It is also true that if n is odd, then $4n$ can be written as a sum of four squares of odd integers. If $n = 23$, we have two such representations

$$92 = 9^2 + 3^2 + 1^2 + 1^2 = 7^2 + 5^2 + 3^2 + 3^2. \tag{14.2.25}$$

If we write $v_i = u^i + u^{n-1}$, $i = 1, \ldots, (n-1)/2$, then

$$W_j = W_j(u) = 1 \pm 2v_i + \cdots \pm 2v_k, \qquad j = 1, 2, 3, 4. \tag{14.2.26}$$

Since putting $u = 1$ makes each $v = 2$, the sign of $W_j(0.1)$ is determined by

$W_j(1) \equiv 1 (\text{mod } 4)$. From the first decomposition of 92 in (14.2.25) we have the possibilities

$$92 = (1 + 2v_a + 2v_b + \cdots)^2 + (1 - 2v_c + \cdots)^2 + (1 + \cdots)^2 + (1 + \cdots)^2, \tag{14.2.27}$$

where the dots are made up of a sum of one or more of $2v_d - 2v_e$, $2v_f - 2v_g$, $2v_h - 2v_i$, $2v_j - 2v_k$, and $a, b, \ldots, k$ are $1, \ldots, 11$ in some order. In this form the problem lends itself to a computer search and it was found by Baumert, Golomb, and Hall [1] that there is a solution

$$92 = (1 + 2v_2 + 2v_6)^2 + (1 - 2v_3 + v_1 - v_{10})^2 + (1 + 2v_5 - v_7)^2 + (1 + 2v_{11} - 2v_8 + 2v_9 - 2v_4)^2. \tag{14.2.28}$$

Thus, there is an H matrix of order 92, which can be constructed from (14.2.28) using (14.2.24) and the form (14.2.1). The other representation, $92 = 7^2 + 5^2 + 3^2 + 3^2$, does not lead to an H matrix. In the same way, Williamson found for $4n = 172$ that if we take

$$\alpha_j = v_{3^j} + v_{3^{7+j}} + v_{3^{14+j}}, \tag{14.2.29}$$

we have

$$172 = (1 + 2\alpha_0 - 2\alpha_2)^2 + (1 + 2\alpha_3 - \alpha_1)^2 + (1 + 2\alpha_4 - 2\alpha_6)^2 + (1 + 2\alpha_5)^2. \tag{14.2.30}$$

This gives the other value 172 in case V of our list. In his original paper [1] Williamson gave a number of solutions of (14.2.22) including (14.2.29), but no solution for $4n = 92$, a value that remained undecided until 1962. L. Baumert [1] has found *all* solutions of (14.2.22) for $4n$, n odd, $n = 3, \ldots, 23$, and also solutions for $n = 25$ and $n = 27$. A machine search by L. Baumert has recently found a solution for $n = 116$:

$$116 = (1 + 2w_{12} + 2w_6 + 2w_2 - 2w_{11} - 2w_9 - 2w_4)^2 + (1 + 2w_{10} + 2w_7 - 2w_8 - 2w_3 - 2w_5)^2 + (1 + 2w_1)^2 + (1 + 2w_{13} + 2w_{14})^2 \tag{14.2.31}$$

There is a paradox associated with this last method of Williamson. There exist H matrices of this type of order $4n$, n odd, for *every* value of n for which a complete search has been made, but no infinite class of H matrices of this type has been found. A proof that this method will always succeed would, of course, resolve the main problem on H matrices, since if H_{4n} exists for n odd, then $H_{2^j n}$ exists for $j > 2$ and so H_m for $m \equiv 0 (\text{mod } 4)$.

14.3 Three Recent Methods

Two more methods may be added to our list for constructing Hadamard matrices. If A, B, C, D are symmetric circulant matrices of order n such that H

as given in (14.2.1) is an H_{4n}, then there is an H_{12n}, given by

$$H_{12n} = \begin{bmatrix} A & A & A & B & -B & C & -C & -D & B & C & -D & -D \\ A & -A & B & -A & -B & -D & D & -C & -B & -D & -C & -C \\ A & -B & -A & A & -D & D & -B & B & -C & -D & C & -C \\ B & A & -A & -A & D & D & D & C & C & -B & -B & -C \\ B & -D & D & D & A & A & A & C & -C & B & -C & B \\ B & C & -D & D & A & -A & C & -A & -D & C & B & -B \\ D & -C & B & -B & A & -C & -A & A & B & C & D & -D \\ -C & -D & -C & -D & C & A & -A & -A & -D & B & -B & -B \\ D & -C & -B & -B & -B & C & C & -D & A & A & A & D \\ -D & -B & C & C & C & B & B & -D & A & -A & D & -A \\ C & -B & -C & C & D & -B & -D & -B & A & -D & -A & A \\ -C & -D & -D & C & -C & -B & B & B & D & A & -A & -A \end{bmatrix} \tag{14.3.1}$$

for we may verify directly that

$$H_{12n}H_{12n}^{T} = (3A^2 + 3B^2 + 3C^2 + 3D^2) \times I_{12} = 12nI_{12n}. \tag{14.3.2}$$

This construction was found by L. Baumert and this author [1]. This adds order 156 to the previously known orders. It may be that the matrix (14.3.1) is an instance of a new kind of composition.

The second new method is due to Karl Goldberg [1]. If H is a skew-type Hadamard matrix of order n, as in (14.1.45), then D is of order $n - 1$ and $D = I_{n-1} = D_1$, where $D_1^T = -D_1$. Define A by

$$A = \tfrac{1}{2}(D + J_{n-1}) - I_{n-1}. \tag{14.3.3}$$

Then an entry of A is 1 if the corresponding entry of D_1 is 1, and other entries of A are all zero. Furthermore,

$$D = I_{n-1} + A - A^T, \qquad A + A^T = J_{n-1} - I_n. \tag{14.3.4}$$

The other relations of (14.1.47) reduce to

$$e_{n-1}A = e_{n-1}A^T, \qquad A^TA = AA^T = \left(\frac{n}{4}\right)I_{n-1} + \frac{(n-4)}{4}J_{n-1}. \tag{14.3.5}$$

Writing (X, Y, Z) for the direct product of the matrices X, Y, Z, we define a matrix $A_{(n-1)^3}$ by

$$A_{(n-1)^3} = (I, A, J) + (J, I, A) + (A, J, I) + (A, A, A) + (A, A^T, A^T) + (A^T, A, A^T) + (A^T, A^T, A). \tag{14.3.6}$$

Then $A_{(n-1)^3}$ is a matrix of zeros and 1's of order $(n - 1)^3$ and satisfies the same relations as A with n replaced by $(n - 1)^3 + 1$, whence if there is a skew type of H matrix of order n, then there is also a skew type of H matrix of order $(n-1)^3+1$.

A third new method is due to H. Ehlich [1].

THEOREM 14.3.1. *If there is a skew-type H matrix of order $n+1$, and if $n-2=m=p^s \equiv 1 \pmod 4$, p a prime, then there is an H matrix of order $(n-1)^2$.*

Proof: ■ We are given an H^* of order $n+1$ with

$$H^* = \begin{bmatrix} 1 & e_n \\ -e_n^T & I_n + G \end{bmatrix}, \qquad G^T = -G, \tag{14.3.7}$$

and $H^*H^{*T} = (n+1)I_{n+1}$.

Also, with $m = p^s \equiv 1 \pmod 4$, from (14.1.29) and (14.1.30), there is a matrix Q of order m such that

$$\begin{gathered} Q = [q_{ij}], \qquad q_{ii} = 0, \qquad q_{ij} = \pm 1, \qquad j \neq i, \\ QQ^T = Q^2 = mI_m - J_m. \end{gathered} \tag{14.3.8}$$

Ehlich constructs

$$K = Q \times G - I_m \times (J_n - I_n) + J_m \times I_n, \tag{14.3.9}$$

and the matrix

$$H = \begin{bmatrix} 1 & e_{mn} \\ e_{mn}^T & K \end{bmatrix} \tag{14.3.10}$$

consists of ± 1's; the preceding relations yield

$$HH^T = (mn+1)I_{mn+1}, \qquad mn + 1 = (n-1)^2, \tag{14.3.11}$$

and our theorem is proved.

15

General Constructions of Block Designs

15.1 Methods of Construction

The known methods for constructing block designs are of two main kinds, direct and recursive. A direct method allows us to construct a design with particular parameters, usually employing finite fields or congruences. A recursive method is a way of constructing designs from smaller ones. Both methods were used in Chapter 13 in the construction of orthogonal Latin squares. Usually, a direct method yields an easier construction, but is applicable only for special values of the parameters, perhaps only when v, the number of objects, is a prime power.

Section 15.2 gives the main recursive theorems of Hanani. Section 15.3 gives a number of direct constructions, most of them due to R. C. Bose [1]. Section 15.4 gives a detailed treatment of triple systems, the designs with $k = 3$. Section 15.5 deals with designs from which $k > 3$, and some results are merely quoted rather than proved in full.

15.2 Basic Definitions. The Hanani Theorems

A balanced incomplete block design has been defined in Section 10.1 as an arrangement of v distinct objects into b blocks such that each block contains exactly k distinct objects, each object occurs in exactly r different blocks, and every (unordered) pair of distinct objects occurs together in exactly λ blocks. The following relations were shown to hold:

$$bk = rv, \tag{15.2.1}$$

$$r(k - 1) = \lambda(v - 1). \tag{15.2.2}$$

We consider only cases with $k \geq 3$, since $k = 1$ and $k = 2$ are trivial.

We shall consider more general block designs in which the number of objects may vary from block to block. Let $K = \{k_1, \ldots, k_n\}$ be a finite set of integers k_i, where $k_i \geq 3$, $i = 1, \ldots, n$. A block design $B[K, \lambda, v]$ is an arrangement of v distinct objects into b blocks, $B_1, \ldots, B_v$, where each block B_j contains k_i objects for some k_i, $i = 1, \ldots, n$, so that every (unordered) pair of distinct objects occurs together in exactly λ blocks. If K consists of a single number k, we write $B[K, \lambda, v] = B[k, \lambda, v]$. Here, each time that an object occurs in a block, it is in a pair with $(k - 1)$ other objects. As it must be paired with each of $(v - 1)$ other objects exactly λ times, the object must be in $\lambda(v - 1)/(k - 1)$ blocks. Thus, the number of times an object occurs in blocks is the same for every object, and so a $B[k, \lambda, v]$ is an incomplete, balanced block design. The class of those v's for which a $B[K, \lambda, v]$ exists is denoted by $B(K, \lambda)$.

A transversal system, or T system, has been introduced by Hanani [1]. This is essentially the same as an orthogonal array.

DEFINITION. *Let a class of m mutually disjoint sets w_i, $i = 0, \ldots, m - 1$, be given, each containing t elements. A system $T_0(m, t)$ consists of t^2 sets Y_j, $j = 1, \ldots, t^2$, each containing m elements such that* (1) *a Y_j has exactly one element in common with each w_i; and* (2) *if $j \neq k$, then Y_j and Y_k have at most one element in common.*

Suppose we have an orthogonal array OA(t, m), and we number the rows from 0 to $m - 1$. The numbers $1, \ldots, t$ in the ith row will correspond to elements $a_{i1}, \ldots, a_{it}$ in a set w_i, $i = 0, \ldots, m - 1$. The t^2 columns of OA(t, m) yield the sets Y_j, $j = 1, \ldots, t^2$. Here, if the jth column of OA(t, m) contains the number u in the ith row, we put the element a_{iu} in Y_j. Thus, each Y_j contains m elements, one from each w_i. The assertion that if $j \neq k$, Y_j and Y_k have at most one element in common is equivalent to the orthogonality of the rows of OA(t, m) saying that between two rows we cannot have a column $\binom{u}{v}$ more than once. Conversely, if we have disjoint sets w_i, $i = 0, \ldots, m$ of t elements each and a system $T_0(m, t)$, let the t elements of w_i be replaced by the numbers $1, 2, \ldots, t$. The sets Y_j now form the columns of a matrix if for our jth column we place u in the ith row if Y_j contains the element of w_i corresponding to the number u. Condition (2) on the Y_j assures the orthogonality of the rows of the matrix so formed, which is therefore an OA(t, m).

The class of numbers t for which a system $T_0(m, t)$ exists is denoted by $T_0(m)$. We may call a set of the transversals Y_j in $T_0(m, t)$ a parallel set if there are t of them and no two of them have an element in common. Thus, given an element, there is exactly one transversal of a parallel set that contains it. If $T_0[m, t]$ contains e (or more) parallel sets of transversals, we denote it as $T_e(m, t)$ and denote as $T_e(m)$ the class of numbers t for which a $T_e(m, t)$ exists.

THEOREM 15.2.1. *If there is a $T_d(m, s)$ and a $T_e(m, t)$, then there is a $T_{de}(m, st)$.*

Proof: ■ We have given sets $v_i = \{a_{i1}, \ldots, a_{is}\}$, $i = 0, \ldots, m - 1$, of s elements and s^2 transversals $X_j, j = 1, \ldots, s^2$ and also sets $w_i = \{b_{i1}, \ldots, b_{it}\}$ $i = 0, \ldots, m - 1$, of t elements and t^2 transversals $Y_k, k = 1, \ldots, t^2$, where there are d parallel sets of X's and e parallel sets of Y's. Let us define elements $c_{i,g,h} = (a_{ig}, b_{ih}), g = 1, \ldots, s$; $h = 1, \ldots, t$, as ordered pairs of a's and b's and put

$$u_i = \{c_{i,1,1}, \ldots, c_{i,g,h}, \ldots\}, \qquad i = 0, \ldots, m - 1.$$

This gives us m u's, each containing st elements. For transversals, define sets $Z_{j,k}$ $j = 1, \ldots, s^2, k = 1, \ldots, t^2$; $c_{i,g,h} \in Z_{jk}$, if and only if $a_{ig} \in X_j$, $b_{ih} \in Y_k$. It is an easy verification that the Z's form a set $T_{de}(m, st)$ if we note that as j runs over a parallel set for X's, k for Y's, then the corresponding $Z_{j,k}$'s form a parallel set.

THEOREM 15.2.2. *There is a system $T_t(m - 1, t)$ if and only if there is a system $T_0(m, t)$.*

Proof: ■ Suppose there is a system $T_0(m, t)$ of transversals to $w_0, w_1, \ldots, w_{m-2}, w_{m-1}$. The t transversals containing a particular element of w_{m-1} have no further elements in common. Hence, if we drop w_{m-1} and its elements, we have t sets of parallels for $w_0, \ldots, w_{m-2}$, one corresponding to each element of w_{m-1}, and so a system $T_t(m - 1, t)$. Conversely, if we have a system $T_t(m - 1, t)$ of transversals to $w_0, \ldots, w_{m-2}$, we may adjoin a new element to every transversal in a parallel set and take the t new elements as a set w_{m-1}, giving a system $T_0(m, t)$.

THEOREM 15.2.3. *If $t = p_1^{e_1} \cdot p_2^{e_2} \cdots p_r^{e_r}$, where the p_i are primes and if $m = min(p_1^{e_1}, \ldots, p_r^{e_r})$, there is a $T_t(m, t)$.*

Proof: ■ For a prime power p^e, by Theorem 13.2.2, there is a complete set of $p^e - 1$ orthogonal squares and an orthogonal array OA$(p^e, p^e + 1)$ and so an OA$(p^e, m + 1)$, and this is equivalent to the existence of a system $T_0(m + 1, p^e)$. By Theorem 15.2.2 we have a system $T_{p^e}(m, p^e)$, and so by repeated application of Theorem 15.2.1, we have a system $T_t(m, t)$.

There is a property of certain block systems $B[K, \lambda, v]$ that is something like the resolvability which Hanani uses. We shall call this "central resolvability."

DEFINITION. *If there is a system $B[K, \lambda, v]$ on the elements of a set E and if there is an element $A \in E$ and a number $m \in K$ such that $m - 1$ divides $v - 1$ and the set $E - A$ can be split into $(v - 1)/(m - 1)$ mutually disjoint subsets E_i, $i = 1, \ldots, (v - 1)/(m - 1)$, each having $(m - 1)$ elements in such a way that each of the sets $E_i \cup A$, $i = 1, \ldots, (v - 1)/(m - 1)$ appears exactly λ times as a block in the system $B[K, \lambda, v]$, then we call this a centrally resolvable system and denote it by $B_m[K, \lambda, v]$ and the class of all numbers v for which systems $B_m[K, \lambda, v]$ exist by $B_m(K, \lambda)$. We call A the center and $E_i \cup A$ the distinguished blocks.*

We can now prove Hanani's two chief recursive theorems for construction of block designs. The first is much like Theorem 13.3.2.

THEOREM 15.2.4. *If $v = (m - 1)u + 1$, where $u \in B(K', \lambda')$, and if for every $k' \in K'$, $(m - 1)k' + 1 \in B_m(K, \lambda'')$, then $v \in B_m(K, \lambda)$, where $\lambda = \lambda'\lambda''$.*

Proof: ■ As our set of elements E, we take a point A and points

$$(x, y), \qquad 0 \le x \le u - 1, \qquad 0 \le y \le m - 2.$$

This gives us $1 + (m - 1)u = v$ elements. Now take subsets

$$(A, i) = \{A, (i, y), \quad 0 \le y \le m - 2\}, \qquad i = 0, \ldots, u - 1.$$

In our system $B[K', \lambda', u]$, let the elements be the numbers $0, 1, \ldots, u - 1$. If B_j' is a block of $B[K', \lambda', u]$, let its elements be $a_1, \ldots, a_r$, where, of course, $r \in K'$. Now take the following $(m - 1)r + 1$ elements as a set

$$E_j': A, (a_i, y), \qquad i = 1, \ldots, r \qquad y = 0, \ldots, m - 2.$$

By assumption there is a system $B_m(K, \lambda'', (m - 1)r + 1)$. Construct a system

$$B_{mj}' = B_m(K, \lambda'', (m - 1)r + 1)$$

with A as center and distinguished blocks (A, a_i), $i = 1, \ldots, r$ on the elements of E_j'. Delete from B_{mj}' the distinguished blocks (A, a_i), leaving a set of blocks $\overline{B_{mj}'}$.

Now we take all blocks of all sets $\overline{B_{mj}'}$ as B_j' runs over all blocks of $B[K', \lambda', u]$ and also each set (A, i) as a block taken λ times. We assert that this set of blocks on the v elements of E is a system $B_m(K, \lambda, v)$. Necessarily, $m \in K$, and every block constructed in every B_{mj}' has a number of elements that is a number of K. The point A occurs only in the distinguished blocks (A, i) and so is paired with every other element exactly λ times. Furthermore, points (i, y_1) and (i, y_2) occur together only in the distinguished blocks of a B_{mj}' and so only in the distinguished blocks (A, i) in the system we have taken. Consider two points (s, y_1) and (t, y_2), where $s \ne t$ but y_1 and y_2 may be equal, s and t occur together in λ' blocks B_j' of $B[K', \lambda', u]$. For each such occurrence the points (s, y_1) and (t, y_2) are in λ'' blocks of the set $\overline{B_{mj}'}$. Hence, (s, y_1) and (t, y_2)

occur together in exactly $\lambda'\lambda''$ blocks of our system, which is therefore a system $B_m(K, \lambda, v)$ that is centrally resolvable with center A and distinguished blocks (A, i).

We give a simple example of Theorem 15.2.4. If $K' = \{4, 3\}$, we have the following blocks of a design $B[K', 1, 10]$ on the numbers $0, \ldots, 9$:

$$\begin{array}{llll} B_0': & 0, 1, 2, 3, & B_6': & 2, 4, 8, \\ B_1': & 0, 4, 5, 6, & B_7': & 2, 5, 9, \\ B_2': & 0, 7, 8, 9, & B_8': & 2, 6, 7, \\ B_3': & 1, 4, 7, & B_9': & 3, 4, 9, \\ B_4': & 1, 5, 8, & B_{10}': & 3, 5, 7, \\ B_5': & 1, 6, 9, & B_{11}': & 3, 6, 8. \end{array} \tag{15.2.3}$$

Let us take $m = 3$. For $1 + (3 - 1)3 = 7$ and for $1 + (3 - 1)4 = 9$ there exists centrally resolvable designs $B_3(3, 1, 7)$ and $B_3(3, 1, 9)$. For $B_3(3, 1, 7)$ on $A, 1, \ldots, 6$, we have

$$\begin{array}{llllll} C_1: & A, 1, 2, & C_4: & 1, 3, 5, & C_6: & 2, 3, 6, \\ C_2: & A, 3, 4, & C_5: & 1, 4, 6, & C_7: & 2, 4, 5. \\ C_3: & A, 5, 6, & & & & \end{array} \tag{15.2.4}$$

For $B_3(3, 1, 9)$ on $A, 1, \ldots, 8$, we have

$$\begin{array}{llllllll} D_1: & A, 1, 2, & D_5: & 1, 3, 5, & D_8: & 2, 3, 8, & D_{11}: & 3, 6, 7, \\ D_2: & A, 3, 4, & D_6: & 1, 4, 7, & D_9: & 2, 4, 6, & D_{12}: & 4, 5, 8. \\ D_3: & A, 5, 6, & D_7: & 1, 6, 8, & D_{10}: & 2, 5, 7, & & \\ D_4: & A, 7, 8, & & & & & & \end{array} \tag{15.2.5}$$

The theorem asserts the existence of a $B_3(3, 1, 21)$ as $21 = 1 + (3 - 1)10$. The 21 elements of E are A and the points (x, y), $x = 0, \ldots, 9, y = 0, 1$. The distinguished blocks are

$$(A, i) = \{A, (i, 0), (i, 1)\} \qquad i = 0, \ldots, 9. \tag{15.2.6}$$

For simplicity in notation, let us write

$$(i, 0) = i, \qquad (i, 1) = 1i, \qquad i = 0, \ldots, 9, \tag{15.2.7}$$

giving our set E as the point A and the numbers $0, 1, \ldots, 19$. The block B_0': 0, 1, 2, 3 in (15.2.3) is used on the set $E_0' = \{A, 0, 1, 2, 3, 10, 11, 12, 13\}$ to yield a design $B_{03}'[3, 1, 9]$ with $i = 0, 1, 2, 3$ in (15.2.6). With the notation of (15.2.7), these blocks are

$$\begin{array}{llll} A, 0, 10; & 0, 1, 2; & 10, 1, 13; & 1, 12, 3; \\ A, 1, 11; & 0, 11, 3; & 10, 11, 12; & 11, 2, 13. \\ A, 2, 12; & 0, 12, 13; & 10, 2, 3; & \\ A, 3, 13; & & & \end{array} \tag{15.2.8}$$

Similarly, the block B_6': 2, 4, 8 is used on the set $E_6' = \{A, 2, 4, 8, 12, 14, 18\}$ to yield a design $B_{63}'[3, 1, 7]$ with appropriate substitutions in (15.2.4) yielding blocks

$$\begin{array}{lll} A, 2, 12; & 2, 4, 8; & 12, 4, 18; \\ A, 4, 14; & 2, 14, 18; & 12, 14, 8. \\ A, 8, 18; & & \end{array} \tag{15.2.9}$$

In this way, using all 12 blocks of (15.2.3), we construct three systems $B_3[3, 1, 9]$ and nine systems $B_3[3, 1, 7]$, and taking all blocks of these together except for using the distinguished blocks A, i, $1i$ only once, we obtain a system $B_3[3, 1, 21]$, which is the balanced, incomplete block design with $v = 21$, $b = 70$, $r = 10$, $k = 3$, $\lambda = 1$. The reader should be able to construct all 70 blocks without difficulty, following the pattern given above. This is not the easiest way of constructing a design with $v = 21$, $b = 3$, $\lambda = 1$, but the example illustrates the way in which the pairwise balanced design of (15.2.3) can be used to construct a design in which K consists of a single number k.

The following theorem is very similar in nature to Theorem 13.3.5.

THEOREM 15.2.5. *Let* s, $s + 1$, $t \in B(K, \lambda)$, *and* $t \in T_q(s)$ *and* $q \in B(K, \lambda)$ *or* $q = 0$ *or* 1; *then* $v = st + q \in B(K, \lambda)$.

Proof: ■ The elements of our set E will be points (x, y) with $0 \leq x \leq t - 1$, $0 \leq y \leq s - 1$ and also (x, s) with $0 \leq x \leq q - 1$. We consider $T_q(s, t)$ to be the t^2 transversals to the sets $w_0, \ldots, w_{s-1}$ where a set w_i consists of the points (x, i), $0 \leq x \leq t - 1$. Our transversals Y_h contain q parallel sets. To each transversal of the jth parallel set we adjoin the point (j, s), $j = 0, \ldots, q - 1$. In this we have qt blocks Y_h^*, each containing $s + 1$ points, and the remaining $(t - q)t$ transversals Y_h are blocks with s points. If $q > 1$, form the set $Z = \{(x, s)\}$, $x = 0, \ldots, q - 1$.

We now form our system $B[K, \lambda, v]$ by taking the blocks of the qt systems $B[K, \lambda, s + 1]$ on the sets from the Y_h^*, the $(t - q)$ systems $B[K, \lambda, s]$ on the sets from the Y_h. We also form the s systems $B[K, \lambda, t]$ on the sets $w_0, \ldots, w_{s-1}$. Finally, if $q > 1$, we form the system $B[K, \lambda, q]$ on the elements of Z. All these blocks together form a system $B[K, \lambda, v]$, for we note first that the sets w_i, Y_h^*, Y_h, Z form a pairwise balanced design with $\lambda = 1$. Hence, constructing a $B[K, \lambda, r]$ with $r = t, s + 1, s, q$ on these in turn yields a $B[K, \lambda, v]$.

The next two theorems are trivial but useful.

Theorem 15.2.6. *If* $v \in B(K', \lambda_1)$ *and every* k' *of* K' *is such that* $k' \in B(K, \lambda_2)$, *then* $v \in B(K, \lambda_1\lambda_2)$.

Proof: ■ Given $B = B[K', \lambda_1, v]$. Take the elements of a block of B containing k' elements and on this set construct the system $B[K, \lambda_2, k']$. All the blocks of all these systems taken together form a system $B(K, \lambda_1\lambda_2, v)$.

THEOREM 15.2.7. *If $v \in B(K, \lambda_1)$ and $v \in B(K, \lambda_2)$, then $v \in B(K, \lambda_1 + \lambda_2)$.*

Proof: ■ Let $B[K, \lambda_1, v]$ and $B[K, \lambda_2, v]$ be represented on the same set E of v elements. Then all the blocks of the two systems taken together form a system $B[K, \lambda_1 + \lambda_2, v]$.

COROLLARY. *If $v \in B(K, \lambda)$, then $v \in B(K, \lambda n)$ for any positive integer n.*

We need a further theorem on transversal systems.

THEOREM 15.2.8. *If $t \in T_s(m)$ and $s \in T_0(m)$, then $st \in T_{s^2}(m)$.*

Proof: ■ We are given systems $T_s(m, t)$ and $T_0(m, s)$. Let the w sets of $T_s(m, t)$ be $w_0, w_1, \ldots, w_{m-1}$, where $w_x = \{b_{x1}, b_{x2}, \ldots, b_{xt}\}$ and the w sets of $T_0(m, s)$ be $w'_0, \ldots, w'_{m-1}$, where $w'_x = \{c_{x1}, \ldots, c_{xs}\}$. We shall identify the elements c_{xj} of w'_x, $x = 0, \ldots, m-1$ with the residues modulo s in some arbitrary but fixed order. It is convenient to think of a transversal Y as a function $Y(x)$, where $x = 0, \ldots, m-1$ and $Y(x)$ is the element belonging to Y in the set w_x. Hence, we are given transversal functions for $T_s(m, t)$:

$$\begin{array}{lllll} Y_{1,1}(x), \ldots, & Y_{1,t}(x), & Y_{2,1}(x), \ldots, & Y_{2,t}(x), \ldots, & \\ Y_{s,1}(x), \ldots, & Y_{s,t}(x), & Y_{st+1}(x), \ldots, & Y_{t^2}(x), & x = 0, \ldots, m-1, \end{array} \tag{15.2.10}$$

where $Y_{i1}, \ldots, Y_{it}$ are the transversals of the ith parallel set, $i = 1, \ldots, s$ and Y_j, $j = st + 1, \ldots, t^2$, are the remaining transversals. For $T_0(m, s)$ let the transversal functions be

$$Y'_1(x), \ldots, \quad Y'_{s^2}(x), \qquad x = 0, \ldots, m-1. \tag{15.2.11}$$

We now define new w sets of ordered pairs

$$w''_x = \{(b_{xi}, c_{xj})\}, \qquad i = 1, \ldots, t; \quad j = 1, \ldots, s; \quad x = 0, \ldots, m-1. \tag{15.2.12}$$

Each of these contains st ordered pairs (b, c). On these w sets we define transversal functions $Y''(x)$ by

$$\begin{aligned} Y''_{e,f,g}(x) &= (Y_{e,f}(x), Y'_g(x) + e), \\ &\qquad e = 1, \ldots, s, \quad f = 1, \ldots, t, \quad g = 1, \ldots, s^2; \\ Y''_{h,g}(x) &= (Y_h(x), Y'_g(x)), \qquad h = st + 1, \ldots, t^2, \qquad g = 1, \ldots, s^2. \end{aligned} \tag{15.2.13}$$

This gives us s^2t^2 transversal functions $Y''(x)$. We must verify that we cannot have

$$Y''_i(x_1) = Y''_j(x_1) \qquad \text{and} \qquad Y''_i(x_2) = Y''_j(x_2)$$

for different functions Y''_i and Y''_i and $x_1 \neq x_2$. This follows directly for the corresponding fact on the functions Y and Y'. Hence, the Y'' form a transversal system $T_0(m, st)$ for the w sets of (15.2.12). For a fixed g, the functions $Y''_{e,f,g}(x)$, $e = 1, \ldots, s, f = 1, \ldots, t$ contain st transversals. Here, if

$$Y''_{e_1,f_1,g}(x_1) = Y''_{e_2,f_2,g}(x_1)$$

for some x_1, then

$$Y_{e_1,f_1}(x_1) = Y_{e_2,f_{21}}(x_1) \qquad \text{and} \qquad Y'_g(x_1) + e_1 = Y'_g(x_1) + e_2 .$$

From the second of these relations we have $e_2 = e_1$. But for e fixed, the functions $Y_{e,f}(x)$ form a parallel set and so $Y_{e_1,f_1}(x) = Y_{e_1,f_2}(x_1)$ is possible only if $f_1 = f_2$ and the transversals are identical. Therefore, for g fixed, the transversals $Y''_{e,f,g}(x)$ form a parallel set. Hence, our system is a system $T_{s^2}(m, st)$, and our theorem is proved.

15.3 Direct Construction Methods

In a long paper published in 1939, R. C. Bose [1] developed a number of direct methods for constructing block designs, the first of which he calls the "method of symmetrically repeated differences." We shall refer to it here simply as the "method of mixed differences."

Suppose that a block design $D(b, v, r, k, \lambda)$ has an Abelian group A of order m as a group of automorphisms. We shall write A additively, and may represent A by residues modulo m if A is cyclic. We say that two objects c, d of D are in the same orbit if there is an automorphism α of A such that $(c)\alpha = d$. Similarly, blocks B_1 and B_2 are in the same orbit if there is an α of A such that $(B_1)\alpha = B_2$. The property of being in the same orbit is an equivalence relation and so the automorphism group A divides the objects and blocks of D into disjoint orbits. Let us represent some arbitrary but fixed object of the ith orbit as $(0)_i$, 0 being the zero of A. Then, for $y \in A$, we write $((0)_i)y = (y)_i$. In general, it may happen that $(y)_i = (z)_i$ even though $y \neq z$; in this case, $(0)_i$ is fixed by some subgroup of A. We shall not have occasion to use this full generality, but only the extreme cases in which $(y)_i \neq (z)_i$ if $y \neq z$ or that in which $(y)_i = (z)_i$ for *all* $y, z \in A$. In this latter alternative we write $(\infty)_i$ for the single object in the orbit. For blocks we shall have occasion to use orbits that are intermediate, having a number of blocks that is some proper divisor of m. If we choose one block from each orbit, the entire design D is determined and such a choice is called a "base." If there is only one orbit of objects and one of blocks, then a base consists of a single block that is a difference set. This case has been treated at some length in Chapter 11.

We give here two instances of designs with cyclic automorphism groups, both yielding designs D with $v = 15, b = 35, r = 7, k = 3, \lambda = 1$. In (15.3.1) the first case, A is the additive group of residues modulo 5; in (15.3.2), the second case, modulo 15.

$$
\begin{array}{lll}
(0_1, 1_2, 4_2) & (0_1, 2_2, 3_2) & \\
(1_1, 2_2, 0_2) & (1_1, 3_2, 4_2) & \\
(2_1, 3_2, 1_2) & (2_1, 4_2, 0_2) & \\
(3_1, 4_2, 2_2) & (3_1, 0_2, 1_2) & \\
(4_1, 0_2, 3_2) & (4_1, 1_2, 2_2) & \\
 & & \\
(0_2, 1_3, 4_3) & (0_2, 2_3, 3_3) & (0_1, 0_2, 0_3) \\
(1_2, 2_3, 0_3) & (1_2, 3_3, 4_3) & (1_1, 1_2, 1_3) \\
(2_2, 3_3, 1_3) & (2_2, 4_3, 0_3) & (2_1, 2_2, 2_3) \\
(3_2, 4_3, 2_3) & (3_2, 0_3, 1_3) & (3_1, 3_2, 3_3) \\
(4_2, 0_3, 3_3) & (4_2, 1_3, 2_3) & (4_1, 4_2, 4_3) \\
 & & \\
(0_3, 1_1, 4_1) & (0_3, 2_1, 3_1) & \\
(1_3, 2_1, 0_1) & (1_3, 3_1, 4_1) & \\
(2_3, 3_1, 1_1) & (2_3, 4_1, 0_1) & \\
(3_3, 4_1, 2_1) & (3_3, 0_1, 1_1) & \\
(4_3, 0_1, 3_1) & (4_3, 1_1, 2_1) &
\end{array}
\qquad (15.3.1)
$$

$$
\begin{array}{rrl}
(0, 1, 4) & (0, 7, 13) & (0, 5, 10) \\
(1, 2, 5) & (1, 8, 14) & (1, 6, 11) \\
(2, 3, 6) & (2, 9, 0) & (2, 7, 12) \\
(3, 4, 7) & (3, 10, 1) & (3, 8, 13) \\
(4, 5, 8) & (4, 11, 2) & (4, 9, 14) \\
(5, 6, 9) & (5, 12, 3) & \\
(6, 7, 10) & (6, 13, 4) & \\
(7, 8, 11) & (7, 14, 5) & \\
(8, 9, 12) & (8, 0, 6) & \\
(9, 10, 13) & (9, 1, 7) & \\
(10, 11, 14) & (10, 2, 8) & \\
(11, 12, 0) & (11, 3, 9) & \\
(12, 13, 1) & (12, 4, 10) & \\
(13, 14, 2) & (13, 5, 11) & \\
(14, 0, 3) & (14, 6, 12) &
\end{array}
\qquad (15.3.2)
$$

Note that in (15.3.1) there are three orbits of objects, and seven of blocks, while in (15.3.2) there is only one orbit of objects and three orbits of blocks, the third orbit containing only five blocks, where a block in the third orbit is fixed by a subgroup of order 3.

We need a theorem to establish conditions under which, given the group A and a set of blocks $\{B\}$, we can form a base for a design $D(v, b, r, k, \lambda)$. If we have a block $B = [(a_1)_{j_1}, (a_2)_{j_2}, \ldots, (a_k)_{j_k}]$, $a_i \in A$, the difference $a_i - a_t$ is of *class* j_i, j_t and is called a *pure* difference if $j_i = j_t$ and a *mixed* difference if $j_i \neq j_t$. Our conditions may be described in terms of pure and mixed differences.

THEOREM 15.3.1. *Given an Abelian group A additively written, of order m, and a set* $\{B\}$ *of blocks of indexed elements of A, a typical block being*

$$B = [(a_1)_{j_1}, (a_2)_{j_2}, \ldots, (a_k)_{j_k}], \qquad a_i \in A.$$

The set $\{B\}$ *will be a base for a design* $D(v, b, r, k, \lambda)$ *having A as a group of automorphisms if* (1) *every block of* $\{B\}$ *contains k objects;* (2) *under the action of A there are b blocks;* (3) *under the action of A every indexed object* $(y)_j$ *occurs r times; and* (4) *in the blocks of* $\{B\}$ *taken together every nonzero pure difference occurs* λ *times. In counting differences, if one of the blocks of* $\{B\}$ *is fixed by a subgroup of A of order w, the counts from it should be divided by w.*

Proof: ■ Let $y \in A$. If $(a_i)_{j_i}$ and $(a_t)_{j_t}$ belong to B, then $(a_i + y)_{j_i}$ and $(a_t + y)_{j_t}$ belong to $(B)y$, whence $(x)_{j_i}$ and $(z)_{j_t}$ belong to block $B(y)$ if and only if $x - z = a_i - a_t$ with y determined by $x = a_i + y$ or $z = a_t + y$ and $(a_i)_{j_i}$ and $(a_t)_{j_t}$ belonging to $B = (B)0$. If B is fixed by a subgroup of order w, the blocks $(B)y$ as y runs over A run over the same set of m/w blocks, each w times, and so the differences arising from B must be divided by w to account for the pairs of objects arising in the m/w blocks that are the images of B under A. This latter situation is illustrated in (15.3.2) where, in the base blocks (0, 5, 10) of the third block orbit, we have $w = 3$ and the differences 5, 10 each occur three times. For a fixed object $(\infty)_i$, the pair $(\infty)_i$ and $(u)_j$ must be regarded as giving every mixed difference of class i, j once.

THEOREM 15.3.2. *With A the additive group of residues modulo* $m = 2t + 1$, *the blocks*

$$\begin{array}{l} [1_1, 2t_1, 0_2], \ldots, [i_1, (2t+1-i)_1, 0_2], \ldots, [t_1, (t+1)_1, 0_2] \\ [1_2, 2t_2, 0_3], \ldots, [i_2, (2t+1-i)_2, 0_3], \ldots, [t_2, (t+1)_2, 0_3] \\ [1_3, 2t_3, 0_1], \ldots, [i_3, (2t+1-i)_3, 0_1], \ldots, [t_3, (t+1)_3, 0_1] \\ \qquad\qquad [0_1, 0_2, 0_3] \end{array} \tag{15.3.3}$$

form a base for a design with $v = 6t + 3$, $b = (3t + 1)(2t + 1)$, $r = 3t + 1$, $k = 3$, $\lambda = 1$.

Proof: ■ All verifications are trivial except for the differences. The zero mixed differences all occur in the block $[0_1, 0_2, 0_3]$. The general block of the first three rows of (15.3.3) is of the form $[x_j, y_j, 0_{j+1}]$, where $x + y = 2t + 1$, and $j \equiv 1, 2, 3 \pmod 3$. To get a pure difference $d \not\equiv 0 \pmod{2t+1}$ of class j, we need to have $x - y \equiv d \pmod{2t+1}$, and as $x + y \equiv 0 \pmod{2t+1}$, this gives $2x \equiv d \pmod{2t+1}$, determining x and so y uniquely. The first row gives every nonzero mixed difference of classes 1 and 2 exactly once. Similarly, the other two rows give each nonzero mixed difference of classes 2 and 3 and of classes 1 and 3. Thus, all conditions are satisfied and the blocks are a base for the design as asserted. The design of (15.3.1) is the special case with $2t + 1 = 5$.

THEOREM 15.3.3. *Suppose* $v = 6t + 3 = 3m$, *where* $m = 2t + 1 \not\equiv 0 \pmod 3$. *Let us determine unordered pairs* (r, s) *modulo* $3m$ *by the conditions* $r \equiv s \equiv 1 \pmod 3$, $r + s \equiv 0 \pmod m$, $r, s \not\equiv 0 \pmod m$. *Then the base blocks*

$$[0, r, s] \pmod{3m} \tag{15.3.4}$$

and the single $[0, m, 2m]$ *of period* m *yield a design with* $v = 6t + 3$, $b = (2t + 1)(3t + 1)$, $r = 3t + 1$, $k = 3$, $\lambda = 1$.

Proof: ■ Here $v = 3m$ and the objects are in a single orbit, and so all differences are pure differences. Let us first consider a difference $d \equiv 0 \pmod 3$, $d \not\equiv 0 \pmod m$. Then the congruences $r - s \equiv d \pmod{3m}$, $r + s \equiv 0 \pmod m$, $r \equiv s \equiv 1 \pmod 3$, require that $2r \equiv d \pmod m$, $r \equiv 1 \pmod 3$, determining r and so s uniquely mod $3m$. If $d \equiv 1 \pmod 3$, $d \not\equiv 0 \pmod m$, then $r - 0 \equiv d \pmod{3m}$ determines r uniquely and so also s with $s \equiv 1 \pmod 3$, $r + s \equiv 0 \pmod m$. If $d \equiv -1 \pmod 3$, $d \equiv 0 \pmod m$, then $0 - r \equiv d \pmod{3m}$ determines an r uniquely and so an s with $s \equiv 1 \pmod 3$, $r + s \equiv 0 \pmod m$. There remain the differences m and $2m$, each of which occurs three times in $[0, m, 2m]$, and as this block is fixed by the subgroup $i \to i + m \pmod{3m}$ of order 3, we divide these counts by 3. This proves our theorem. The example (15.3.2) is an instance of this theorem with $3m = 15$.

For the next theorem we require that $v = p^n$, p prime, and we use the finite field $GF(p^n)$. Here A, our automorphism group, is the additive group of $GF(p^n)$.

THEOREM 15.3.4. *Let* $v = 6t + 1 = p^n$, p *a prime. In the field* $GF(p^n)$, *let* x *be a primitive root. Then the blocks*

$$(x^0, x^{2t}, x^{4t}), \ldots, (x^i, x^{2t+i}, x^{4t+i}), \ldots, (x^{t-1}, x^{3t-1}, x^{5t-1}) \tag{15.3.5}$$

are base blocks with respect to A, *the additive group of* $GF(p^n)$ *of a design* D *with* $v = 6t + 1$, $b = 6t^2 + t$, $r = 3t$, $k = 3$, $\lambda = 1$.

Proof: ■ Since x is a primitive root of $GF(p^n)$, $x^{6t} = 1$. Here

$$(x^{3t} - 1)(x^{3t} + 1) = 0,$$

and as $x^{3t} \neq 1$, we have $x^{3t} + 1 = 0$. Also $x^{2t} - 1 \neq 0$ and so let us determine

s by $x^{2t} - 1 = x^s$. Then the six differences arising from (x^0, x^{2t}, x^{4t}) are $\pm(x^{2t} - 1)$, $\pm(x^{4t} - 1)$, $\pm(x^{4t} - x^{2t})$. As $-1 = x^{3t}$,

$$-(x^{2t} - 1) = x^{s+3t},$$

$$x^{s+4t} = x^{4t}(x^{2t} - 1) = 1 - x^{4t} = -(x^{4t} - 1).$$

Hence, the six differences are x^s, x^{s+t}, x^{s+2t}, x^{s+3t}, x^{s+4t}, x^{s+5t} and the differences of the base blocks are given by

$$x^{s+i},\ x^{s+i+t},\ x^{s+i+2t},\ x^{s+i+3t},\ x^{s+i+4t},\ x^{s+i+5t}, \qquad i = 0, \ldots, t-1. \quad (15.3.6)$$

These together give all x^j, $j = 0, \ldots, 6t - 1$, and so every nonzero difference of A, the additive group of $\mathrm{GF}(p^n)$, and our theorem is proved.

THEOREM 15.3.5. *If $v = p^n = 12t + 1$, p prime, and if x is a primitive root of* $\mathrm{GF}(p^n)$ *such that $x^{4t} - 1 = x^q$ where q is odd, the blocks*

$$(0, x^0, x^{4t}, x^{8t}), \ldots, (0, x^{2i}, x^{2i+4t}, x^{2i+8t}), \ldots, (0, x^{2t-2}, x^{6t-2}, x^{10t-2}) \quad (15.3.7)$$

are a base with respect to A, the additive group of $\mathrm{GF}(p^n)$, *of a design D with $v = 12t + 1$, $b = t(12t + 1)$, $r = 4t$, $k = 4$, $\lambda = 1$.*

Proof: ∎ Noting that

$$x^{6t} = -1, \qquad x^{4t} - 1 = x^q \qquad \text{and} \qquad x^{8t+q} = x^{8t}(x^{4t} - 1) = 1 - x^{8t},$$

the twelve differences of the first block become

$$x^0,\ x^{6t},\ x^{4t},\ x^{10t},\ x^{8t},\ x^{2t},\ x^q,\ x^{q+6t},\ x^{q+2t},\ x^{q+8t},\ x^{q+4t},\ x^{q+10t}. \quad (15.3.8)$$

The differences of the other blocks are obtained by multiplying these by x^{2i}, $i = 1, \ldots, t - 1$, and so if q is odd, we obtain every nonzero difference in $\mathrm{GF}(p^n)$ exactly once, and our theorem is proved.

THEOREM 15.3.6. *Let $4t + 1 = p^n$, p a prime, and let x be a primitive root of* $\mathrm{GF}(p^n)$. *Then there exists a pair of odd integers c, d such that $(x^c + 1)/(x^c - 1) = x^d$. Then the blocks*

$$\begin{array}{ll} (x_1^{2i}, x_1^{2t+2i}, x_2^{2i+c}, x_2^{2t+2i+c}), & \\ (x_2^{2i}, x_2^{2t+2i}, x_3^{2i+c}, x_3^{2t+2i+c}), & \\ (x_3^{2i}, x_3^{2t+2i}, x_1^{2i+c}, x_1^{2t+2i+c}), & i = 0, \ldots, t-1 \\ (\infty, 0_1, 0_2, 0_3), & \end{array} \quad (15.3.9)$$

form a base with respect to A, the additive group of $\mathrm{GF}(p^n)$, *of a design with $v = 12t + 4$, $b = (3t + 1)(4t + 1)$, $r = 4t + 1$, $k = 4$, $\lambda = 1$.*

Proof: ∎ As x is a primitive root of $\mathrm{GF}(p^n)$, $x^{2t} = -1$. Hence, for $c \neq 0$, $2t$, $0 < c \leq 4t - 1$, $x^c + 1 \neq 0$ and $x^c - 1 \neq 0$, and so $(x^c + 1)/(x^c - 1) = x^d$ for some d. Here, d is uniquely determined by c. Of the $(4t - 2)$ values of c, $1, \ldots, 2t - 1, 2t + 1, \ldots, 4t - 1$, $2t - 2$ are even and $2t$ are odd. Also, if $x^d = \pm 1$, we have $(x^c + 1) = \pm(x^c - 1)$, giving either $2 = 0$ or $2x^c = 0$, which conflicts with the fact that p is odd. Hence, also $d \neq 0, 2t$,

$$0 < d \leq 4t - 1,$$

and there are $(2t - 2)$ even values and $2t$ odd values for d. Hence, for at least two choices, both c and d are odd.

The last block gives all mixed differences of the fixed element ∞ with the other three classes and also the mixed difference zero for the three possible pairs. As $x^{2t} = -1$, the pure differences of the first class, given by the first and third rows, are $\pm 2x^{2i}$ and $\pm 2x^{2i+c}$ with $i = 0, \ldots, t$, or alternately, $2x^{2i}$, $2x^{2i+2t}$, $2x^{2i+c}$, $2x^{2i+c+2t}$. As c is odd, this gives all nonzero pure differences of the first class, and similarly we have all nonzero pure differences of the second and third classes. For mixed differences of type [2, 1]—of the second and first classes—we have, remembering that $x^{2t} = -1$,

$$x^{2i}(x^c - 1), \quad x^{2i}(x^c + 1), \quad x^{2i+2t}(x^c + 1), \quad x^{2i+2t}(x^c - 1). \qquad (15.3.10)$$

Since $x^c - 1 = x^d(x^c + 1)$, these become

$$(x^c - 1)x^{2i}, \quad (x^c - 1)x^{2i+d}, \quad (x^c - 1)x^{2i+d+2t}, \quad (x^c - 1)x^{2i+2t},$$
$$i = 0, \ldots, t - 1. \qquad (15.3.11)$$

But these run over all nonzero elements of GF(p^n), each exactly once, since d is odd. The same argument applies to the other mixed differences, and our theorem is proved.

We list more base blocks for designs based on finite fields GF(p^n). In every case x is a primitive root, and the additional conditions, if any, are given. The proof in every case is essentially the same as that in the preceding theorems.

$$v = 20t + 1 = p^n, \quad b = t(20t + 1), \quad r = 5t, \quad k = 5, \quad \lambda = 1,$$
$$x^{4t} + 1 = x^q, \quad q \text{ odd}, \qquad (15.3.12)$$
$$(x^{2i}, x^{4t+2i}, x^{8t+2i}, x^{12t+2i}, x^{16t+2i}), \quad i = 0, \ldots, t - 1.$$

$$v = 20t + 5, \quad b = (5t + 1)(4t + 1), \quad r = 5t + 1, \quad k = 5, \quad \lambda = 1,$$
$$4t + 1 = p^n, \quad \frac{x^c + 1}{x^c - 1} = x^d, \quad c, d \text{ both odd};$$

$$\begin{aligned}
&(x_1^{2i}, x_1^{2i+2t}, x_3^{2i+c}, x_3^{2i+c+2t}, 0_2)\\
&(x_2^{2i}, x_2^{2i+2t}, x_4^{2i+c}, x_4^{2i+c+2t}, 0_3)\\
&(x_3^{2i}, x_3^{2i+2t}, x_5^{2i+c}, x_5^{2i+c+2t}, 0_4) \qquad i = 0, 1, \ldots, t - 1.\\
&(x_4^{2i}, x_4^{2i+2t}, x_1^{2i+c}, x_1^{2i+c+2}, 0_5)\\
&(x_5^{2i}, x_5^{2i+2t}, x_2^{2i+c}, x_2^{2i+c+2t}, 0_1)\\
&(0_1, 0_2, 0_3, 0_4, 0_5)
\end{aligned}$$
$$(15.3.13)$$

$$v = 6t + 1 = p^n, \quad b = t(6t + 1), \quad r = 4t, \quad k = 4, \quad \lambda = 2,$$
$$(0, x^i, x^{2t+i}, x^{4t+i}), \quad i = 0, \ldots, t - 1. \tag{15.3.14}$$

$$v = 4t + 1 = p^n, \quad b = t(4t + 1), \quad r = 4t, \quad k = 4, \quad \lambda = 3.$$
$$(x^i, x^{t+i}, x^{2t+i}, x^{3t+i}), \quad i = 0, 1, \ldots, t - 1. \tag{15.3.15}$$

$$v = p^n > k, \quad b = (p^n - 1)p^n, \quad r = k(p^n - 1), \quad k = k, \quad \lambda = k(k - 1),$$
$$(x^i, x^{i+1}, \ldots, x^{i+k-1}), \quad i = 0, \ldots, p^n - 2, \tag{15.3.16}$$

$$v = 6t + 6, \quad b = 2(t + 1)(6t + 5), \quad r = 6t + 5, \quad k = 3, \quad \lambda = 2,$$

A: additive group mod $6t + 5$

$$\begin{aligned} &(\infty, 0, 3t + 2) \\ &(0, i, 2t + 3 - i), \quad i = 1, \ldots, t + 1 \\ &(0, 2i, 3t + 3 + i), \quad i = 1, \ldots, t. \end{aligned} \tag{15.3.17}$$

$$v = 6t + 4, \quad b = 2(2t + 1)(3t + 2), \quad r = 6t + 3, \quad k = 3, \quad \lambda = 2,$$

A: additive group mod $6t + 3$

$$\begin{aligned} &(\infty, 0, 3t + 1) \\ &(0, i, 2t + 1 - i), \quad i = 1, \ldots, t \\ &(0, 2i, 3t + 1 + i), \quad i = 1, \ldots, t \\ &(0, 2t + 1, 4t + 2) \text{ of period } 2t + 1. \end{aligned} \tag{15.3.18}$$

15.4 Triple Systems

A block design with $k = 3$ is called, reasonably enough, a "triple system." The necessary conditions on the parameters for a design $D(v, b, r, 3, \lambda)$ are

$$3b = rv, \quad 2r = \lambda(v - 1). \tag{15.4.1}$$

We may rewrite these in terms of v and λ:

$$r = \frac{\lambda(v - 1)}{2}, \quad b = \frac{\lambda v(v - 1)}{6}. \tag{15.4.2}$$

The conditions that r and b be integers may be expressed as congruences on v and λ:

$$\lambda(v - 1) \equiv 0 \pmod 2, \quad \lambda v(v - 1) \equiv 0 \pmod 6. \tag{15.4.3}$$

The chief result of this section will be the proof that these necessary conditions on the parameters are also sufficient for the existence of the designs. This will be proved by use of Hanani's recursive methods.

A triple system with $\lambda = 1$ is called a Steiner triple system. Here a necessary condition for the existence of a Steiner triple system is that $v \equiv 1, 3 \pmod 6$. Steiner [1] posed as a problem in 1853 whether these necessary conditions were sufficient for its existence. This was solved affirmatively by Reiss [1] in 1859. Both papers appeared in *Crelle's Journal.* These writers were not aware that the problem had been posed and solved by Kirkman [1] in 1847 in an article in the *Cambridge and Dublin Mathematical Journal,* nor until recently did anyone else seem to be aware of Kirkman's work.

The Steiner triple systems of orders $v = 3, 7, 9$ are unique to within equivalence, two systems being considered equivalent if one can be obtained from the other by substitution on the elements and permutation of the blocks:

$$\begin{array}{ccccccccc}
v = 3 & \multicolumn{2}{c}{v = 7} & \multicolumn{4}{c}{v = 9} \\
123 & 123 & 246 & 123 & & & \\
 & 145 & 257 & 145 & 249 & 348 & \\
 & 167 & 347 & 168 & 256 & 357 & 467 \\
 & & 356 & 179 & 278 & 369 & 589
\end{array} \qquad (15.4.4)$$

For $v = 13$ there are exactly two nonisomorphic solutions. In both solutions we may include the triples

$$\begin{array}{llllll}
1, 2, 3, & 2, 4, 6, & & & & \\
1, 4, 5, & 2, 5, 7, & 4, 3, 8, & 7, 3, 11, & & \\
1, 6, 7, & 2, 8, 10, & 4, 7, 9, & 7, 8, 13, & 8, 5, 11, & 6, 9, 11, \\
1, 8, 9, & 2, 9, 12, & 4, 10, 13, & 7, 10, 12. & 8, 6, 12. & 3, 5, 12. \\
1, 10, 11, & 2, 11, 13, & 4, 11, 12. & & & \\
1, 12, 13. & & & & &
\end{array} \qquad (15.4.5)$$

For the first solution we also take the triples

$$\begin{array}{ll}
3, 6, 10, & 5, 6, 13, \\
3, 9, 13. & 5, 9, 10.
\end{array} \qquad (15.4.6)$$

For the second solution we take

$$\begin{array}{ll}
3, 6, 13, & 5, 6, 10, \\
3, 9, 10. & 5, 9, 13.
\end{array} \qquad (15.4.7)$$

Cole, White, and Cummings [1] found 80 distinct Steiner triple systems on 15 letters. Fisher [1] listed the systems on 15 letters, finding 79 systems, missing one on Cole's list. The author and J. D. Swift [1] made a systematic search and verified the correctness of Cole's list.

THEOREM 15.4.1. *If there are Steiner triple systems of orders v_1 and v_2, there is a Steiner triple system of order $v = v_1v_2$ containing subsystems isomorphic to those of orders v_1 and v_2.*

Proof: ■ Let $A = S(v_1)$ and $B = S(v_2)$ be Steiner triple systems of orders v_1 and v_2, respectively, and let (a_i, a_j, a_k) be any triple of A, and (b_r, b_s, b_u) any of B. Form a new system C with elements c_{ij}, $i = 1, \ldots, v_1$, $j = 1, \ldots, v_2$. Then (c_{ir}, c_{js}, c_{ku}) is taken as a triple of system C if (1), $r = s = u$ and (a_i, a_j, a_k) is a triple of A; or (2), $i = j = k$ and (b_r, b_s, b_u) is a triple of B; or (3), (a_i, a_j, a_k) is a triple of A and (b_r, b_s, b_u) is a triple of B. It is easily checked that these rules make C a Steiner triple system. Those triples with $r = s = u = 1$ form a subsystem of C isomorphic to A and those with $i = j = k = 1$ form a subsystem isomorphic to B. Thus, our theorem is proved.

We have an alternate proof. From a Steiner triple system S we form a quasi-group S^* in the following way: The elements of S^* are the elements of S. A product xy is defined by the rules (1), $xx = x$ for all $x \in S^*$, and (2), $xy = z$ if $x \neq y$ and (x, y, z) is a triple of S. The quasi-group S^* may be characterized by the idempotent property $x^2 = x$, and the properties that $yx = xy$ and $xy = z$ imply $yz = x$, and such a quasi-group determines a Steiner triple system. Here, if S_1^* and S_2^* are Steiner quasi-groups on v_1 and v_2 elements, respectively, then a new system S^* may be determined whose elements are ordered pairs (x_1, x_2), $x_1 \in S_1^*$, $x_2 \in S_2^*$ and the product rule $(x_1, x_2)(y_1, y_2) = (x_1y_1, x_2y_2)$. Here, S^* has the same properties as S_1^* and S_2^* and determines a Steiner triple system S on v_1v_2 elements.

A recursive method of constructing Steiner triple systems, due to Moore [1], is as follows.

THEOREM 15.4.2. *If there is a Steiner triple system of order v_2 containing a subsystem of order v_3 (or $v_3 = 1$), we can construct a system of order $v = v_3 + v_1(v_2 - v_3)$ containing v_1 subsystems of order v_2 and one of order v_1 and order v_3.*

Proof: ■ We construct an array of $v = v_3 + v_1(v_2 - v_3)$ elements listed in $(v_1 + 1)$ sets:

$$\begin{array}{llllll} S_0: & a_1 & a_2 & \cdots & a_{v_3} & \\ S_1: & b_{11} & \cdots & \cdots & b_{1_s} & \\ S_2: & b_{21} & \cdots & \cdots & b_{2_s}, & s = v_2 - v_3. \\ \cdots & \cdots & \cdots & \cdots & \cdots & \\ S_{v_1}: & b_{v_1 1} & \cdots & \cdots & b_{v_{1_s}} & \end{array} \qquad (15.4.8)$$

We form triples of these v elements by three rules:

1. We associate the elements of S_0 with the system of order v_3 and take a triple (a_i, a_j, a_k) if these correspond to a triple of that system.

2. We make S_0 and S_i, $i = 1, \ldots, v_1$ correspond to the system of order v_2, making S_0 correspond to its subsystem of order v_3. Triples of a's are already determined by rule (1). Other triples contain at most one a_i, and we construct triples (a_m, b_{ij}, b_{ik}), (b_{ij}, b_{ik}, b_{ir}) corresponding to the remaining triples of the system of order v_2.
3. Writing the system of order v_1 on the numbers $1, \ldots, v_1$, if (j, k, r) is a triple of this system, form all triples (b_{jx}, b_{ky}, b_{rz}) for subscripts x, y, z satisfying $x + y + z \equiv 0 \pmod{s}$.

These rules taken together give the triples of a system of order v. The first rule gives the triples in S_0 and we note that a triple containing two a's will have as its third element another a. The second rule gives the triples containing an a and a b, the third element being another b in the same row, and also the triples containing two b's from the same row, the third element being either an a or another b in the same row. For b's in different rows b_{jx}, b_{ky}, we determine r from the triple (j, k, r) of the system with v_1 elements and z from $x + y + z \equiv 0 \pmod{s}$. Thus, any pair of our $v = v_3 + v_1(v_2 - v_3)$ elements determines a unique triple containing it. Triples from S_0 form a system of order v_3, those from S_0 and any S_i, $i = 1, \ldots, v_1$, a subsystem of order v_2, and the triples from $b_{1s}, b_{2s}, \ldots, b_{v_1 s}$ a subsystem of order v_1. In each case the subsystems are isomorphic to the given subsystems. The proof of the theorem is now complete.

We may use this recursive theorem to construct Steiner triple systems of all orders v of the form $v = 6t + 1$, $v = 6t + 3$, showing the necessary condition to be sufficient.

THEOREM 15.4.3. *If* $v = 6t + 1$ *or* $v = 6t + 3$, *there is a Steiner triple system of order* v.

Proof: ∎ We use several specializations of Theorem 15.4.2, listing them as recursive rules:

(A) $v_1 = v'$, $v_2 = 3$, $v_3 = 1$, $v = 2v' + 1$, $v' \geq 3$.
(B) $v_1 = 3$, $v_2 = v'$, $v_3 = 1$, $v = 3v' - 2$, $v' \geq 3$.
(C) $v_1 = 3$, $v_2 = v'$, $v_3 = 3$, $v = 3v' - 6$, $v' \geq 7$.
(D) $v_1 = v'$, $v_2 = 9$, $v_3 = 3$, $v = 6v' + 3$, $v' \geq 3$.
(E) $v_1 = 3$, $v_2 = v'$, $v_3 = 7$, $v = 3v' - 14$, $v' \geq 15$.
(F) $v_1 = v'$, $v_2 = 7$, $v_3 = 1$, $v = 6v' + 1$, $v' \geq 3$.

We note that whenever we have a system of order v', we may use any rule above, except that to use rule E, the system of order v' must have a subsystem of order 7. Since the system of order $v = v_3 + v_1(v_2 - v_3)$ has subsystems of orders v_1, v_2, v_3 isomorphic to those used in its construction, the property of having a subsystem of order 7 carries over in the construction. We construct systems recursively according to the residue of v modulo 36, as in Table 15.1.

TABLE 15.1 SUBSYSTEMS OF v

Form of v	Rule	Value of v'
$36t + 1$	(B)	$12t + 1$
$36t + 3$	(A)	$18t + 1$
$36t + 7$	(F)	$6t + 1$
$36t + 9$	(D)	$6t + 1$
$36t + 13$	(E)	$12t + 9$
$36t + 15$	(A)	$18t + 7$
$36t + 19$	(F)	$6t + 3$
$36t + 21$	(D)	$6t + 3$
$36t + 25$	(B)	$12t + 9$
$36t + 27$	(A)	$18t + 13$
$36t + 31$	(A)	$18t + 15$
$36t + 33$	(C)	$12t + 13$

Alternate rules may be used in some cases, and we shall always prefer a rule that assures the existence of a subsystem of order 7. Rule F always gives a subsystem of order 7. We may, of course, also make use of Theorem 15.4.1 and, in particular, taking the triple system of order 3 have the rule $v = 3v'$, which applies in the following form:

$$\begin{aligned} 36t + 3 &= 3(12t + 1), \\ 36t + 9 &= 3(12t + 3), \\ 36t + 21 &= 3(12t + 7), \\ 36t + 27 &= 3(12t + 9). \end{aligned} \tag{15.4.9}$$

Applying these rules and some other applications of Theorems 15.4.1 and 15.4.2, we can find Steiner triple systems of all orders of the form $v = 6t + 1$ or $6t + 3$, and find one with a subsystem of order 7 except for orders 1, 3, 9, 13, 25, 27, 33, 37, 67, 69, 75, 81, 97, 109, 201, 289, and 321. We need to supplement the general rules of Tables 15.1 and 15.2 and (15.4.9) with the

TABLE 15.2 SUBSYSTEMS OF ORDER 7

Form of v	Rule	Value of v'
$36t + 3$	(C)	$12t + 3$
$36t + 15$	(C)	$12t + 7$
$36t + 21$	(C)	$12t + 9$
$36t + 25$	(E)	$12t + 13$
$36t + 31$	(E)	$12t + 15$

following special cases, the form $v = v_1 v_2$ or $v = v_3 + v_1(v_2 - v_3)$ indicating an application of Theorem 15.4.1 or 15.4.2.

$$\begin{aligned}
49 &= 7^2, & 229 &= 1 + 19(13 - 1), \\
73 &= 3 + 7(13 - 3), & 285 &= 15 \cdot 19, \\
85 &= 1 + 7(13 - 1), & 325 &= 3 + 7(49 - 3), \\
103 &= 3 + 25(7 - 3), & 589 &= 19 \cdot 31, \\
105 &= 7 \cdot 15, & 861 &= 7 \cdot 123, \\
193 &= 3 + 19(13 - 3), & 865 &= 7 + 13(73 - 7), \\
195 &= 13 \cdot 15, & 949 &= 13 \cdot 73, \\
225 &= 15^2, & 961 &= 31^2.
\end{aligned} \tag{15.4.10}$$

If we begin with initial Steiner triple systems $S(v)$ of order v for $v = 3, 7, 9, 13$, as listed in (15.4.4), (15.4.5), (15.4.6), and (15.4.7), the rules of Table 15.1 are sufficient to derive all values of v of the form $6t + 1$ or $6t + 3$ from these inductively. But application of rule E requires that we begin with an $S(v')$ containing an $S(7)$. The nature of our constructions in Theorems 15.4.1 and 15.4.2 is such that the property of containing an $S(7)$ is hereditary—that is, if an $S(v_1)$, $S(v_2)$, or $S(v_3)$ contains an $S(7)$, so does our constructed $S(v)$, where $v = v_1 v_2$ or $v = v_3 + v_1(v_2 - v_3)$. Rule F always yields an $S(v)$ containing an $S(7)$. Let us call a value of v *hereditarily sound* if we have an $S(v)$ containing an $S(7)$, and *hereditarily unsound* if we do not have an $S(v)$ containing an $S(7)$. Our theorem will be proved if we can show that only the values 1, 3, 9, 13, 25, 27, 33, 37, 67, 69, 75, 81, 97, 109, 201, 289, and 321 are unsound and that all other numbers of the form $6t + 1$ or $6t + 3$ are sound. An unsound value v can arise only by application of one of the rules A, . . . , E to an unsound v'. If we make use of the alternate choices afforded by Table 15.2, (15.4.9), or (15.4.10), we find alternate expressions for all other numbers. Thus, for $v' = 321$, we have

$$\begin{aligned}
&(\text{A}) & 2 \cdot 321 + 1 &= 643 = 3 \cdot 219 - 14, \\
&(\text{B}) & 3 \cdot 321 - 2 &= 961 = 31^2, \\
&(\text{C}) & 3 \cdot 321 - 6 &= 957 = 3 \cdot 319, \\
&(\text{D}) & 6 \cdot 321 + 3 &= 1929 = 3 \cdot 643, \\
&(\text{E}) & 3 \cdot 321 - 14 &= 949 = 13 \cdot 73,
\end{aligned} \tag{15.4.11}$$

and all successors of 321 are sound because of their alternate expressions. Thus our theorem is proved.

It would have been possible to prove our theorem by directly constructing Steiner triple systems of orders 25, 27, 33, and 37 containing $S(7)$'s.

Kirkman has posed a further problem on triple systems whose general solution is unknown, at least to the writer. This is the "Kirkman schoolgirl problem." The original problem asks for a walking schedule for 15 schoolgirls. The girls are to walk in five groups of three in an afternoon. The problem is to arrange a walking schedule so that in seven afternoons every girl will have walked in a group of three once with every other girl. This amounts to asking for a Steiner triple system with $v = 15$, $b = 35$, $r = 7$, $k = 3$, $\lambda = 1$, where the triples are resolvable into seven sets of five triples each, such that each of these is a replication containing every element exactly once. A solution of this is given here, with automorphism group A the additive group modulo 7, with base blocks as follows:

$$\begin{aligned} &(\infty, 0_1, 0_2), \\ &(1_1, 2_1, 4_1), \\ &(5_1, 1_2, 6_2), \\ &(3_1, 2_2, 5_2), \\ &(6_1, 3_2, 4_2). \end{aligned} \tag{15.4.12}$$

Here the base blocks form a complete replication, and adding 1, 2, . . . , 6(mod 7) to the values in (15.4.12) gives six more replications.

The general Kirkman schoolgirl problem is to find a triple system in which the triples can be divided into r complete replications. If this is possible, then necessarily the number v of elements, being the number in a replication of triples, must be a multiple of 3 and so v must be of the form $v = 6t + 3$. For $v = 9$, the solution is easy and is given by the lines of the affine plan of order 3 divided into four sets of parallels. For $v = 15$, the solution is given in the preceding paragraph. For $v = 21$, we have the following solution, with A the additive group modulo 7. Base blocks are:

$$v = 21, \qquad b = 70, \qquad r = 10, \qquad k = 3, \qquad \lambda = 1,$$

$$\begin{array}{ll} (0_1, 1_1, 3_1) & \\ (0_2, 1_2, 3_2) & (0_1, 1_3, 3_2) \\ (0_3, 1_3, 3_3) & (0_2, 1_1, 3_3) \\ (2_1, 4_2, 5_3) & (0_3, 1_2, 3_1) \\ (2_2, 4_3, 5_1) & \\ (2_3, 4_1, 5_2) & \\ (6_1, 6_2, 6_3) & \end{array} \tag{15.4.13}$$

The first seven of these are a complete replication, and applying A gives us six more complete replications. Each of the last three base blocks under the action of A gives a complete replication. Thus, (15.4.13) yields the desired system on 21 elements resolvable into 10 complete replications. Solutions of this schoolgirl problem are known for a number of specific values of $v = 6t + 3$, but no solution for the general case is known to the writer.

For a triple system with $\lambda = 2$, the necessary conditions (15.4.3) on v reduce to $v \equiv 0 \pmod 3$ or $v \equiv 1 \pmod 3$. We can now easily prove the necessary conditions to be sufficient.

THEOREM 15.4.4 (BHATTACHARYA). *If $v \geq 3$ is of the form $v = 3m$ or $v = 3m + 1$, there is a triple system with v elements and $\lambda = 2$.*

Proof: ■ If v is of the form $v = 6t + 1$ or $v = 6t + 3$, there is a Steiner triple system of order v. Taking its blocks twice, we have a triple design of order v with $\lambda = 2$. The other values to be considered are those v's of the form $v = 6t + 6$ or $v = 6t + 4$. The designs determined by the base blocks of (15.3.17) and (15.3.18), due to Bhattacharya [1], cover these cases, and our theorem is proved.

THEOREM 15.4.5 (HANANI). *A necessary and sufficient condition for the existence of a balanced incomplete block design with $k = 3$ and any λ is that*

$$\lambda(v - 1) \equiv 0 \pmod 2 \qquad \textit{and} \qquad \lambda v(v - 1) \equiv 0 \pmod 6. \qquad (15.4.14)$$

Proof: ■ In a block design $D(v, b, r, 3, \lambda)$, the basic conditions (15.2.1) and (15.2.2) give $r = \lambda(v - 1)/2$ and $b = \lambda v(v - 1)/6$. Hence, for r and b to be integers, the conditions (15.4.14) are clearly necessary. The main problem is, of course, to prove the sufficiency of these conditions, and this will be proved using Hanani's recursive theorems given in Section 15.2.

We begin by restating the conditions (15.4.9) in the form:

$$\begin{array}{ll} \text{If } \lambda \equiv 1 \text{ or } 5 \pmod 6, & v \equiv 1 \text{ or } 3 \pmod 6; \\ \text{If } \lambda \equiv 2 \text{ or } 4 \pmod 6, & v \equiv 0 \text{ or } 1 \pmod 3; \\ \text{If } \lambda \equiv 3 \pmod 6, & v \equiv 1 \pmod 2; \\ \text{If } \lambda \equiv 0 \pmod 6, & \text{no restrictions on } v. \end{array} \qquad (15.4.15)$$

From the corollary to Theorem 15.2.7, the sufficiency of (15.4.16) will follow if it can be shown that for $v \geq 3$,

$$v \equiv 1 \text{ or } 3 \pmod 6 \quad \text{implies } v \in B(3, 1), \qquad (15.4.16a)$$

$$v \equiv 0 \text{ or } 1 \pmod 3 \quad \text{implies } v \in B(3, 2), \qquad (15.4.16b)$$

$$v \equiv 1 \pmod 2 \quad \text{implies } v \in B(3, 3), \qquad (15.4.16c)$$

$$\text{For every } v \geq 3, \quad v \in B(3, 6), \qquad (15.4.16d)$$

Here (15.4.16a) is the statement of Theorem 15.4.3 and (15.4.16b) is the statement of Theorem 15.4.4. But we shall follow Hanani's argument and make no use of either of these theorems. We rely on the recursive Theorems

15.2.4 and 15.2.5 and explicit constructions for certain initial values of v. For convenience of reference we quote the statements of these theorems.

THEOREM 15.2.4. *If* $v = (m - 1)u + 1$ *where* $u \in B(K', \lambda')$ *and if for every* $k' \in K'$, $(m - 1)k' + 1 \in B_m(K, \lambda'')$, *then* $v \in B_m(K, \lambda)$, *where* $\lambda = \lambda'\lambda''$.

THEOREM 15.2.5. *Let* $s, s + 1 \in B(K, \lambda)$ *and* $t \in T_q(s)$ *and* $q \in B(K, \lambda)$ *or* $q = 0$ *or* 1; *then* $v = st + q \in B(K, \lambda)$.

Our proof proceeds through several lemmas.

LEMMA 15.4.1. *If* $u \equiv 0$ *or* $1(\text{mod } 3)$ *and* $u \geq 3$, *then* $u \in B(K_3^1, 1)$, *where* $K_3^1 = \{3, 4, 6\}$.

Proof: ■ By Theorem 15.2.3, $t \in T_t(3)$ for $t \geq 3$ and $t \equiv 0, 1, 3(\text{mod } 4)$. Since $2 \in T_0(3)$, and for n odd, $n > 1$, $n \in T_2(3)$, from Theorem 15.2.8, $t = 2n \in T_4(3)$, whence for $t \equiv 2(\text{mod } 4)$, $t \geq 6$, $t \in T_4(3)$. Trivially $3, 4, 6 \in B(K_3^1, 1)$. In (15.4.4) the Steiner triple systems with $v = 7, 9$ show that $7, 9 \in B(K_3^1, 1)$, For $u > 9$, remembering that $3 \in T_3(3)$ and $t \in T_4(3)$ for $t \geq 4$, we apply Theorem 15.2.5 recursively, taking $s = 3$ in all cases and choosing q and t by the rules, depending on the residue of $u(\text{mod } 9)$:

$$\begin{aligned}
&u \equiv 0(\text{mod } 9), && q = 0, && t = \frac{u}{3},\\
&u \equiv 1(\text{mod } 9), && q = 1, && t = \frac{u-1}{3},\\
&u \equiv 3(\text{mod } 9), && q = 0, && t = \frac{u}{3},\\
&u \equiv 4(\text{mod } 9), && q = 1, && t = \frac{u-1}{3},\\
&u \equiv 6(\text{mod } 9), && q = 3, && t = \frac{u-3}{3},\\
&u \equiv 7(\text{mod } 9), && q = 4, && t = \frac{u-4}{3}.
\end{aligned} \qquad (15.4.17)$$

This proves Lemma 15.4.1.

LEMMA 15.4.2. *If* $u \geq 3$, *then* $u \in B(K_3^2, 1)$, *where* $K_3^2 = \{3, 4, 5, 6, 8, 11, 14\}$.

Proof: ■ For $u \equiv 0$ or $1(\text{mod } 3)$, this is already covered by Lemma 15.4.1, since K_3^1 is a subset of K_3^2. The values $u = 8, 11, 14$ are trivial, being included in K_3^2. Hence, the first value for which proof is needed is $u = 17$. Our proof is inductive on the value of u, and we choose the values of q, s, and t according to the value of u by the following rules:

$$\begin{array}{llll}
u = 17, & q = 1, & s = 4, & t = 4, \\
u = 18, 19, 20, & q = u - 15, & s = 3, & t = 5, \\
u = 21, 22, & q = u - 21, & s = 3, & t = 7, \\
u = 23, & q = 3, & s = 4, & t = 5, \\
u = 24, \ldots, 28, & q = u - 21, & s = 3, & t = 7, \\
u = 29, & q = 1, & s = 4, & t = 7, \\
u = 30, \ldots, 36, & q = u - 27, & s = 3, & t = 9, \\
u = 37, \ldots, 44, & q = u - 33, & s = 3, & t = 11, \\
u = 45, \ldots, 50, & q = u - 39, & s = 3, & t = 13, \\
u \geq 51, & q \equiv u (\mathrm{mod}\ 12), & 3 \leq q \leq 14, & t = (u - q)/3. \\
 & & s = 3, &
\end{array} \tag{15.4.18}$$

Here, by construction, $u = st + q$ in every case. Also, s, $s + 1 \in B(K_3^2, 1)$, since $s = 3$ or 4 in every case. By induction also, $t \in B(K_3^2, 1)$. We must check $t \in T_q(s)$ in every case. We note that $t \in T_t(3)$ and $T_t(4)$ for $t = 4, 5, 7, 9, 11, 13$, as these are prime powers. For $u \geq 51$, the choice of t is such that $t \equiv 0 (\mathrm{mod}\ 4)$ and $t > q$. But here $t \in T_t(3)$ by Theorem 15.2.3, and as $t > q$, $t \in T_q(3)$. Thus our lemma is proved.

We have now used the transversal systems (essentially orthogonal arrays) and the recursive methods of Theorem 15.2.5 to construct the pairwise balanced designs

$$B[\{3, 4, 6\}, 1, u] \quad \text{with} \quad u \equiv 0 \quad \text{or} \quad 1 (\mathrm{mod}\ 3), \quad u \geq 3,$$

and

$$B[\{3, 4, 5, 6, 8, 11, 14\}, 1; u] \quad \text{for all} \quad u \geq 3.$$

Our proof of the main theorem will be completed by appealing to Theorem 15.2.4 and K_3^1 or K_3^2 in the role of K', or to the more elementary Theorem 15.2.6, which we quote for reference.

THEOREM 15.2.6. *If $v \in B(K', \lambda_1)$ and every $k' \in K'$ is such that $k' \in B(K, \lambda_2)$, then $v \in B(K, \lambda_1\lambda_2)$.*

Thus, at this stage our proof is reduced to a finite number of constructions involving the finite set of numbers k' in K'.

LEMMA 15.4.3. *If $v \equiv 1$ or $3 (\mathrm{mod}\ 6)$, then $v \in B(3, 1)$.*

Proof: ∎ If $v = 6t + 1$, take $u = 3t$, and if $v = 6t + 3$, take $u = 3t + 1$. In both cases, $v = 2u + 1$ and $u \in B(\{3, 4, 6\}, 1)$ by Lemma 15.4.1. In Theorem 15.2.4, take $m = 3$ and $K = \{3\}$. The Steiner triple systems of orders 7, 9 and

13, given in (15.4.4), (15.4.5), (15.4.6), and (15.4.7), show that $7, 9, 13 \in B_3(\{3\}, 1)$, and so our Lemma is proved. This is (15.4.16a) and is, of course, equivalent to Theorem 15.4.3.

LEMMA 15.4.4. *If* $v \equiv 0$ *or* $1 (\text{mod } 3)$, $v \geq 3$, *then* $v \in B(3, 2)$.

Proof: ■ By Lemma 15.4.1, $v \in B(\{3, 4, 6\}, 1)$. We appeal to Theorem 15.2.6, with $\lambda_1 = 1$, $\lambda_2 = 2$, and our lemma will be proved if we can show that 3, 4, $6 \in B(3, 2)$. For 3, this is trivial. For 4, we construct the design $B(4, 4, 3, 3, 2)$—that is, $4 \in B(3, 2)$:

$$\begin{array}{c} 1, 2, 3, \\ 1, 2, 4, \\ 1, 3, 4, \\ 2, 3, 4. \end{array} \tag{15.4.19}$$

For 6, we construct the design $B(6, 10, 5, 3, 2)$, so that $6 \in B(3, 2)$:

$$\begin{array}{cc} 1, 2, 3, & 2, 3, 6, \\ 1, 2, 4, & 2, 4, 5, \\ 1, 3, 5, & 2, 5, 6, \\ 1, 4, 6, & 3, 4, 5, \\ 1, 5, 6, & 3, 4, 6. \end{array} \tag{15.4.20}$$

These constructions complete the proof of the lemma, which is (15.4.16b) and equivalent to Theorem 15.4.4.

LEMMA 15.4.5. *If* $v \equiv 1 (\text{mod } 2)$, *then* $v \in B(3, 3)$.

Proof: ■ This is (15.4.16c). Here, for $v \equiv 1$ or $3 (\text{mod } 6)$ as $v \in B(3, 1)$, we may take the design $B[3, 1, v]$ three times and have $v \in B_3(3, 3)$. If we apply Theorem 15.2.4, with $m = 3$, $K' = K_3^2$, $\lambda' = 1$, $\lambda'' = 1$, since by Lemma 15.4.2 if $u \geq 3$, then $u \in B(K_3^2, 1)$, and we conclude that $v = 2u + 1 \in B_3(3, 3)$ if $2k' + 1 \in B_3(3, 3)$ for $k' \in K_3^2$. Here, $v = 2u + 1$, includes all odd numbers greater than or equal to 7. Hence, to prove our lemma, we must show that $3, 5 \in B(3, 3)$ and that $7, 9, 11, 13, 17, 23, 29 \in B_3(3, 3)$, these being the values of $2k' + 1$ for $k' \in K_3^2$. Since $3, 7, 9, 13 \in B(3, 1)$, if we take the corresponding design three times, we have $3, 7, 9, 13 \in B_3(3, 3)$. The remaining values 5, 11, 17, 23, 29 must be handled separately.

Sublemma 1: ■ $5 \in B(3, 3)$. Base blocks mod 5.

$$(0, 1, 4)\,; \qquad (0, 2, 3).$$

Sublemma 2: ■ $11 \in B_3(3, 3)$. Apply Theorem 5.2.4, with $m = 3$, $K' = \{3\}$, $\lambda' = 3$, writing $11 = (3 - 1)5 + 1$, where

$$5 \in B(3, 3), \qquad K = \{3\}, \qquad \lambda'' = 1, \qquad \text{as } 7 \in B(3, 1).$$

Sublemma 3: ■ $17 \in B_3(3, 3)$. We first show that $8 \in B(4, 3)$. Here we take elements (i, j), $i \equiv 0, 2, 3 (\mathrm{mod}\ 4)$, $j \equiv 0, 1 (\mathrm{mod}\ 2)$. Blocks are

$$\{(0, b_0), (1, b_1), (2, b_2), (3, b_3)\} \sum b_i \equiv 0 (\mathrm{mod}\ 2),$$
$$\{(i, 0), (i, 1), (i', 0)(i', 1)\}, \qquad i < i'.$$

Now apply Theorem 15.2.4, with $m = 3$, $u = 8$, $K' = \{4\}$, $\lambda' = 3$, $K = \{3\}$, $\lambda'' = 1$. Here, $(m - 1)k' + 1 = 9 \in B_3(3, 1)$, and as $8 \in B(4, 3)$, we may conclude that $17 = (3 - 1)8 + 1 \in B_3(3, 3)$.

Sublemma 4: ■ $23 \in B_3(3, 3)$. Apply Theorem 15.2.4, with $m = 3$, $u = 11$, $K' = \{3\}$, $\lambda' = 3$, $K = \{3\}$, $\lambda'' = 1$. Here, $(m - 1)k' + 1 = 7 \in B_3(3, 1)$, and as $11 \in B_3(3, 3)$, by Sublemma 2 we may conclude that $23 = (3 - 1)11 + 1 \in B_3(3, 3)$.

Sublemma 5: ■ $29 \in B_3(3, 3)$. We first show $14 \in B(\{3, 4\}, 3)$; elements i, $i \equiv 0, \ldots, 12 (\mathrm{mod}\ 13)$ and ∞; base blocks modulo 13(1, 2, 6, 12); (2, 4, 12, 11):

$$(\infty, 1, 3, 9), \qquad (2, 6, 5).$$

Now apply Theorem 15.2.4, with $m = 3$, $u = 14$, $K' = \{4, 3\}$, $\lambda' = 3$, $K = \{3\}$, $\lambda'' = 1$. Here, $7 \in B_3(3, 1)$ and $9 \in B(3, 1)$, and we conclude that

$$29 = (3 - 1)14 + 1 \in B_3(3, 3).$$

These five sublemmas cover all the specific cases needed to prove our lemma.

LEMMA 15.4.6. *If $v \geq 3$, then $v \in B(3, 6)$.*

Proof: ■ This is (15.4.16d) and is the last lemma needed to prove the main Theorem 15.4.5. Here we appeal to Theorem 15.2.6, with $K' = K_3^2 = \{3, 4, 5, 6, 8, 11, 14\}$, $\lambda_1 = 1$, and $K = \{3\}$, $\lambda_2 = 6$. By Lemma 15.4.2 we know that $v \in B(K', 1)$. Thus, to prove our lemma, we must show that 3, 4, 5, 6, 8, 11, $14 \in B(3, 6)$. From the corollary to Theorem 15.2.7 and Lemmas 15.4.3, 15.4.4, and 15.4.5, it follows that 3, 4, 5, 6, $11 \in B(3, 6)$. It remains to be shown that $8 \in B(3, 6)$ and $14 \in B(3, 6)$.

Sublemma 6: ■ $8 \in B(3, 6)$. In $GF(2^3)$ let x be a primitive root satisfying $x^3 = x + 1$. We take base blocks (x^i, x^{i+1}, x^{i+2}), $i = 0, \ldots, 6$ over the additive group of $GF(2^3)$. This is a special case of (15.3.16).

Sublemma 7: ■ $14 \in B(3, 6)$. Take residues modulo 13 and ∞. Base blocks are

$$(1, 3, 9), \quad \text{taken 5 times},$$
$$(2, 6, 5), \quad \text{taken 6 times},$$
$$(\infty, 1, 12), \quad (\infty, 3, 10), \quad (\infty, 4, 9).$$

This completes the proof of our lemma, and so in turn the proof of the main theorem.

15.5 Block Designs with k Greater than 3

Hanani [1] has shown that for $k = 4$, the situation is exactly the same as that for $k = 3$, namely, that the basic necessary conditions are also sufficient. For $k = 5$, this is not true, and it may be that for every $k \geq 5$, there is at least one set of parameters satisfying the basic relations $bk = vr$, $r(k - 1) = \lambda(v - 1)$ of (10.1.1), for which no block design exists. Nevertheless, the writer is inclined to support a conjecture that this is, in some sense, a relatively unusual situation. It is not known who first advanced the following conjecture:

CONJECTURE. *Let k and λ be given. Then, with a finite number of exceptions, there is a block design for every set of parameters v, b, r, k, λ satisfying $bk = vr$, $r(k - 1) = \lambda(v - 1)$.*

We shall give an outline of Hanani's results for $k = 4$. For details the reader is referred to the original paper, Hanani [1].

THEOREM 15.5.1. *A necessary and sufficient condition for the existence of a balanced incomplete block design with $v \geq 4$ elements with $k = 4$ and any λ is that*

$$\lambda(v - 1) \equiv 0 (\text{mod } 3) \quad \text{and} \quad \lambda v(v - 1) \equiv 0 (\text{mod } 12). \tag{15.5.1}$$

Note that the conditions are trivially necessary as a consequence of (10.1.1). These conditions are equivalent to

$$\begin{array}{ll} \text{If } \lambda \equiv 1 \text{ or } 5 (\text{mod } 6), & v \equiv 1 \text{ or } 4 (\text{mod } 12); \\ \text{If } \lambda \equiv 2 \text{ or } 4 (\text{mod } 6), & v \equiv 1 (\text{mod } 3); \\ \text{If } \lambda \equiv 3 (\text{mod } 6), & v \equiv 0 \text{ or } 1 (\text{mod } 4); \\ \text{If } \lambda \equiv 0 (\text{mod } 6), & \text{there are no restrictions on } v. \end{array} \tag{15.5.2}$$

These in turn will be satisfied if it can be shown that, with $v \geq 4$,

$$v \equiv 1 \text{ or } 4 (\text{mod } 12) \quad \text{implies } v \in B(4, 1); \tag{15.5.3a}$$

$$v \equiv 1 (\text{mod } 3) \quad \text{implies } v \in B(4, 2); \tag{15.5.3b}$$

$$v \equiv 0 \text{ or } 1 (\text{mod } 4) \quad \text{implies } v \in B(4, 3); \tag{15.5.3c}$$

$$v \geq 4 \quad \text{implies } v \in B(4, 6). \tag{15.5.3d}$$

The lemmas required are

LEMMA 15.5.1. *If $u \equiv 0$ or $1 (\text{mod } 4)$ and $u \geq 4$, then $u \in B(K_4^1, 1)$, where* $K_4^1 = \{4, 5, 8, 9, 12\}$.

LEMMA 15.5.2. *If $u \geq 4$, then $u \in B(K_4^2, 1)$, where*

$$K_4^2 = \{4, 5, 6, 7, 8, 9, 10, 11, 12, 14, 15, 18, 19, 22, 23\}.$$

LEMMA 15.5.3. *If* $v \equiv 1$ *or* $4 \pmod{12}$, $v \geq 4$, *then* $v \in B(3, 1)$.

This is proved by putting $v = 3u + 1$, $u \equiv 0$ or $1 \pmod 4$ and applying Theorem 15.2.4, with $m = 4$, since by Lemma 15.5.1, $u \in B(K_4^1, 1)$. This requires proving $v \in B(4, 1)$ for the specific values $v = 4, 13, 16, 25, 28, 37$. This is (15.5.3a).

LEMMA 15.5.4. *If* $v \equiv 1 \pmod 3$, *then* $v \in B(4, 2)$.

Here, Theorem 15.2.4 is applied with $m = 4$ and $v = 3u + 1$, where for $u \geq 4$, by Lemma 15.5.2, $u \in B(K_4^2, 1)$. This requires showing directly that $v \in B_2(4, 2)$ for the set $v = 3u + 1$, $u \in K_4^2$. This is (15.5.3b).

LEMMA 15.5.5. *If* $v \equiv 0$ *or* $1 \pmod 4$, *then* $v \in B(4, 3)$.

By Lemma 15.5.1, $v \in B(K_4^1, 1)$. We apply Theorem 15.2.6 and need to show only $4, 5, 8, 9, 12 \in B(4, 3)$. This is (15.5.3c).

LEMMA 15.5.6. *If* $v \geq 4$, *then* $v \in B(4, 6)$.

By Lemma 15.5.2, $v \in B(K_4^2, 1)$. Application of Theorem 15.2.6 proves the lemma by showing that for the 15 numbers $v \in K_4^2$, we have $v \in B(4, 6)$. This is (15.5.3d). These lemmas together establish the theorem.

For $k \geq 5$, the situation is different. The following theorem will be proved in the next chapter.

THEOREM 15.5.2. *Suppose we have a design D with parameters*

$$v = v_1 - k_1, \quad b = v_1 - 1, \quad r = k_1, \quad k = k_1 - \lambda, \quad \lambda = \lambda,$$

where $k_1(k_1 - 1) = \lambda(v_1 - 1)$ *and* $\lambda = 1$ *or* 2. *Then D can be embedded in a symmetric block design* D^* *with parameters* v_1, k_1, λ. *D is a residual design of* D^*.

A consequence of this theorem is that the nonsymmetric design D^* exists. In particular, if there were a design D, $v = 15, b = 21, r = 7, k = 5, \lambda = 2$, by Theorem 15.5.2, it could be embedded in a symmetric design D^* with $v_1 = 22$, $k_1 = 7$, $\lambda = 2$. By Theorem 10.3.1, the design D^* does not exist because v_1 is even, but $n = k_1 - \lambda = 5$ is not a square. Hence, we conclude that D does not exist. Similarly, since by Theorem 10.3.1 there is no design with $v_1 = 43$, $k_1 = 7$, $\lambda = 1$, we conclude from Theorem 15.5.2 that there is node sign with $v = 36$, $b = 32$, $r = 7$, $k = 6$, $\lambda = 1$. Indeed, for most values of $k \geq 5$, application of Theorems 10.3.1 and 15.5.2 shows that there are some parameters satisfying the basic relations (10.1.1) for which no design exists.

For $k = 5$, Hanani gives partial results. He needs to prove

$$\text{If } v \equiv 1 \text{ or } 5 (\text{mod } 20), \quad v \in B(5, 1); \tag{15.5.4a}$$

$$\text{If } v \equiv 1 \text{ or } 5 (\text{mod } 20), \quad v \in B(5, 2); \tag{15.5.4b}$$

$$\text{If } v \equiv 0 \text{ or } 1 (\text{mod } 5), \quad v \in B(5, 4); \tag{15.5.4c}$$

$$\text{If } v \equiv 1 (\text{mod } 4), \quad v \in B(5, 5); \tag{15.5.4d}$$

$$\text{If } v \equiv 1 (\text{mod } 2), \quad v \in B(5, 10); \tag{15.5.4e}$$

$$\text{If } v \geq 5, \quad v \in B(5, 20). \tag{15.5.4f}$$

He uses two basic sets:

$$K_5^1 = \{5, 6, 10, 11, 15, 16, 20, 35, 36, 40, 70, 71, 75, 76\}, \tag{15.5.5}$$

$$K_5^2 = \{5, 6, \ldots, 20, 22, 23, 24, 27, 28, 29, 32, 33, 34, 38, 39\}. \tag{15.5.6}$$

He proves that for $u \geq 5$ satisfying $u \equiv 0$ or $1 (\text{mod } 5)$, $u \in B(K_5^1, 1)$ and, even omitting 35 from K_5^1, the same result for all other $u \equiv 0$ or $1 (\text{mod } 5)$. From this he can prove (15.5.4a) with the possible exception of $141 = 4 \cdot 35 + 1$. He also proves that for $u \geq 5$, $u \in B\{K_5^2, 1\}$. From this he is able to prove (15.5.4c) and (15.5.4f). He remarks that (15.5.4d) is true if $4u + 1 \in B_5(5, 5)$ for every $u \in K_5^2$.

Although we have observed that there is no design with $v = 15$, $k = 5$, $\lambda = 2$, nevertheless there is a design with $v = 15$, $k = 5$, $\lambda = 4$ given here:

$$v = 15, \quad b = 42, \quad r = 14, \quad k = 5, \quad \lambda = 4.$$

Elements: (i, j), $i \equiv 0, 2, 3, 4 (\text{mod } 5)$, $j \equiv 0, 1, 2 (\text{mod } 3)$.

Blocks:
$\{(i, j), (i + 2, j)(i + 3, j)(i, j + 1)(i + 4, j + 2)\}$
$\{(i, j), (i + 1, j)(i, j + 1), (i + 2, j + 1), (i, j + 2)\}$
$\{(i, 0), (i + 2, 1)(i + 3, 1)(i + 4, 1), (i + 1, 2)\}$
$\{(i, 0), (i + 1, 0)(i + 2, 1)(i + 3, 2), (i + 4, 2)\}$
$\{(0, 0), (1, 0), (2, 0), (3, 0), (4, 0)\}$
$\{(0, 2)(1, 2), (2, 2), (3, 2)(4, 2)\}$

Construction of the design $D(15, 63, 21, 5, 6)$ would show that for $v = 15$, $k = 5$, and $\lambda = 0 (\text{mod } 2)$, all λ's except $\lambda = 2$ are possible, since any even number ≥ 4 can be written as a sum of 4's and 6's. But it may be remarked that the set of all $\binom{15}{5} = 3003$ combinations of 15 elements, 5 at a time, is a design $D(15, 3003, 1001, 5, 286)$, and so in any event, with $v = 15$, $k = 5$, and $\lambda \equiv 0 (\text{mod } 2)$, $\lambda \geq 284$, there is a design.

For any design $D(v, b, r, k, \lambda)$ to each block of k elements, there is a complementary set of the remaining $(v - k)$ elements, and all these complementary sets are the blocks of a design $D'(v, b, b - r, v - k, b - 2r + \lambda)$. Each of the designs is the complement of the other, and in listing designs it is simpler to list only the one with the smaller block size; so without loss of generality, we may take $k \leq v/2$. For a given value of r, since $r \geq k$, there are only a finite number of possibilities for k, and from $r(k - 1) = \lambda(v - 1)$, in turn, only a finite number of possibilities for λ and v. Given these values, b is determined by $bk = vr$. In the Appendix, designs are listed for $r = 3, \ldots, 15$ with $k \leq v/2$ and ordering on r, v, k, respectively.

16

Theorems on Completion and Embedding

16.1 Connor's Method

Let $D(v, b, r, k, \lambda)$ be a block design. It has been observed in Section 10.1 that D is determined by its incidence matrix

$$A = (a_{ij}), \qquad i = 1, \ldots, v, \qquad j = 1, \ldots, b,$$

corresponding to objects $a_1, \ldots, a_v$ and blocks $B_1, \ldots, B_b$, where

$$\begin{aligned} a_{ij} &= 1, &\quad &\text{if } a_i \in B_j, \\ a_{ij} &= 0, &\quad &\text{if } a_i \notin B_j. \end{aligned} \tag{16.1.1}$$

Then the matrix A satisfies

$$\begin{aligned} AA^T &= B = (r - \lambda)I_v + \lambda J_v, \\ w_v A &= k w_b, \qquad A w_b^T = r w_v^T. \end{aligned} \tag{16.1.2}$$

Here, J_v is the $v \times v$ matrix of all 1's and w_v and w_b are the row vectors of v and b 1's respectively. We associate indeterminates $x_1, \ldots, x_v$ with the v objects and define linear forms L_j by

$$L_j = \sum_{i=1}^{v} a_{ij} x_i, \qquad j = 1, \ldots, b. \tag{16.1.3}$$

Then (16.1.2) is equivalent to

$$\begin{aligned} L_1^2 + \cdots + L_b^2 &= (r - \lambda)(x_1^2 + \cdots + x_v^2) + \lambda(x_1 + \cdots + x_v)^2 \\ &= Q(x_1, \ldots, x_v). \end{aligned} \tag{16.1.4}$$

The underlying idea of Connor's method [1] is the observation that for given parameters v, b, r, k, λ, the form $Q = Q(x_1, \ldots, x_v)$ of (16.1.4) is known, and that if we choose the first t blocks of our design $B_1, \ldots, B_t$, the forms $L_1, \ldots, L_t$ are known. Hence, if we write $Q^* = Q^*(x_1, \ldots, x_v)$, then

$$Q^* = Q - L_1^2 - \cdots - L_t^2 = L_{t+1}^2 + \cdots + L_b^2. \tag{16.1.5}$$

Here, if v, b, r, k, λ are given, and $B_1, \ldots, B_t$ are chosen, then Q and $L_1, \ldots, L_t$ are known and so Q^* is known. Hence, from (16.1.5), if $L_{t+1}, \ldots, L_b$ exist, then necessarily $Q^* = L_{t+1}^2 + \cdots + L_b^2$ must be a positive semidefinite form. Thus, a necessary condition for the existence of a block design $D(v, b, r, k, \lambda)$ with initial blocks $B_1, \ldots, B_t$ is that the form Q^* of (16.1.5) be positive semidefinite. A virtue of Connor's method is that he can test Q^* for being positive semidefinite by evaluating a $t \times t$ determinant.

We begin with

LEMMA 16.1.1. *If $|K|$ is the determinant defined by*

$$|K| = \begin{vmatrix} a & b & \cdots & b & e_{1,v+1} & \cdots & e_{1,v+t} \\ b & a & \cdots & b & e_{2,v+1} & \cdots & e_{2,v+t} \\ \cdot & \cdot & & \cdot & \cdot & & \\ \cdot & \cdot & & \cdot & \cdot & & \\ \cdot & \cdot & & \cdot & \cdot & & \\ b & b & \cdots & a & e_{v,v+1} & \cdots & e_{v,v+t} \\ e_{v+1,1} & e_{v+1,2} & \cdots & e_{v+1,v} & e_{v+1,v+1} & \cdots & e_{v+1,v+t} \\ \cdot & \cdot & & \cdot & \cdot & & \cdot \\ \cdot & \cdot & & \cdot & \cdot & & \cdot \\ \cdot & \cdot & & \cdot & \cdot & & \cdot \\ e_{v+t,1} & e_{v+t,2} & \cdots & e_{v+t,v} & e_{v+t,v+1} & \cdots & e_{v+t,v+t} \end{vmatrix}, \tag{16.1.6}$$

then

$$|K| = (a + (v-1)b)^{-t+1}(a - b)^{v-t-1}|B_t|, \tag{16.1.7}$$

where B_t is of order $t \times t$ and the elements of B_t are

$$b_{ju} = (a + (v-1)b)(a - b)e_{v+j,v+u} - (a + (v-1)b)\sum_{i=1}^{v} e_{i,v+u}e_{v+j,i} + b\sum_{i=1}^{v} e_{i,v+u}\sum_{i=1}^{v} e_{v+j,i}. \tag{16.1.8}$$

Proof: ■ We carry out the following operations on $|K|$:

1. Multiply the last t columns by $[a + (v - 1)b][a - b]$ and write an offsetting factor outside.
2. Add rows $1, 2, \ldots, v - 1$ to row v.
3. Take the factor $[a + (v - 1)b]$ out of row v.
4. Multiply row v by b and subtract this product from rows $1, 2, \ldots, v - 1$.
5. Take the factor $(a - b)$ out of rows $1, 2, \ldots, v - 1$.
6. Subtract rows $1, 2, \ldots, v - 1$ from row v.
7. Subtract suitable multiples of columns $1, 2, \ldots, v$ from column $v + 1$, $v + 2, \ldots, v + t$ so as to make the elements that are both in the first v rows and in the last t columns equal to zero, and the lemma follows.

If $A = (a_{ij})$, $i = 1, \ldots, v$, $j = 1, \ldots, b$, is the incidence matrix for a design $D(v, b, r, k, \lambda)$, let $A_{11} = (a_{ij})$, $i = 1, \ldots, v$, $j = 1, \ldots, t$, be the incidence matrix for the first t blocks, $B_1, \ldots, B_t$, which we assume chosen. Now define C_t by

$$C_t = r(r - \lambda)I_t + \lambda k J_t - r A_{11}^T A_{11}. \tag{16.1.9}$$

We call the matrix C_t the *characteristic matrix* of the t chosen blocks $B_1, \ldots, B_t$.

THEOREM 16.1.1. *If C_t is the characteristic matrix of any set of t blocks chosen from a balanced, incomplete block design with parameters v, b, r, k, λ, then*

1. $|C_t| \geq 0$, *if* $t < b - v$.
2. $|C_t| = 0$, *if* $t > b - v$.
3. $k(r)^{-b+v+1}(r - \lambda)^{v-b-1}\,|C_{b-v}|$ *is a perfect integral square.*

Proof: ■ We define the $v + t \times b$ matrix A_1 by

$$A_1 = \begin{bmatrix} A_{11} & A_{12} \\ I_t & 0 \end{bmatrix}, \tag{16.1.10}$$

where

$$\begin{aligned} A &= (A_{11}, A_{12}), \\ A_{11} &= (a_{ij}), \qquad i = 1, \ldots, v, \qquad j = 1, \ldots, t, \\ A_{12} &= (a_{ij}), \qquad i = 1, \ldots, v, \qquad j = t + 1, \ldots, b. \end{aligned} \tag{16.1.11}$$

Then, by block multiplication,

$$B_1 = A_1 A_1^T = \begin{bmatrix} A_{11}A_{11}^T + A_{12}A_{12}^T & A_{11} \\ A_{11}^T & I_t \end{bmatrix}$$

$$= \begin{bmatrix} B & A_{11} \\ A_{11}^T & I_t \end{bmatrix} \tag{16.1.12}$$

where $B = (r - \lambda)I_v + \lambda J_v$, as given in (16.1.2).

Applying Lemma 16.1.1, we find

$$|B_1| = |A_1A_1^T| = kr^{-t+1}(r-\lambda)^{v-t-1}|C_t|. \qquad (16.1.13)$$

From the theory of quadratic forms it follows that for a real matrix A_1, the determinant $|A_1A_1^T| \geq 0$, and as $k, r, r-\lambda$ are positive, this proves condition 1 of the theorem, $|C_t| \geq 0$. Also rank $A_1A_1^T \leq$ rank A_1, whence if $t > b - v$, as $A_1A_1^T$ is singular, then $|A_1A_1^T| = 0$, and so $|C_t| = 0$. If $t = b - v$, then A_1 is a square matrix and $|A_1A_1^T| = |A_1|^2$. As the entries of A_1 are integers, then $|A_1|$ is an integer, whence part (3) of our theorem follows.

The formulation of Theorem 16.1.1 is essentially the same as the requirement that $Q^* = Q - L_1^2 - \cdots - L_t^2$ be semidefinite. The matrix B_1 of (16.1.12) is associated with the quadratic form

$$\begin{aligned} Q_1 &= Q_1(x_1, \ldots, x_v, y_1, \ldots, y_t) \\ &= (L_1 + y_1)^2 + \cdots + (L_t + y_t)^2 + L_{t+1}^2 + \cdots + L_b^2, \end{aligned} \qquad (16.1.14)$$

where $y_1, \ldots, y_t$ are further indeterminates. For any real values of the x's we may choose $y_1, \ldots, y_t$ so that $L_1 + y_1 = \cdots = L_t + y_t = 0$, and so Q_1 is positive semidefinite if and only if $L_{t+1}^2 + \cdots + L_b^2 = Q - L_1^2 - \cdots - L_t^2 = Q^*$ is.

Let us define the structure matrix S_t by

$$S_t = A_{11}^TA_{11} = (s_{ij}), \qquad i, j = 1, \ldots, t. \qquad (16.1.15)$$

Here, $s_{ii} = k$ and s_{ij}, $i \neq j$, is the number of elements common to blocks B_i and B_j. By Theorem 10.2.2, if D is a symmetric design, we must have $s_{ij} = \lambda$ if $i \neq j$. In case D is symmetric and we have $s_{ij} = \lambda$ for $i \neq j$, the matrix C_t is identically zero. There appears to be no further information obtainable from a closer study of the zero matrix.

In Section 10.1 the derived and residual designs of a symmetric design were defined. Here, if v_1, k_1, λ_1 are the parameters of a symmetric design D satisfying

$$k_1(k_1 - 1) = \lambda_1(v_1 - 1), \qquad (16.1.16)$$

the residual design $\bar{D}$, obtained by deleting a block B_0 of D and all its elements, has parameters

$$v = v_1 - k_1, \quad b = v_1 - 1, \quad r = k_1, \quad k = k_1 - \lambda_1, \quad \lambda = \lambda_1 \qquad (16.1.17)$$

and the derived design D' has parameters

$$v = k_1, \quad b = v_1 - 1, \quad r = k_1 - 1, \quad k = \lambda_1, \quad \lambda = \lambda_1 - 1. \qquad (16.1.18)$$

A question that arises naturally is whether a design with parameters of a residual design as given by (16.1.17) subject to (16.1.16) can be embedded into a symmetric design by adjoining k_1 new elements as a new block and adjoining λ_1 of these appropriately to each of the $v_1 - 1$ blocks. This is not always possible, as the following example, due to Bhattacharya [2], shows:

$$v = 16, \qquad b = 24, \qquad r = 9, \qquad k = 6, \qquad \lambda = 3,$$

$$\begin{array}{lll}
(1, 2, 7, \ 8, 14, 15) & (3, 5, 7, \ 8, 11, 13) & (2, 3, 8, \ 9, 13, 16) \\
(3, 5, 8, \ 9, 12, 14) & (1, 6, 7, \ 9, 12, 13)^* & (2, 5, 7, 10, 13, 15) \\
(3, 4, 7, 10, 12, 16) & (3, 4, 6, 13, 14, 15) & (4, 5, 7, \ 9, 12, 15) \\
(2, 4, 9, 10, 11, 13) & (3, 6, 7, 10, 11, 14) & \underline{(1, 2, 3, \ 4, \ 5, \ 6)} \\
(1, 4, 7, \ 8, 11, 16) & (2, 4, 8, 10, 12, 14) & (5, 6, 8, 10, 15, 16) \\
(1, 6, 8, 10, 12, 13)^* & (1, 2, 3, 11, 12, 15) & (2, 0, 7, \ 9, 14, 16) \\
(1, 4, 5, 13, 14, 15) & (2, 5, 6, 11, 12, 16) & (1, 3, 9, 10, 15, 16) \\
(4, 6, 8, \ 9, 11, 15) & (1, 5, 9, 10, 11, 14) & \underline{(11, 12, 13, 14, 15, 16)}
\end{array} \qquad (16.1.19)$$

Here, the two starred blocks have four elements in common, and the two underlined blocks have no elements in common. If it were possible to adjoin nine more elements as a new block and three of these to each of the above 24 blocks to form a symmetric design D with $v_1 = 25, k_1 = 9, \lambda_1 = 3$, then any two blocks of D would have exactly three elements in common. But the starred blocks already have four elements in common, and so this is impossible.

We can formulate conditions under which the embedding is possible.

THEOREM 16.1.2. *A design D with parameters satisfying*

$$r = k + \lambda, \qquad v\lambda = k(k + \lambda - 1), \qquad b\lambda = (k + \lambda)(k + \lambda - 1) \quad (16.1.20)$$

can be embedded as a residual design in a symmetric design if and only if we can find in D sets of blocks $S_j, j = 1, \ldots, k + \lambda$, *such that*

1. *Each* S_j *consists of* $(k + \lambda - 1)$ *blocks of D.*
2. *The blocks of an* S_j *together contain each element* λ *times.*
3. *Any two distinct* S_i, S_j *have exactly* $(\lambda - 1)$ *blocks in common.*
4. *Any block of D is in exactly* λ *sets* S_j.

Proof: ■ Let us adjoin to D new elements $x_1, \ldots, x_{k+\lambda}$ and a new block B_0 containing these elements. Also adjoin the new element x_j to every block of S_j. Properties 1 to 4 of the theorem show immediately that block B_0 and the blocks B_i^*, $i = 1, \ldots, b$, where B_i^* is the block B_i of D together with the new elements x_j adjoined to B_i, form a symmetric design with $v_1 = v + \lambda$, $k_1 = r$, $\lambda_1 = \lambda$. Conversely, if from a symmetric design with parameters v_1, k_1, λ_1 we remove a block B_0 and its elements, we are left with a design D whose parameters satisfy (16.1.20), and if $x_1, \ldots, x_{k_1}$ are the suppressed elements of B_0, those blocks from which the element x_j was removed form a set S_j and the sets $S_1, \ldots, S_{k_1}$ have the properties of the theorem.

Although the Bhattacharya example shows that the embedding is not possible in every case, we shall prove that the embedding is possible when $\lambda = 1$ or $\lambda = 2$.

THEOREM 16.1.3. *Let v_1, k_1, λ_1 satisfy $k_1(k_1 - 1) = \lambda_1(v_1 - 1)$ and suppose we are given a design D with parameters $v = v_1 - k_1$, $b = v_1 - 1$, $r = k_1$, $k = k_1 - \lambda_1$, $\lambda = \lambda_1$, and that $\lambda = 1$ or 2. Then D can be embedded as a residual design in a symmetric design with parameters v_1, k_1, λ_1.*

Proof: ∎ For $\lambda = 1$ this is the embedding of a finite affine plane in a finite projective plane, and this result has already been proved in Theorem 12.3.3. For $\lambda = 2$, the proof is considerably more difficult and leans heavily on applications of Theorem 16.1.1. Our proof will depend on establishing

LEMMA 16.1.2. *Every block design D with parameters satisfying*

$$v = \frac{k(k+1)}{2}, \qquad b = \frac{(k+1)(k+2)}{2}, \qquad r = k + 2, \qquad \lambda = 2, \quad (16.1.21)$$

satisfies the conditions of Theorem 16.1.2 and can be embedded in a symmetric design. This embedding is unique up to isomorphism.

Proof: ∎ If $k = 2$, our parameters are $v = 3$, $b = 6$, $r = 4$, $k = 2$, $\lambda = 2$, and the design is the trivial one:

$$\begin{matrix} (1,2) & (1,3) & (2,3) \\ (1,2) & (1,3) & (2,3) \end{matrix} \qquad (16.1.22)$$

Our new elements will be x_1, x_2, x_3, x_4, and our essentially unique embedding is

$$v = b = 7, \qquad r = k = 4, \qquad \lambda = 2,$$

$$\begin{matrix} (x_1, x_2, x_3, x_4) & \\ (x_1, x_2, 1, 2) & (x_2, x_4, 1, 3) \\ (x_3, x_4, 1, 2) & (x_1, x_4, 2, 3) \\ (x_1, x_3, 1, 3) & (x_2, x_3, 2, 3) \end{matrix} \qquad (16.1.23)$$

Here, the given design is trivial and the symmetric design is the complement of PG(2, 2) with $v = 7$, $k = 3$, $\lambda = 1$. Henceforth, in (16.1.21) we assume $k \geq 3$.

Our proof begins by examining the intersection pattern of a given block B_0 with the remaining $(b - 1)$ blocks of D. Let there be n_i blocks of $B_1, \ldots, B_{b-1}$, which have exactly i elements in common with B_0. Then the following relations hold:

$$\sum_{i=0}^{k} n_i = b - 1 = \frac{(k^2 + 3k)}{2}. \tag{16.1.24}$$

$$\sum_{i=0}^{k} in_i = (r - 1)k = k^2 + k. \tag{16.1.25}$$

$$\sum_{i=0}^{k} i(i - 1)n_i = (\lambda - 1)(k^2 - k) = k^2 - k. \tag{16.1.26}$$

The first of these merely counts the remaining $(b - 1)$ blocks. If

$$B_0 = \{a_1, \ldots, a_k\},$$

the second of these says that each of $a_1, \ldots, a_k$ occurs $(r - 1)$ times in the remaining blocks, and a block B_j that has i elements in common with B_0 accounts for i of these occurrences. Furthermore, if B_j has i elements in common with B_0, then it accounts for $i(i - 1)/2$ of the unordered pairs (a_m, a_n), $m, n = 1, \ldots, k$, $m \neq n$, which must occur together $(\lambda - 1)$ times in the remaining blocks. We may combine these relations to obtain

$$\sum_{i=0}^{k} [i(i - 1) - 2i + 2]n_i = k^2 - k - 2(k^2 + k) + k^2 + 3k = 0, \tag{16.1.27}$$

or

$$\sum_{i=0}^{k} (i - 1)(i - 2)n_i = 0. \tag{16.1.28}$$

For $i = 0$ or $i > 2$, the coefficient $(i - 1)(i - 2) > 0$, and so (16.1.28) shows that $n_i = 0$ for $i \neq 1, 2$. The relations (16.1.24) show that $n_1 = 2k$, $n_2 = (k^2 - k)/2$. Let us call two blocks "first associates" if they have one element in common and "second associates" if they have two elements in common. We may state our findings as a lemma in these terms.

LEMMA 16.1.3. *In our design D, given a block B_0, the $b - 1 = (k^2 + 3k)/2$ remaining blocks consist of $2k$ first associates and $(k^2 - k)/2$ second associates.*

We now consider the intersection pattern of two blocks with the remaining $(b - 2)$ blocks. Let the blocks be B_1, B_2. We write

$$\begin{aligned} B_1 &= \{z_1, z_t, a_1, \ldots, a_{k-t}\}, \qquad t = 1 \text{ or } 2, \\ B_2 &= \{z_1, z_t, b_1, \ldots, b_{k-t}\}. \end{aligned} \tag{16.1.29}$$

Here, z_1 or z_1, z_2 are the elements common to B_1 and B_2, according as they are first or second associates. The remaining blocks intersect each of B_1 and B_2 in one or two elements, but we note that since $\lambda = 2$ if $t = 2$, there cannot be another block containing both z_1 and z_2. Thus, the patterns of intersection are given by

$$\begin{array}{llll} x_1: & \{z_h, a_i, b_s \cdots\} & y_1: & \{a_i, a_j, b_s, b_u \cdots\} \\ x_2: & \{z_h, a_i \cdots\} & y_2: & \{a_i, a_j, b_s \cdots\} \\ x_3: & \{z_h, b_s \cdots\} & y_3: & \{a_i, b_s, b_u \cdots\} \\ x_4: & \{z_h, \cdots\} & y_4: & \{a_i, b_s \cdots\} \end{array} \qquad (16.1.30)$$

Here, $x_1, \ldots, x_4, y_1, \ldots, y_4$ denote the number of blocks having precisely the indicated intersection pattern with B_1 and B_2. These numbers satisfy the following relations:

$$x_1 + x_2 + x_3 + x_4 = tk, \qquad y_1 + y_2 + y_3 + y_4 = \frac{(k^2 + 3k - 2)}{2} - tk,$$

$$x_1 + x_2 = t(k - t), \qquad y_1 + y_2 = \frac{(k - t)(k - t - 1)}{2},$$

$$x_1 + x_3 = t(k - t), \qquad y_1 + y_3 = \frac{(k - t)(k - t - 1)}{2},$$

$$x_1 + 4y_1 + 2y_2 + 2y_3 + y_4 = 2(k - t)^2. \qquad (16.1.31)$$

Of these, $x_1 + x_2 + x_3 + x_4$ counts the remaining occurrences of z_1 and z_t; $y_1 + y_2 + y_3 + y_4$ counts the remaining blocks. Also, $x_1 + x_2$ counts the remaining pairs $z_h a_i$, $h = 1, t$; $x_1 + x_3$ counts the pairs $z_h b_s$, $h = 1, t$; $y_1 + y_2$, pairs a_i, a_j; $y_1 + y_3$, pairs $b_2 b_u$; and the last counts the pairs a_i, b_s, i, $s = 1, \ldots, k - t$, each pair occurring twice in the remaining blocks.

Solutions are

$$t = 1$$

$$\begin{array}{ll} x_1 = k - 2 + x_4, & y_1 = \dfrac{(k \;\; - 5k + 6)}{2} - x_4, \\ x_2 = 1 - x_4, & y_2 = k - 2 + x_4, \\ x_3 = 1 - x_4, & y_3 = k - 2 + x_4, \\ x_4 = x_4, & y_4 = k - x_4. \end{array} \qquad (16.1.32)$$

$$t = 2$$

$$\begin{array}{ll} x_1 = 2k - 8 + x_4, & y_1 = \dfrac{(k \;\; - 9k + 2)}{2} - x_4, \\ x_2 = 4 - x_4, & y_2 = 2k - 8 + x_4, \\ x_3 = 4 - x_4, & y_3 = 2k - 8 + x_4, \\ x_4 = x_4, & y_4 = 4 - x_4. \end{array}$$

Note that although the value of x_4 is undetermined, we do know the values of $x_4 + y_4$, the number of remaining blocks that are first associates of both B_1 and B_2. Similarly, $x_3 + y_3$, $x_2 + y_2$, and $x_1 + y_1$ are known, and give exactly the number of remaining blocks whose type of association with B_1 and B_2 is prescribed. We state these conclusions as

LEMMA 16.1.4. *If B_1 and B_2 are first associates, there are k blocks that are first associates of both, $(k - 1)$ first associates of B_1 and second of B_2, $(k - 1)$ second of B_1 and first of B_2, and $(k^2 - 3k + 2)/2$ blocks that are second associates of both. If B_1 and B_2 are second associates, these numbers are respectively* 4, $2k - 4$, $2k - 4$, *and* $(k^2 - 5k + 6)/2$.

In the matrix C_t of (16.1.9), the entries have the values

$$\begin{aligned} c_{jj} &= r(r - \lambda) + \lambda k - rk = (r - k)(r - \lambda), \qquad j = 1, \ldots, t, \\ c_{ju} &= \lambda k - rs_{ju}, \qquad j \neq u, \qquad j, u = 1, \ldots, t, \end{aligned} \tag{16.1.33}$$

where s_{ju} is the corresponding entry in S_t. In our case, $r = k + 2$, $\lambda = 2$, and by Lemma 16.1.3, $s_{ju} = 1$ or 2. Hence, $c_{jj} = 2k$ and $c_{ju} = k - 2$ if $s_{ju} = 1$, $c_{ju} = -4$ if $s_{ju} = 2$.

We apply Theorem 16.1.1 in three special cases

$$S_5^{(1)} = \begin{bmatrix} k & 1 & 1 & 2 & 2 \\ 1 & k & 1 & 1 & 1 \\ 1 & 1 & k & 1 & 1 \\ 2 & 1 & 1 & k & s_{45} \\ 2 & 1 & 1 & s_{45} & k \end{bmatrix}, \qquad S_5^{(2)} = \begin{bmatrix} k & 1 & 1 & 1 & 2 \\ 1 & k & 1 & 2 & 1 \\ 1 & 1 & k & 2 & 1 \\ 1 & 2 & 2 & k & s_{45} \\ 2 & 1 & 1 & s_{45} & k \end{bmatrix},$$

$$S_6 = \begin{bmatrix} k & 1 & 1 & 1 & 1 & 2 \\ 1 & k & 1 & 2 & 2 & 1 \\ 1 & 1 & k & 2 & 2 & 1 \\ 1 & 2 & 2 & k & s_{45} & 2 \\ 1 & 2 & 2 & s_{45} & k & 2 \\ 2 & 1 & 1 & 2 & 2 & k \end{bmatrix}. \tag{16.1.34}$$

LEMMA 16.1.5. *In $S_5^{(1)}$ we must have $S_{45} = 1$; in $S_5^{(2)}$ we must have $s_{45} = 2$; and in S_6 we must have $s_{45} = 1$.*

Proof: ■ We evaluate the determinants $|C_t|$ corresponding to these structure matrices, using (16.1.33), and find

$$\begin{aligned} |C_5^{(1)}| &= 4(k + 2)^2(2k - c_{45})[(k - 1)c_{45} + 2(k - 4)], \\ |C_5^{(2)}| &= 4(k + 2)^2[-(k - 1)c_{45}^2 - (k - 2)(k + 8)c_{45} \\ &\qquad + (k - 2)(k^2 - k - 18)], \\ |C_6| &= 4(k + 2)^3(2k - c_{45})[(k - 2)c_{45} + 2(k - 6)]. \end{aligned} \tag{16.1.35}$$

We note that if $c_{45} = -4$ in $|C_5^{(1)}|$ or $|C_6|$, or $c_{45} = k - 2$ in $|C_5^{(2)}|$, the determinant is negative. From Theorem 16.1.1 the conclusion follows.

Let any block be B_1 and let B_2 be a first associate of B_1. Then, by Lemma 16.1.4, the first two rows of the structure matrix of the whole design may be put in the form

$$\begin{array}{cc|cccc|ccc|ccc|ccc} k & 1 & 1 & 1 & \cdots & 1 & 1 & \cdots & 1 & 2 & \cdots & 2 & 2 & \cdots & 2 \\ 1 & k & 1 & 1 & \cdots & 1 & 2 & \cdots & 2 & 1 & \cdots & 1 & 2 & \cdots & 2 \end{array}, \tag{16.1.36}$$

the columns corresponding left to right to $B_1, B_2, \ldots, B_b$; we have ordered and divided these into 2, k, $k - 1$, $k - 1$, and $(k^2 - 3k + 2)/2$ columns according to the association of the blocks with B_1 and B_2, as indicated. With B_3 as any block that is a first associate of both B_1 and B_2, the pattern of the first three rows is given by

$$\left|\begin{array}{ccc|ccc|ccc|ccc|ccc|ccc|ccc|ccc|ccc} k & 1 & 1 & 1 & \cdots & 1 & 1 & \cdots & 1 & 1 & \cdots & 1 & 1 & \cdots & 1 & 2 & \cdots & 2 & 2 & \cdots & 2 & 2 & \cdots & 2 & 2 & \cdots & 2 \\ 1 & k & 1 & 1 & \cdots & 1 & 1 & \cdots & 1 & 2 & \cdots & 2 & 2 & \cdots & 2 & 1 & \cdots & 1 & 1 & \cdots & 1 & 2 & \cdots & 2 & 2 & \cdots & 2 \\ 1 & 1 & k & 1 & \cdots & 1 & 2 & \cdots & 2 & 1 & \cdots & 1 & 2 & \cdots & 2 & 1 & \cdots & 1 & 2 & \cdots & 2 & 1 & \cdots & 1 & 2 & \cdots & 2 \end{array}\right|. \tag{16.1.37}$$

In this pattern, since B_3 is a first associate of both B_1 and B_2, Lemma 16.1.4 applies. Let the number of columns $\begin{bmatrix} 1 \\ 1 \\ 1 \end{bmatrix}$ in (16.1.37) be j. Then the partition (16.1.37) is into 3, j, $k - 1 - j$, $k - 1 - j$, j, $k - 1 - j$, j, j, and $(k^2 - 3k + 2)/2 - j$ columns. We consider the following blocks: B_1, B_2, B_3, the j blocks headed $\begin{bmatrix} 1 \\ 2 \\ 2 \end{bmatrix}$ and the $(k - 1 - j)$ blocks headed $\begin{bmatrix} 2 \\ 1 \\ 1 \end{bmatrix}$. We form their structural matrix:

$$S_{k+2} = \left[\begin{array}{ccc|ccc|ccc} k & 1 & 1 & 1 & \cdots & 1 & 2 & \cdots & 2 \\ 1 & k & 1 & 2 & \cdots & 2 & 1 & \cdots & 1 \\ 1 & 1 & k & 2 & \cdots & 2 & 1 & \cdots & 1 \\ \hline 1 & 2 & 2 & & & & & & \\ & \cdots & & & F & & & G & \\ 1 & 2 & 2 & & & & & & \\ \hline 2 & 1 & 1 & & & & & & \\ & \cdots & & & G^T & & & H & \\ 2 & 1 & 1 & & & & & & \end{array}\right] \tag{16.1.38}$$

Considering B_1, B_2, B_3 and any two blocks headed $\begin{bmatrix} 2 \\ 1 \\ 1 \end{bmatrix}$, we may conclude from $S_5^{(1)}$ in Lemma 16.1.5 that every off-diagonal element of H is a 1. Similarly, $S_5^{(2)}$ shows that every element in G and G^T is a 2. Similarly, from S_6, if $k - 1 - j > 0$, we can conclude that every nondiagonal element in F is 1. As

$0 \leq j \leq k - 1$, this excludes only the case $j = k - 1$. If we evaluate the determinant $|C_{k+2}|$, we find

$$|C_{k+2}| = (k - 6)j(j - k + 2)(k + 2)^{k+1}. \tag{16.1.39}$$

By Theorem 16.1.1, since $k + 2 > b - v$, we must have $|C_{k+2}| = 0$. We conclude that if $k \neq 6$,

$$j = 0, \qquad k - 2, \qquad \text{or} \qquad k - 1. \tag{16.1.40}$$

The last value, $j = k - 1$, corresponds to the case in which S_6 is not applicable. We shall exclude $k = 6$ from further consideration. A special proof for this case was given in Connor's first paper [1].

Let us consider the part of the structure matrix of our design D consisting of the first associates of B_1:

$$\begin{array}{cc|cccc|cccc}
B_1 & B_2 & B_3 & B_4 & \cdots & B_{k+2} & B_{k+3} & & \cdots & B_{2k+1} \\
\hline
k & 1 & 1 & 1 & \cdots & 1 & 1 & 1 & \cdots & 1 \\
1 & k & 1 & 1 & \cdots & 1 & 2 & 2 & \cdots & 2 \\
\hline
1 & 1 & & & & & & & & \\
1 & 1 & & & & & & & & \\
\cdot & \cdot & & & A & & & & B & \\
\cdot & \cdot & & & & & & & & \\
\cdot & \cdot & & & & & & & & \\
\hline
1 & 1 & & & & & & & & \\
1 & 2 & & & & & & & & \\
1 & 2 & & & & & & & & \\
\cdot & \cdot & & & B^T & & & & C & \\
\cdot & \cdot & & & & & & & & \\
\cdot & \cdot & & & & & & & & \\
1 & 2 & & & & & & & &
\end{array} \tag{16.1.41}$$

Here B has k rows and $(k - 1)$ columns. Since $B_{k+3}, \ldots, B_{2k+1}$ are first associates of B_1, every one of the columns $k + 3, \ldots, 2k + 1$ in (16.1.41) contains, besides the 1 and 2 in rows 1 and 2, exactly k 1's and $(k - 2)$ 2's. Thus, every column of the submatrix B contains at least two 1's, and as $2(k - 1) > k$, there must be at least one row of B containing at least two 1's. Let the blocks be numbered so that one such row is the third row and corresponds to B_3. In the notation of (16.1.37), for such a choice of B_3 we have $(k - 1 - j)$ columns $\begin{bmatrix} 1 \\ 2 \\ 1 \end{bmatrix}$, and for this choice, $k - 1 - j \geq 2$ or $j \leq k - 3$.

From (16.1.40) this is possible only if $j = 0$. Hence, (16.1.41) now has the form

$$
\begin{array}{c|ccc|ccc|ccc}
 & B_1 & B_2 & B_3 & B_4 & \cdots & B_{k+2} & B_{k+3} & \cdots & B_{2k+1} \\
B_1 & k & 1 & 1 & 1 & \cdots & 1 & 1 & \cdots & 1 \\
B_2 & 1 & k & 1 & 1 & \cdots & 1 & 2 & \cdots & 2 \\
B_3 & 1 & 1 & k & 2 & \cdots & 2 & 1 & \cdots & 1 \\
\hline
B_4 & 1 & 1 & 2 & & & & & & \\
 & \cdot & \cdot & \cdot & & & & & & \\
 & \cdot & \cdot & \cdot & & A & & & B & \\
 & \cdot & \cdot & \cdot & & & & & & \\
B_{k+2} & 1 & 1 & 2 & & & & & & \\
\hline
B_{k+3} & 1 & 2 & 1 & & & & & & \\
 & \cdot & \cdot & \cdot & & & & & & \\
 & \cdot & \cdot & \cdot & & B^T & & & C & \\
 & \cdot & \cdot & \cdot & & & & & & \\
B_{2k+1} & 1 & 2 & 1 & & & & & &
\end{array}
\qquad (16.1.42)
$$

If we consider the structure matrix for B_3, B_1, B_2, B_i, B_j for any $i, j, 4 \leq i$, $j \leq k + 2$, we may appeal to Lemma 16.1.5 as applicable to $S_5^{(1)}$ and conclude that $s_{ij} = 1$. Hence, all off-diagonal elements in A are 1's. Since each row of $B_4, \ldots, B_{k+2}$ contains exactly $(b + 1)$ 1's, we conclude that each row of B has exactly one 1 and the rest have 2's. This shows that our choice of B_3 as some row with at least two 1's in columns $k + 3, \ldots, 2k + 1$, was in fact a unique choice. If we take B_2, B_1, B_3, B_i, B_j with $k + 3 \leq i, j \leq 2k + 1$, we have the structure matrix $S_5^{(1)}$, and by Lemma 16.1.5 we conclude that every off-diagonal element in C is a 1, whence it follows also that there is exactly one 1 in every column of B. We may, if we choose, renumber $B_{k+3}, \ldots, B_{2k+1}$ so that the 1's that appear in B are on its main diagonal, since there is exactly one 1 in each row and in each column of B.

We now find that

$$S_1 = \{B_1, B_2, B_4, B_5, \ldots, B_{k+2}\}$$

are all first associates of each other and also

$$S_2 = \{B_1, B_3, B_{k+3}, B_{k+4}, \ldots, B_{2k+1}\}.$$

Our conclusion will be formulated as

LEMMA 16.1.6. *Any block B_1 and its $2k$ first associates uniquely comprise two sets S_1 and S_2 of $(k + 1)$ blocks, which pairwise are first associates and have only the block B_1 in common.*

Proof: ■ All that remains to be proved for this lemma is the uniqueness of the subdivision of $B_1, B_2, \ldots, B_{2k+1}$ into the sets S_1 and S_2. If $B_{k+3}, \ldots, B_{2k+1}$

are numbered so that the 1's in B of (16.1.42) are on its main diagonal, then B_i and B_{i+k-1} for $i = 4, \ldots, k+2$ are first associates, as are B_2 and B_3, but otherwise any $B_i \neq B_1$ of S_1 and $B_j \neq B_1$ of S_2 are second associates. Thus, a set of $(k+1)$ blocks that are pairwise first associates is uniquely determined by B_1 and any additional first associate of B_1, and such a set is either S_1 or S_2.

We are now able to proceed to the proof of Theorem 16.1.3, and it is sufficient to establish the properties of the sets S_j as given in Theorem 16.1.2. By Lemma 16.1.6, we have found sets S_j satisfying properties 1 and 4. Let n be the number of sets $S_1, \ldots, S_n$. Since there are $b = (k+1)(k+2)/2$ blocks and each of these is in exactly two sets containing $(k+1)$ blocks, the number n must satisfy $2b = n(k+1)$, whence $n = k+2$ as required by Theorem 16.1.2. Furthermore, any block is the unique block common to the two sets S containing it. As $n = k+2$, this means that the $n(n-1)/2 = b$ set intersections consist of each block used exactly once. This is property 3 of Theorem 16.1.2.

To prove 2 of Theorem 16.1.2, let m_i be the number of elements appearing i times in a set S_j. Then the following relations hold:

$$\sum_{i=0}^{k+1} m_i = v = \frac{k(k+1)}{2},$$

$$\sum_{i=0}^{k+1} i m_i = k(k+1), \tag{16.1.43}$$

$$\sum_{i=0}^{k+1} i(i-1) m_i = k(k+1).$$

The first of these merely says that each element is counted exactly once in some m, the second counts the total number of elements in the $(k+1)$ blocks of S_j, and the third counts the intersections of the $(k+1)$ blocks, noting that any two blocks of S_j are first associates. These relations lead to

$$\sum_{i=0}^{k+1} (i-2)^2 m_i = 0. \tag{16.1.44}$$

Since, trivially, $m_i \geq 0$ for any i, this gives $m_j = 0$ for $j \neq 2$ and $m_2 = v$. This is property 2 and the proof of our theorem is now complete. By the uniqueness of the construction of the S_j's, the uniqueness of the embedding of Theorem 16.1.3 follows.

16.2 Copositive and Completely Positive Quadratic Forms

Connor's method does not use the full amount of information implicit in the equation (16.1.5) which we rewrite as

$$Q^* = Q - L_1^2 - \cdots - L_t^2 = L_{t+1}^2 + \cdots + L_b^2, \tag{16.2.1}$$

where Q and $L_1, \ldots, L_t$ are known, but $L_{t+1}, \ldots, L_b$ are unknown. Connor uses the fact the Q^* must be positive semidefinite. Of course the Bruck-Ryser-Chowla methods rely on the rationality of $L_{t+1}, \ldots, L_b$. But another kind of information is available, namely, that $L_{t+1}, \ldots, L_b$ are linear forms with nonnegative coefficients (namely, coefficients zero or 1). A natural question arises as to how restrictive this information is. Naturally, Q^* is positive semidefinite and has nonnegative coefficients. But for quadratic forms in five or more variables, these two properties are not sufficient to guarantee that Q^* can be written as a sum of squares of linear forms, where the linear forms have nonnegative coefficients.

Consider the following example:

$$\begin{aligned} Q(x_1, \ldots, x_5) &= x_1^2 + x_2^2 + x_3^2 + x_4^2 + x_5^2 + x_1x_2 \\ &\quad + x_1x_5 + x_2x_3 + \tfrac{3}{2}x_3x_4 + x_4x_5 \\ &= (x_2 + \tfrac{1}{2}x_1 + \tfrac{1}{2}x_3)^2 + (x_5 + \tfrac{1}{2}x_1 + \tfrac{1}{2}x_4)^2 \\ &\quad + \tfrac{1}{2}(x_1 - \tfrac{1}{2}x_3 - \tfrac{1}{2}x_4)^2 + \tfrac{5}{8}(x_3 + x_4)^2. \end{aligned} \tag{16.2.2}$$

From the first form, Q has positive coefficients, and from the second, Q is semidefinite. But we shall show directly that Q cannot be written as a sum of squares of nonnegative linear forms. For suppose we have

$$Q = L_1^2 + \cdots + L_r^2 + L_{r+1}^2 + \cdots + L_n^2, \tag{16.2.3}$$

where the linear forms L have nonnegative coefficients and are numbered so that $L_1, \ldots, L_r$ are those in which both x_3 and x_4 have positive coefficients. Since in Q there are no terms in x_1x_3, x_2x_4, and x_3x_5, the forms $L_1, \ldots, L_r$ must have zero coefficients for x_1, x_2, and x_5, or $L_i = u_{i3}x_3 + u_{i4}x_4$, $u_{i3} > 0$, $u_{i4} > 0$, $i = 1, \ldots, r$. Hence,

$$L_1^2 + \cdots + L_r^2 = Ax_3^2 + \tfrac{3}{2}x_3x_4 + Bx_4^2. \tag{16.2.4}$$

Let us write

$$Q_1 = Q_1(x_1, \ldots, x_5) = L_{r+1}^2 + \cdots + L_n^2. \tag{16.2.5}$$

Then (16.2.3) becomes

$$Q(x_1, \ldots, x_5) = Ax_3^2 + \tfrac{3}{2}x_3x_4 + Bx_4^2 + Q_1(x_1, \ldots, x_5). \tag{16.2.6}$$

Put

$$x_1 = \frac{x_3 + x_4}{2}, \qquad x^2 = \frac{-(3x_3 + x_4)}{4}, \qquad x_5 = -\frac{(x_3 + 3x_4)}{4}. \tag{16.2.7}$$

This substitution leaves (16.2.4) unchanged, and from (16.2.2) Q reduces to $\frac{5}{8}(x_3 + x_4)^2$, whence (16.2.6) becomes

$$\tfrac{5}{8}(x_3 + x_4)^2 = Ax_3^2 + \tfrac{3}{2}x_3x_4 + Bx_4^2 + Q_1. \tag{16.2.8}$$

Under the substitution, Q_1 loses the property of being completely positive, but retains the property of being semidefinite, so that $0 \le A \le \frac{5}{8}$, $0 \le B \le \frac{5}{8}$. Now, in (16.2.4) put $x_3 = 1$, $x = -1$, giving

$$(u_{13} - u_{14})^2 + \cdots + (u_{r3} - u_{r4})^2 = A - \tfrac{3}{2} + B \le \tfrac{5}{8} - \tfrac{3}{2} + \tfrac{5}{8} = -\tfrac{1}{4}. \quad (16.2.9)$$

This is a conflict. Hence, the assumption (16.2.3) that we could write Q as a sum of squares of nonnegative linear forms is false. Hence, in general, the property of Q^* in (16.2.1) of being expressible as a sum of squares of nonnegative linear forms is stronger than the property of being positive semidefinite and having nonnegative coefficients.

To proceed to our general problem, we use the following notations:

$$Q = Q(x_1, \ldots, x_n) = \sum_{i,j=1}^{n} a_{ij}x_ix_j, \qquad a_{ij} = a_{ji}, \qquad (16.2.10)$$

and where no ambiguity arises, we can abbreviate (16.2.10) to $Q = \sum a_{ij}x_ix_j$.

$L_i \ge 0$ means

$$L_i = u_{1i}x_1 + \cdots + u_{ni}x_n, \qquad u_{si} \ge 0, \qquad s = 1, \ldots, n, \quad (16.2.11)$$

and $x = (x_1, \ldots, x_n) \ge 0$ means

$$x_i \ge 0, \qquad i = 1, \ldots, n. \qquad (16.2.12)$$

DEFINITION. *A quadratic form $Q = Q(x_1, \ldots, x_n)$ is called completely positive if Q can be expressed in the form*

$$Q = \sum_{i=1}^{N} L_i^2, \qquad L_i \ge 0, \qquad i = 1, \ldots, N. \qquad (16.2.13)$$

DEFINITION. *A quadratic form $Q = Q(x_1, \ldots, x_n)$ is called copositive if $Q(x_1, \ldots, x_n) \ge 0$ whenever $(x_1, \ldots, x_n) \ge 0$.*

We shall associate a quadratic form $Q(x_1, \ldots, x_n) = \sum a_{ij}x_ix_j$ with a point in Euclidean space E_m, where $m = (n^2 + n)/2$. Our correspondence is

$$Q = \sum_{i,j=1}^{n} a_{ij}x_ix_j \rightleftharpoons (a_{11}, a_{21}, \ldots, a_{nn}, \sqrt{2}a_{12}, \ldots, \sqrt{2}a_{1n}, \ldots, \sqrt{2}a_{n-1,n}). \qquad (16.2.14)$$

Under this correspondence, every quadratic form in n variables corresponds to a unique point in E_m, and conversely. In these terms, a family of quadratic forms corresponds to a set of points in E_m. The terminology and theorems on convex spaces given in Chapter 8 are applicable here.

THEOREM 16.2.1. *The spaces of copositive quadratic forms in n variables and of completely positive quadratic forms in n variables are both closed convex cones. Each is the dual cone of the other.*

Proof: ■ If Q_1 and Q_2 are copositive quadratic forms, it is immediate from the definition that $Q_1 + Q_2$ and bQ_1, $b \geq 0$ are copositive. Thus, the copositive quadratic forms are a convex cone. Let a quadratic form Q be the limit of a sequence of copositive forms Q_k, $k = 1, 2, \ldots$ (where our limit is in terms of the Euclidean metric in E_m). Then, for a point $x = (x_1, \ldots, x_n) \geq 0$ in E_n, each $Q_k(x_1, \ldots, x_n) \geq 0$, and as quadratic forms are continuous functions of their coefficients, it follows that $Q(x_1, \ldots, x_n) \geq 0$. Hence, Q is copositive, and it follows that the space of copositive quadratic forms is closed.

Similarly, if Q_1 and Q_2 are completely positive, then from the definition, $Q_1 + Q_2$ is completely positive. If $b \geq 0$ when $Q_1 = L_1^2 + \cdots + L_N^2$, $L_i \geq 0$, and if we choose $c \geq 0$ so that $c^2 = b$, then $bQ_1 = (cL_1)^2 + \cdots + (cL_N)^2$, where also $cL_i \geq 0$. Hence, the completely positive quadratic forms are a convex cone. The closure of this cone is a little less obvious. For this we appeal to

LEMMA 16.2.1. *If $Q = Q(x_1, \ldots, x_n) = L_1^2 + \cdots + L_N^2$ with $L_i \geq 0$, $i = 1, \ldots, N$, then Q has a representation of this kind with $N < 2^n$.*

Proof: ■ Suppose two of the L_i have positive coefficients for exactly the same variables x_j. By appropriate renumbering we take these to be L_1 and L_2 and the variables $x_1, \ldots, x_s$. Then we have

$$\begin{aligned} L_1 &= c_1x_1 + \cdots + c_sx_s, \qquad c_i > 0, \qquad i = 1, \ldots, s, \\ L_2 &= d_1x_1 + \cdots + d_sx_s, \qquad d_j > 0, \qquad i = 1, \ldots, s. \end{aligned} \tag{16.2.15}$$

Now, for any angle θ, we have the identity

$$\begin{aligned} L_1^2 + L_2^2 &= (L_1 \cos\theta + L_2 \sin\theta)^2 + (-L_1 \sin\theta + L_2 \cos\theta)^2 \\ &= L_1'^2 + L_2'^2. \end{aligned} \tag{16.2.16}$$

For any θ in the first quadrant, L_1' has positive coefficients, and for θ sufficiently small, so does L_2', though for $Q = \pi/2$, $L_2' = -L_1$ has all coefficients negative. Hence, as θ increases from 0 to $\pi/2$, there will be a value of θ that makes at least one coefficient in L_2' zero and leaves the nonzero coefficients in L_2' positive. For this value of θ, L_1' has positive coefficients for the same x's as L_1, and L_2' has positive coefficients for fewer x's. Thus, in $Q = L_1^2 + \cdots + L_N^2$, with $L_i \geq 0$, we can assume that no L_i is identically zero and that no two L's have positive coefficients for exactly the same variables. Hence, N is at most the number of distinct nonvacuous subsets of n, and so $N < 2^n$. Appeal to a general theorem on convex cones, not proved here, would give the limit $N \leq m = (n^2 + n)/2$, the dimensionality of the space, but the limit $N < 2^n$ is sufficient for our purposes.

Hence, if $Q = Q(x_1, \ldots, x_n)$ is the limit of a sequence $Q_k (k = 1, 2, \ldots)$ of completely positive quadratic forms, then each Q_k may be expressed as a sum of squares of N nonnegative linear forms, with $N < 2^n$. If the coefficients of Q_k for $k \geq k_0$ differ from those of Q by less than $\frac{1}{2}$, then the coefficients of the N linear forms are uniformly bounded, and for a subsequence of the Q_k the forms $L_1, \ldots, L_N$ have limits that are nonnegative linear forms, the sum of whose squares is Q. Thus, Q is completely positive and the completely positive forms are closed.

If $(c_1, \ldots, c_n) \geq 0$, then by definition

$$Q_2 = (c_1 x_1 + \cdots + c_n x_n)^2 = \sum c_i c_j x_i x_j = \sum b_{ij} x_i x_j \tag{16.2.17}$$

is a completely positive form. Let $Q_1 = \sum a_{ij} x_i x_j$ be a quadratic form. Then, if q_1, q_2 are the points of E_m corresponding to Q_1 and Q_2, respectively, the inner product (q_1, q_2) as given by (16.2.14) has the value

$$\begin{aligned}(q_1, q_2) &= a_{11} b_{11} + \cdots + a_{nn} b_{nn} + 2a_{12} b_{12} + \cdots + 2a_{n-1,n} b_{n-1,n} \\ &= \sum_{i,j} a_{ij} b_{ij}, \qquad i, j = 1, \ldots, n.\end{aligned} \tag{16.2.18}$$

Here, with $b_{ij} = c_i c_j$ as in (16.2.17),

$$(q_1, q_2) = \sum a_{ij} c_i c_j. \tag{16.2.19}$$

Now suppose that Q_1 has the property $(q_1, q_2) \geq 0$ for every Q_2 of the form

$$Q_2 = L^2 = (c_1 x_1 + \cdots + c_n x_n)^2, \qquad (c_1, \ldots, c_n) \geq 0.$$

Then, (16.2.19) becomes

$$\sum a_{ij} c_i c_j \geq 0 \qquad \text{for} \qquad (c_1, \ldots, c_n) \geq 0. \tag{16.2.20}$$

But this is the assertion that Q_1 is copositive. If, on the other hand,

$$Q_2 = L_1^2 + \cdots + L_t^2 = Q_2^{(1)} + \cdots + Q_2^{(t)}, \qquad L_i \geq 0, \quad i = 1, \ldots, t,$$

then from $(q_1, q_2^{(i)}) \geq 0$, $i = 1, \ldots, t$, and the bilinearity of the inner product follows, $(q_1, q_2) \geq 0$. This shows that if H is the cone of completely positive quadratic forms, then H^* is the cone of copositive forms. Since H is a closed convex cone, it follows from Theorem 8.1.5 that $(H^*)^* = H$. This proves that the cones of copositive and completely positive forms are the duals of each other. Now all parts of Theorem 16.2.1 have been proved.

For sets A and B in E_m, we note

$$A \supseteq B \qquad \text{implies} \qquad A^* \subseteq B^*. \tag{16.2.21}$$

This is an immediate consequence of the definition of the dual, since if $(x, y) \geq 0$ for a given x and all $y \in A$, then a fortiori we have $(x, y) \geq 0$ for all $y \in B$. Thus, $x \in A^*$ implies $x \in B^*$.

For any two sets A and B, let $A \cap B$ be the set of points common to A and B and $A + B$ the set of all points $a + b$, $a \in A$, $b \in B$.

LEMMA 16.2.2. *If C_1 and C_2 are two closed convex cones, then $C_1 \cap C_2$ and $C_1 + C_2$ are closed convex cones and $(C_1 + C_2)^* = C_1^* \cap C_2^*$, $(C_1 \cap C_2)^* = C_1^* + C_2^*$.*

Proof: ■ It is an easy verification that $C_1 \cap C_2$ and $C_1 + C_2$ are closed convex cones. Consider $(C_1 + C_2)^*$. Since $C_1 \subseteq C_1 + C_2$ and $C_2 \subseteq C_1 + C_2$ from (16.2.11), it follows that $(C_1 + C_2)^* \subseteq C_1^*$ and $(C_1 + C_2)^* \subseteq C_2^*$, whence $(C_1 + C_2)^* \subseteq C_1^* \cap C_2^*$. Let $y \in C_1^* \cap C_2^*$. Any $x \in C_1 + C_2$ can be put in the form $x = x_1 + x_2$, $x_1 \in C_1$, $x_2 \in C_2$. Thus,

$$(x, y) = (x_1 + x_2, y) = (x_1, y) + (x_2, y).$$

Hence as $(x_1, y) \geq 0$ and $(x_2, y) \geq 0$, we have $(x, y) \geq 0$, and so $y \in (C_1 + C_2)^*$. Thus, $C_1^* \cap C_2^* \subset (C_1 + C_2)^*$ and so from our two inclusions, $(C_1 + C_2)^* = C_1^* \cap C_2^*$. Applying this,

$$(C_1^* + C_2^*)^* = (C_1^*)^* \cap (C_2^*)^* = C_1 \cap C_2$$

by Theorem 8.1.5. Then

$$(C_1 \cap C_2)^* = (C_1^* + C_2)^{**} = C_1^* + C_2^*,$$

our other relation.

Let us denote the class of completely positive quadratic forms by B, using this designation because of their relation to block designs. We denote the class of copositive forms by C. Two other classes are related to these, the class P of forms with positive coefficients, and the class S of semidefinite quadratic forms.

LEMMA 16.2.3. *Each of the classes P and S is self-dual.*

Proof: ■ In P, choose Q_{rs} with $b_{rs} = 1$, $b_{ij} = 0$ otherwise. Then, if $Q = \sum a_{ij}x_ix_j$ and $Q \in P^*$,

$$(q, q_{rs}) = a_{rs} \qquad \text{or} \qquad 2a_{rs} \geq 0.$$

Hence, $Q \in P$. The closure of P is trivial and so $P = P^*$. In S, let

$$Q_0 = (c_1x_1 + \cdots + c_nx_n)^2$$

be any form of rank 1. If $Q = \sum a_{ij}x_ix_j$ is in S^*, then

$$(q, q_0) = \sum a_{ij}c_ic_j \geq 0.$$

This says that Q is nonnegative for all real values of $x_1, \ldots, x_n$, whence $Q \in S$ and $S = S^*$, the closure of S being trivial.

Since every positive form and every semidefinite form is copositive, we have

$$C \supseteq P + S. \tag{16.2.22}$$

From Theorem 16.2.1, $C^* = B$ and $B^* = C$, and Lemmas 16.2.2 and 16.2.3 yield

$$B = C^* \subseteq (P + S)^* = P^* \cap S^* = P \cap S. \tag{16.2.23}$$

Equality in either of these implies equality in the other. Diananda [1] has proved $C = P + S$ for forms in at most four variables. But our example of (16.2.2) shows that for five variables, $B \subset P \cap S$.

The determination of the extreme points (extreme forms) for the classes *P*, *S*, and *B* is easy.

THEOREM 16.2.2. *The extreme forms for the class P of positive forms, S of positive semidefinite forms, and B of completely positive forms are:*

1. *For class P,*

$$ax_i^2, \quad a > 0, \quad bx_ix_j, \quad b > 0 \quad i \neq j, \quad i, j = 1, \ldots, n.$$

2. *For class S, the forms of rank* 1:

$$(c_1x_1 + \cdots + c_nx_n)^2,$$

c_i real not all zero.

3. *For class B, the forms of rank* 1:

$$(c_1x_1 + \cdots + c_nx_n)^2, \quad c_i \geq 0, \quad i = 1, \ldots, n,$$

c_i not all zero.

Proof: ■ For class *P*, the verification is immediate. For the class *S*, every form can be written

$$Q = L_1^2 + \cdots + L_t^2.$$

Here, if $Q = (c_1x_1 + \cdots + c_nx_n)^2 \neq 0$, then unless each L_i is a constant multiple of $L_0 = c_1x_1 + \cdots + c_nx_n$, we may find numbers $x_1, \ldots, x_n$ not all zero such that $L_0 = 0$ and some $L_i \neq 0$. But in the relation $L_0^2 = L_1^2 + \cdots + L_t^2$, this yields a conflict. The same argument holds for class *B*. Since the forms of the classes *P*, *S*, *B* can be expressed as a sum of a finite number of the extreme forms listed, it follows that these are all the extreme forms of the respective classes.

The extreme forms of the class *C* are still unknown. The following theorem gives some of them.

THEOREM 16.2.3. *The extreme forms of the class C of copositive forms in $x_1, \ldots, x_n$ include the following: ax_i^2, $a > 0$, $i = 1, \ldots, n$, bx_ix_j, $b > 0$,*

$i \neq j$, $i, j = 1, \ldots, n$ and $(U - V)^2$ where

$$U = \sum_{i=1}^{s} a_i u_i, \qquad V = \sum_{i=1}^{s} b_i v_i, \qquad a_i > 0, \qquad b_i > 0, \qquad r \geq 1, \qquad s \geq 1,$$

and the u's and v's are disjoint subsets of $x_1, \ldots, x_n$. These are the extreme forms of $P + S$.

Proof: ■ If $Q = \sum a_{ij} x_i x_j$ is copositive, by taking $x_i = 1$, $x_j = 0$, $j \neq i$, we see that $Q = a_{ii} \geq 0$. Also, in

$$Q = a_{ii} x_i^2 + 2a_{ij} x_i x_j = x_i(a_{ii} x_i + 2a_{ij} x_j)$$

with $a_{ii} \geq 0$ and $a_{ij} < 0$, if we take $x_j = 1$ and $x_i = \epsilon > 0$ sufficiently small, then $Q < 0$. Hence, in a copositive form if $a_{ii} = 0$ or $a_{jj} = 0$, then $a_{ij} \geq 0$.

Now suppose that $a_{rr} x_r^2 = Q_1 + Q_2$ with both Q_1 and Q_2 copositive:

$$Q_1 = \sum b_{ij} x_i x_j, \qquad Q_2 = \sum c_{ij} x_i x_j.$$

Then, for $j \neq r$,

$$b_{jj} \geq 0, \qquad c_{jj} \geq 0, \qquad \text{and} \qquad b_{jj} + c_{jj} = 0,$$

whence $b_{jj} = c_{jj} = 0$. Also, $b_{ij} \geq 0$, $c_{ij} \geq 0$, $b_{ij} + c_{ij} = 0$ in all cases with $i \neq j$, and so $b_{ij} = c_{ij} = 0$. Hence,

$$Q_1 = b_{rr} x_r^2, \qquad Q_2 = c_{rr} x_r^2, \qquad c_{rr} \geq 0$$

and so $Q = a_{rr} x_r^2$ is extreme. Similarly, if $Q = a_{rs} x_r x_s$, $r \neq s$ and $Q = Q_1 + Q_2$, $Q_1 = \sum b_{ij} x_i x_j$, $Q_2 = \sum c_{ij} x_i x_j$, with Q_1 and Q_2 copositive. Here, for all i, $b_{ii} \geq 0$, $c_{ii} \geq 0$, and $b_{ii} + c_{ii} = 0$, whence all $b_{ii} = c_{ii} = 0$. Hence, from the preceding remark, $b_{ij} \geq 0$, $c_{ij} \geq 0$ for all $i \neq j$. Then, except for $i = r$, $j = s$, we have

$$b_{ij} = c_{ij} = 0 \qquad \text{and} \qquad Q_1 = 2b_{rs} x_r x_s, \qquad Q_2 = 2c_{rs} x_r x_s,$$

and thus $a_{rs} x_r x_s$ is extreme.

To show that $(U - V)^2$ is extreme where

$$U = a_1 u_1 + \cdots + a_r u_r, \qquad V = b_1 v_1 + \cdots + b_s v_s,$$
$$a_i > 0, \qquad b_i > 0, \qquad r \geq 1, \qquad s \geq 1,$$

where the u's and v's are disjoint subsets of $x_1, \ldots, x_n$, it is sufficient to replace $a_i u_i$ by y_i and $b_j v_j$ by z_j and show that

$$(y_1 + \cdots + y_r - z_1 - \cdots - z_s)^2, \qquad r \geq 1, \qquad s \geq 1,$$

is an extreme form in variables y_i, z_j. First consider $(y_1 - z_1)^2 = Q_1 + Q_2$, where Q_1 and Q_2 are copositive in y_1 and z_1. For $y_1 = z_1$ we have $Q_1 = Q_2 = 0$, whence

$$Q_1 = (y_1 - z_1)(ay_1 - bz_1), \qquad Q_2 = (y_1 - z_1)(cy_1 - dz_1), \quad (16.2.24)$$

where a, b, c, d are nonnegative and $a + c = 1$, $b + d = 1$. If $a > b$, we can put $z_1 = y_1 + \epsilon$, and with y_1, z_1 positive and $\epsilon > 0$ sufficiently small, we have $Q_1 < 0$, a conflict. Similarly, if $a < b$, we can put $z_1 = y_1 - \epsilon$, and with $\epsilon > 0$ sufficiently small, $Q_1 < 0$, again a conflict. Hence, $a = b$ and

$$Q_1 = a(y_1 - z_1)^2, \qquad Q_2 = (1 - a)(y_1 - z_1)^2,$$

and so Q is extreme. Next consider $Q = (y_1 + y_2 - z_1)^2 = Q_1 + Q_2$. Putting y_1 and y_2 respectively equal to zero, the previous case yields

$$\begin{aligned} Q_1 &= a(y_1^2 + y_2^2 + z_1^2 - 2y_1z_1 - 2y_2z_1) + ky_1y_2, \\ Q_2 &= (1 - a)(y_1^2 + y_2^2 + z_1^2 - 2y_1z_1 - 2y_2z_1) + ty_1y_2, \end{aligned} \quad (16.2.25)$$

with $0 \le a \le 1$ and, of course, $k + t = 2$. But from this we have

$$\begin{aligned} Q_1 &= a(y_1 + y_2 - z_1)^2 + (k - 2a)y_1y_2, \\ Q_2 &= (1 - a)(y_1 + y_2 - z_1)^2 + (t - 2a - 2)y_1y_2. \end{aligned} \quad (16.2.26)$$

Here, if $k - 2a < 0$, we can put $y_1 = y_2 = 1$, $z_1 = 2$, and $Q_1 < 0$, whence we must have $k - 2a \ge 0$. Similarly, we must have $t + 2a - 2 \ge 0$. As $k + t = 2$,

$$0 = (k - 2a) + (t + 2a - 2), \quad (16.2.27)$$

whence $k = 2a$, $t = 2 - 2a$, $Q_1 = a(y_1 + y_2 - z_1)^2$ and

$$Q_2 = (1 - a)(y_1 + y_2 - z_1)^2,$$

and $Q = (y_1 + y_2 - z_1)^2$ is extreme. The same argument shows that

$$(y_1 - z_1 - z_2)^2$$

is extreme.

Now generally, suppose that

$$Q = (y_1 + \cdots + y_r - z_1 - \cdots - z_s)^2 = Q_1 + Q_2$$

with Q_1 and Q_2 copositive. Let

$$Q_1 = ay_1^2 + \cdots \qquad \text{and} \qquad Q_2 = (1 - a)y_1^2 + \cdots.$$

Putting appropriate y's and z's equal to zero, we can use the previous cases to determine all coefficients of Q_1, and we find that $Q_1 = aQ$, $Q_2 = (1 - a)Q$, with $0 \le a \le 1$, whence Q is extreme. We note that

$$(U + V)^2 = (U - V)^2 + 2UV$$

is never extreme. Since every form in $P + S$ is the sum of a finite number of these forms, these are all the extreme forms of $P + S$.

There are other forms in C that are not contained in $P + S$. The following is the "Horn form," discovered by A. Horn:

$$Q = (x_1 + x_2 + x_3 + x_4 + x_5)^2 - 4x_1x_2 - 4x_2x_3 - 4x_3x_4 - 4x_4x_5 - 4x_5x_1 . \tag{16.2.28}$$

We may express Q in the following two ways:

$$\begin{aligned} Q &= (x_1 - x_2 + x_3 + x_4 - x_5)^2 + 4x_2x_4 + 4x_3(x_5 - x_4), \\ &= (x_1 - x_2 + x_3 - x_4 + x_5)^2 + 4x_2x_5 + 4x_1(x_4 - x_5) . \end{aligned} \tag{16.2.29}$$

With $x_5 \geq x_4$, the first expression shows that $Q \geq 0$, whereas if $x_4 \geq x_5$, the second expression yields $Q \geq 0$. Hence, Q is copositive. Furthermore, Q is an extreme form, for if $Q = Q_1 + Q_2$ with Q_1 and Q_2 copositive, putting $x_4 = x_5 = 0$, Q reduces to the extreme form $(x_1 - x_2 + x_3)^2$, whence

$$\begin{aligned} Q_1 &= a(x_1 - x_2 + x_3)^2 + x_4L + x_5M , \\ Q_2 &= (1 - a)(x_1 - x_2 + x_3)^2 + x_4U + x_5V. \end{aligned} \tag{16.2.30}$$

Similarly, putting $x_i = x_{i+1} = 0$, $i = 1, 2, 3$, and $x_5 = x_1 = 0$, Q becomes extreme and these cases together show that $Q_1 = aQ$ and $Q_2 = (1 - a)Q$, whence Q is extreme. Clearly, $Q \notin P$ and $Q \notin S$, and as Q is extreme, $Q \notin P + S$.

If we take Q_1 as the quadratic form of (16.2.2) and Q_2 as the Horn form of (16.2.28), then q_1 and q_2 being the corresponding points of E_{15}, we find

$$(q_1, q_2) = -\tfrac{1}{2} . \tag{16.2.31}$$

Since $Q_2 \in C$, this shows that $Q_1 \notin B$, as we showed directly in (16.2.3) through (16.2.9). Since $Q_1 \in P \cap S$, this shows how copositive forms not in $P + S$ can be used to show that a form does not belong to B even though it may belong to $P \cap S$.

Recently, Baumert has shown the existence of further extreme copositive forms Q in five variables. If $Q = \sum a_{ij}x_ix_j$ is normalized so that $a_{11} = a_{22} = \cdots = a_{55} = 1$, then Q has zeros of the form

$$\begin{aligned} u &= (u_1, u_2, 1, 0, 0), \\ v &= (0, v_2, v_3, 1, 0), \\ w &= (0, 0, w_3, w_4, 1), \\ y &= (1, 0, 0, y_4, y_5), \\ z &= (z_1, 1, 0, 0, z_5), \end{aligned} \tag{16.2.32}$$

where no number $u, \ldots, z_5$ is zero. These numbers determine the coefficients a_{ij}.

One such choice of zeros is

$$u = \left(\frac{1}{8}, \frac{7+\sqrt{8^4-15}}{64}, 1, 0, 0\right) = (0.125, 1.1075+, 1, 0, 0),$$
$$v = \left(0, \frac{1}{8}, \frac{\sqrt{8^4-15}+\sqrt{8^6-15}}{512}, 1, 0\right) = (0, 0.125, 1.1247+, 1, 0),$$
$$w = \left(0, 0, \frac{1}{8}, \frac{\sqrt{8^6-15}+\sqrt{8^8-15}}{4096}, 1\right) = (0, 0, 0.125, 1.1249+, 1),$$
$$y = \left(1, 0, 0, \frac{1}{8}, \frac{\sqrt{8^8-15}+\sqrt{8^{10}-15}}{8^5}\right) = (1, 0, 0, 0.125, 1.1249+),$$
$$z = \left(\frac{7+\sqrt{8^{10}-15}}{8}, 1, 0, 0, 8^4\right) = (4096.874+, 1, 0, 0, 4096). \tag{16.2.33}$$

Although we do not know the extreme forms of copositive forms, we do have a test to decide whether or not a given form is copositive. The formulation given here is due to A. Garsia.

TEST FOR COPOSITIVE QUADRATIC FORMS

Let $Q = Q(x_1, \ldots, x_n) = \sum a_{ij}x_ix_j$ be a quadratic form in n variables and suppose that putting any one of x_i, $i = 1, \ldots, n$ equal to zero gives us a copositive form in $(n-1)$ variables. With $A = [a_{ij}]$, put $A(\epsilon) = A + \epsilon I$ and let $D(\epsilon)$ be the determinant of $A(\epsilon)$, and $E_1(\epsilon), \ldots, E_n(\epsilon)$ be the cofactors of the last row of $A(\epsilon)$. Then Q is not copositive if and only if for sufficiently small positive values of ϵ the quantities $E_1(\epsilon)D(\epsilon), \ldots, E_n(\epsilon)D(\epsilon)$ are all positive. This may be checked by expanding the determinants as polynomials in ϵ and examining the terms with the lowest power of ϵ in each case. In case $E_n(0) \neq 0$, the stronger result holds: Q is not copositive if and only if the quantities $E_1(0)D(0), \ldots, E_n(0)D(0)$ are all positive.

16.3 Rational Completions of Incidence Matrices

It was noted in Section 16.1 that for a symmetric design $D(v, k, \lambda)$, the Connor condition requires $s_{ij} = \lambda$, $i \neq j$ in (16.1.5), as was shown necessary in Theorem 10.2.2. A natural question is whether requiring $L_{t+1}, \ldots, L_b$ in (16.1.5) to be rational as well as real is an added restriction, and if it is, in what way this limits choices of initial blocks for symmetric designs. Somewhat surprisingly, it turns out that if there is *any* rational $v \times v$ matrix A satisfying

$$AA^T = (k-\lambda)I + \lambda J, \tag{16.3.1}$$

rational completion is possible, and indeed with $s_{ij} = \lambda$, there is even a rational normal completion. This result is due to the author and H. J. Ryser [2].

Consider parameters v, k, λ satisfying

$$k(k-1) = \lambda(v-1). \tag{16.3.2}$$

The proof of Theorem 10.3.1 proves more than the statement of the theorem. The Bruck-Ryser-Chowla conditions are

1. *If v is even, $k - \lambda$ is a square.*
2. *If v is odd, $z^2 = (k-\lambda)x^2 + (-1)^{(v-1)/2}\lambda y^2$ has a solution in integers x, y, z not all zero.*

It was proved that the existence of rational linear forms

$$L_j = \sum_i a_{ij}x_i, \qquad j = 1, \ldots, v,$$

satisfying (10.3.1) implied the Bruck-Ryser-Chowla conditions. But (10.3.1) is equivalent to saying that the matrix $A = (a_{ij})$ satisfies (16.3.1). In Section 10.4 it was shown conversely, using the deep Hasse-Minkowski theory, that conditions 1 and 2 implied the existence of rational L_j satisfying (10.3.1), and so a rational A satisfying (16.3.1). In Theorem 10.2.3 it was shown that either of the further conditions

$$AJ = kJ, \qquad JA = kJ, \tag{16.3.3}$$

implied the other and the normality of A, which is the statement $AA^T = A^TA$, whence

$$AA^T = A^TA = (k-\lambda)I + \lambda J. \tag{16.3.4}$$

The main theorem of this section is

THEOREM 16.3.1. *Given positive integers v, k, λ satisfying $k(k-1) = \lambda(v-1)$ and suppose that the appropriate condition 1 or 2 is satisfied. Suppose further that we are given a rational $t \times v$ matrix A_1 $(0 \leq t \leq v)$ satisfying*

$$A_1A_1^T = (k-\lambda)I_t + \lambda J_t \qquad \textit{and} \qquad A_1J_v = kJ_{t,v}. \tag{16.3.5}$$

Then there exists a rational $v \times v$ matrix A having A_1 as its first t rows such that

$$AA^T = A^TA = (k-\lambda)I_v + \lambda J_v, \qquad AJ_v = J_vA = kJ_v. \tag{16.3.6}$$

Here, J_t, J_v are square matrices of all 1's and $J_{t,v}$ is the $t \times v$ matrix of all 1's.

Comment: ■ With obvious modifications we could have assumed the first t columns given. If we are given t initial blocks of k elements such that $s_{ij} = \lambda$, $i \neq j$, in its structural matrix, this gives the requirement of (16.3.5) expressed for t initial columns. But the statement of the theorem is entirely in terms of rational matrices and no explicit reference to block designs is made. Nevertheless, the conclusions are in sharp contrast to those of Connor's with respect to block designs. It says that if the basic conditions 1 and 2 are satisfied, any choice of t initial blocks subject to the condition $s_{ij} = \lambda$ can be completed not only in the real field but even in the rational field to a rational matrix A, satisfying not only (13.3.1) but also the other relations of (16.3.6). In a sense, this is negative information, since it means that no t initial blocks of k elements with $s_{ij} = \lambda$ can be excluded. A rational completion will exist even for certain initial choices of blocks that are clearly impossible, for a pair of elements may occur together in more than λ of an initial set of t blocks without invalidating the condition $s_{ij} = \lambda$, which refers only to the intersections of blocks.

The proof of Theorem 16.3.1 depends on some preliminary results, which will be given as theorems rather than lemmas because of their independent importance.

We follow some standard notation relating quadratic forms and symmetric matrices, as given in B. W. Jones [1]. Let

$$Q_1 = \sum_{i,j=1}^{n} a_{ij}x_ix_j, \qquad Q_2 = \sum_{i,j=1}^{n} b_{ij}y_iy_j, \qquad a_{ij} = a_{ji}, \qquad b_{ij} = b_{ji}. \tag{16.3.7}$$

Then Q_1 and Q_2 correspond to the symmetric matrices A and B:

$$A = (a_{ij}), \qquad a_{ji} = a_{ij}, \qquad B = (b_{ij}), \qquad b_{ji} = b_{ij}. \tag{16.3.8}$$

If

$$x_i = \sum_{j=1}^{n} c_{ij}y_j, \qquad i = 1, \ldots, n, \tag{16.3.9}$$

is such that when these values of x's are substituted into Q_1 we obtain Q_2, then, writing

$$C = (c_{ij}), \tag{16.3.10}$$

we find

$$C^TAC = B. \tag{16.3.11}$$

We say that Q_1 *represents* Q_2. If C is nonsingular, then Q_2 also represents Q_1, and we say that Q_1 and Q_2 are *congruent* and that the symmetric matrices A and B are *congruent* and write $A \overset{c}{=} B$. From (16.3.11), if A and B are nonsingular, C is also nonsingular.

THEOREM 16.3.2 (WITT [1]). *Suppose that* $ax_0^2 + Q_1$ *represents* $ax_0^2 + Q_2$, *where these are quadratic forms over a field* F *of characteristic different from* 2, *and that* Q_1 *and* Q_2 *are forms in* $x_1, \ldots, x_n$. *Then* Q_1 *represents* Q_2.

Proof: ■ Let $Q_1 = \sum a_{ij}x_ix_j$ and $Q_2 = \sum b_{ij}x_ix_j$, $A = (a_{ij})$, $B = (b_{ij})$. We are given the existence of a matrix C:

$$C = \begin{bmatrix} c_0 & c_1 \\ c_2 & C_0 \end{bmatrix} \tag{16.3.12}$$

such that

$$\begin{bmatrix} c_0 & c_2^T \\ c_1^T & C_0^T \end{bmatrix} \begin{bmatrix} a & 0 \\ 0 & A \end{bmatrix} \begin{bmatrix} c_0 & c_1 \\ c_2 & C_0 \end{bmatrix} = \begin{bmatrix} a & 0 \\ 0 & B \end{bmatrix}. \tag{16.3.13}$$

This equation is equivalent to the three equations

$$\begin{aligned} c_0^2 a + c_2^T A c_2 &= a\,, \\ c_0 a c_1 + c_2^T A C_0 &= 0\,, \\ c_1^T a c_1 + C_0^T A C_0 &= B\,. \end{aligned} \tag{16.3.14}$$

Choose the sign in $c_0 \pm 1$ so that it is not zero; write $u = (c_0 \pm 1)^{-1}$ and define

$$S = C_0 - u c_2 c_1\,. \tag{16.3.15}$$

Then

$$\begin{aligned} S^T A S &= (C_0^T - u c_1^T c_2^T) A (C_0 - u c_2 c_1) \\ &= C_0^T A C_0 - u c_1^T c_2^T A C_0 - u C_0^T A c_2 c_1 + u^2 c_1^T c_2^T A c_2 c_1\,. \end{aligned} \tag{16.3.16}$$

Using (16.3.14) gives

$$\begin{aligned} S^T A S &= C_0^T A C_0 + u c_1^T c_0 a c_1 + u c_1^T a c_0 c_1 + u^2 c_1^T (a - c_0^2 a) c_1 \\ &= C_0^T A C_0 + u c_1^T c_1 a \{c_0 + c_0 + u(1 - c_0^2)\} \\ &= C_0^T A C_0 + a c_1^T c_1 \\ &= B\,. \end{aligned} \tag{16.3.17}$$

This proves the theorem.

Since a quadratic form $Q(x_1, \ldots, x_r)$ over a field F of characteristic different from 2 is congruent to a diagonal form $d_1 x_1^2 + \cdots + d_r x_r^2$, we have the important

Corollary. *If $Q_1(x_1, \ldots, x_r) + Q_2(x_{r+1}, \ldots, x_n)$ represents*

$$Q_1(x_1, \ldots, x_r) + Q_3(x_{r+1}, \ldots, x_n)\,,$$

then $Q_2(x_{r+1}, \ldots, x_n)$ represents $Q_3(x_{r+1}, \ldots, x_n)$.

We write $A \oplus B$ for the direct sum of matrices A and B, where this is the matrix defined by

$$A \oplus B = \begin{bmatrix} A & 0 \\ 0 & B \end{bmatrix}, \tag{16.3.18}$$

the 0's standing for matrices of zeros.

Theorem 16.3.3. *Suppose that $AA^T = D_1 \oplus D_2$, where D_1 is nonsingular of order r, D_2 is nonsingular of order s, and $r + s = n$. Let X be an arbitrary $r \times n$*

matrix such that $XX^T = D_1$. *Then there exists an* $n \times v$ *matrix* Z *having* X *as its first* r *rows such that* $ZZ^T = D_1 \oplus D_2$.

Proof: ■ X is of rank r, since D_1 is nonsingular. Hence, the vectors $W = (w_1, \ldots, w_n)$ such that the inner product $(x_i, w) = 0$ for every row x_i, $i = 1, \ldots, r$ of X, form a space of dimension $s = n - r$. Let W be an $s \times n$ matrix whose rows form a basis for this space. Then, writing

$$Y = \begin{bmatrix} X \\ W \end{bmatrix}, \tag{16.3.19}$$

we have

$$YY^T = D_1 \oplus D_3. \tag{16.3.20}$$

We assert that Y is nonsingular. Since D_1 is nonsingular, there exists a nonsingular C such that

$$C^T D_1 C = D, \qquad D = \text{diag}[d_1, \ldots, d_r], \qquad d_i \neq 0. \tag{16.3.21}$$

Then

$$\begin{bmatrix} C^T, X \\ W \end{bmatrix} [X^T C, W^T] = \begin{bmatrix} D & 0 \\ 0 & D_3 \end{bmatrix}. \tag{16.3.22}$$

Let the row vectors of

$$K = \begin{bmatrix} C^T, X \\ W \end{bmatrix}$$

be denoted by $\alpha_1, \ldots, \alpha_r, \beta_1, \ldots, \beta_s$. Here, $\alpha_1, \ldots, \alpha_r$ are linear combinations of the rows of X, and so

$$(\alpha_i, \beta_k) = 0, \qquad i = 1, \ldots, r, \qquad k = 1, \ldots, s.$$

Also, since D is diagonal,

$$(\alpha_i, \alpha_j) = 0, \qquad i \neq j \qquad \text{and} \qquad (\alpha_i, \alpha_i) = d_i \neq 0.$$

If we have a relation

$$a_1\alpha_1 + \cdots + a_r\alpha_r + b_1\beta_1 + \cdots + b_s\beta_s = 0, \tag{16.3.23}$$

taking the inner product with α_i gives

$$a_i d_i = 0, \qquad i = 1, \ldots, r, \tag{16.3.24}$$

and as $d_i \neq 0$, $i = 1, \ldots, r$, this implies

$$a_1 = a_2 = \cdots = a_r = 0. \tag{16.3.25}$$

Since the β's were chosen to be independent, it follows that also

$$b_1 = \cdots = b_s = 0. \tag{16.3.26}$$

Hence, the rows of K are independent, and as C is nonsingular, the same is true of Y and so Y is nonsingular. Now we may write

$$\begin{aligned}(AY^{-1})(D_1 \oplus D_3)(AY^{-1})^T &= AY^{-1}(YY^T)(Y^T)^{-1}A^T \\ &= AA^T = D_1 \oplus D_2 .\end{aligned} \tag{16.3.27}$$

Hence, by Witt's theorem, a nonsingular matrix E exists with $E^T D_3 E = D_2$, and writing

$$Z = \begin{bmatrix} X \\ E^T W \end{bmatrix}, \tag{16.3.28}$$

we have

$$ZZ^T = \begin{bmatrix} D_1 & 0 \\ 0 & E^T D_3 E \end{bmatrix} = \begin{bmatrix} D_1 & 0 \\ 0 & D_2 \end{bmatrix}. \tag{16.3.29}$$

This is the statement of the theorem.

THEOREM 16.3.4. *Let $x = (x_1, \ldots, x_n)$ and $y = (y_1, \ldots, y_n)$ be vectors such that*

$$x_1^2 + \cdots + x_n^2 = y_1^2 + \cdots + y_n^2 = c \neq 0 . \tag{16.3.30}$$

Then there exists an orthogonal matrix O such that $xO = y$. This holds in any field F of characteristic $\neq 2$.

Proof: ■ There exists an $(n-1) \times n$ matrix W of rank $n-1$ such that

$$X = \begin{bmatrix} x \\ W \end{bmatrix}, \qquad XX^T = \begin{bmatrix} c & 0 \\ 0 & D_2 \end{bmatrix}. \tag{16.3.31}$$

By Theorem 16.3.3, there exists a matrix Y:

$$Y = \begin{bmatrix} y \\ R \end{bmatrix}, \qquad YY^T = \begin{bmatrix} c & 0 \\ 0 & D_2 \end{bmatrix}. \tag{16.3.32}$$

Here, X and Y are nonsingular and $O = X^{-1}Y$ is orthogonal; so

$$(1, 0, \ldots, 0)XO = (1, 0, \ldots, 0)Y \tag{16.3.33}$$

or

$$xO = y . \tag{16.3.34}$$

This is the conclusion of our theorem.

Theorem 16.3.4 asserts the existence of a rotation taking a vector of given length into any other vector of the same length. This is, of course, familiar if F is the real field, but we wish to apply this when F is the rational field, where the result is not obvious.

THEOREM 16.3.5. *Let $k(k-1) = \lambda(v-1)$ and let A be a rational nonsingular matrix, such that $AA^T = B = (k-\lambda)I + \lambda J$. Then a rational matrix C exists such that $CC^T = C^TC = B$, $CJ = JC = kJ$.*

Proof: ∎ As was shown in Section 10.4, a rational A exists if the Bruck-Ryser-Chowla condition in (10.3.3) holds. Hence, if s_i is the sum of the ith column of A, we have

$$\begin{aligned} JAA^TJ &= (s_1^2 + \cdots + s_v^2)J \\ &= JBJ = (kv - \lambda v + \lambda v^2)J \\ &= k^2 vJ. \end{aligned} \tag{16.3.35}$$

Hence, comparing coefficients of J,

$$s_1^2 + \cdots + s_v^2 = k^2 v. \tag{16.3.36}$$

Thus, by Theorem 16.3.4, there exists a rational orthogonal matrix O such that

$$(s_1, \ldots, s_v)O = (k, \ldots, k). \tag{16.3.37}$$

Hence,

$$JAO = \begin{bmatrix} s_1 & s_2 & \cdots & s_v \\ \cdots & \cdots & \cdots & \cdots \\ \cdots & \cdots & \cdots & \cdots \\ s_1 & s_2 & \cdots & s_v \end{bmatrix} O = kJ. \tag{16.3.38}$$

If we now put $C = AO$, then as $CC^T = I$, we have

$$CC^T = ACC^TA^T = AA^T = B = (k-\lambda)I + \lambda J, \tag{16.3.39}$$

and also from (16.3.38),

$$JC = JAO = kJ. \tag{16.3.40}$$

From Theorem 10.2.3, the conditions (16.3.39) and (16.3.40) imply that also

$$C^TC = B, \qquad CJ = kJ. \tag{16.3.41}$$

This completes the proof of our theorem.

We are now in a position to prove our main theorem.

Proof of Theorem 16.3.1: ∎ We are given a $t \times v$ matrix A_1 such that

$$A_1A_1^T = B_1 = (k-\lambda)I_t + \lambda J_t, \qquad A_1J_v = kJ_{t,v}. \tag{16.3.42}$$

Since we have also assumed the Bruck-Ryser-Chowla conditions 1 and 2 of Theorem 10.3.1, by Theorem 16.9.5 a rational $v \times v$ matrix C exists such that

$$\begin{aligned} CC^T &= C^TC = B = (k-\lambda)I_v + \lambda J_v, \\ CJ_v &= J_vC = kJ_v. \end{aligned} \tag{16.3.43}$$

Consider the $t \times v$ matrix A_1C^{-1}. It satisfies

$$\begin{aligned}(A_1C^{-1})(A_1C^{-1})^T &= A_1C^{-1}(C^{-1})^TA_1^T \\ &= A_1B^{-1}A_1^T \\ &= A_1[(k-\lambda)^{-1}(I_v - \lambda k^{-2}J_v)]A_1^T \\ &= (k-\lambda)^{-1}B_1 - \lambda(k-\lambda)J_t. \end{aligned} \tag{16.3.44}$$

It is easy to check that B^{-1} has the value used here. We have used both properties of A_1 given in (16.3.42). From the value of B_1, (16.3.44) simplifies to

$$(A_1C^{-1})(A_1C^{-1})^T = I_t. \tag{16.3.45}$$

Hence, by Theorem 16.3.3, there exists a rational $(v-t) \times v$ matrix Q such that

$$Y = \begin{bmatrix} A_1C^{-1} \\ Q \end{bmatrix}, \qquad YY^T = I_v. \tag{16.3.46}$$

From this,

$$I_v = Y^TY = [(C^{-1})^TA_1^T, C^T]\begin{bmatrix} A_1C^{-1} \\ Q \end{bmatrix}. \tag{16.3.47}$$

This yields

$$C^TI_vC = [A_1^T, C^TQ^T]\begin{bmatrix} A_1 \\ QC \end{bmatrix} = C^TC = B. \tag{16.3.48}$$

Let r_i denote the sum of the ith row of QC. We assert that

$$r_1^2 + \cdots + r_s^2 = k^2s, \qquad s = v - t, \tag{16.3.49}$$

for (16.3.48) gives

$$A_1^TA_1 + (QC)^TQC = B, \tag{16.3.50}$$

whence

$$J_vA_1^TA_1J_v + J_v((QC)^TQC)J_v = J_vBJ_v = k^2vJ_v. \tag{16.3.51}$$

Using (16.3.35) and (16.3.42), we find

$$J_v((QC)^TQC)J_v = k^2(v-t)J_v. \tag{16.3.52}$$

But this yields the relation (16.3.49). By Theorem 16.3.4, there is a rational $s \times s$ orthogonal matrix O_s^T such that

$$(r_1, \ldots, r_s)O_s^T = (k, \ldots, k). \tag{16.3.53}$$

Now define the matrix A by

$$A = \begin{bmatrix} A_1 \\ O_sQC \end{bmatrix}. \tag{16.3.54}$$

From (16.3.50) we have

$$A^T A = B. \tag{16.3.55}$$

Also, we find

$$AJ_v = kJ_v, \tag{16.3.56}$$

since from (16.3.5) the row sums of A_1 are k, and from (16.3.53) the row sums of $O_s QC$ are also k. From Theorem 10.2.3, (16.3.55) and (16.3.56) imply that also

$$AA^T = B, \qquad J_v A = kJ_v. \tag{16.3.57}$$

Thus, we have finally proved Theorem 16.3.1.

This theorem has been generalized by Eugene Johnsen [1]. The proof is similar to the preceding proof, but somewhat more complicated.

THEOREM 16.3.6 (JOHNSEN [1]). *Suppose the conditions* 1 *and* 2 *of Theorem* 10.3.1 *hold for v, k, λ, where $k(k-1) = \lambda(v-1)$. Let us be given the first r rows and the first s columns of a $v \times v$ matrix where A_r is the $r \times v$ matrix of the first r rows and A_s is the $v \times s$ matrix of the first s columns. Then we are given that A_r and A_s satisfy*

$$\begin{aligned} A_r A_r^T &= (k-\lambda)I_r + \lambda J_r, \qquad & A_r J_v &= kJ_{r,v}, \\ A_s^T A_s &= (k-\lambda)I_s + \lambda J_s, \qquad & J_v A_s &= kJ_{v,s}. \end{aligned} \tag{16.3.58}$$

Then there exists a rational $v \times v$ matrix A whose first r rows are A_r and whose first s columns are A_s, which satisfies the equations of (16.3.6).

16.4 Integral Solutions of the Incidence Equation

Let $A = (a_{ij})$ be a $v \times v$ matrix where the a_{ij} are rational integers, and suppose the incidence equation holds:

$$AA^T = B = (k-\lambda)I + \lambda J, \qquad k(k-1) = \lambda(v-1). \tag{16.4.1}$$

In this section we are concerned with knowing whether such matrices A exist, and if they do, whether or not A is necessarily the incidence matrix of a symmetric block design with parameters v, k, λ. The first theorem is easy, but not quite trivial.

THEOREM 16.4.1. *If every a_{ij} in an A satisfying* (16.4.1) *is zero or* 1, *then A is the incidence matrix of a symmetric block design with parameters v, k, λ.*

Proof: ■ Let the columns of A correspond to blocks $B_1, \ldots, B_v$ and the rows to elements $a_1, \ldots, a_v$. Let us define an incidence system S, saying $a_i \in B_j$

if $a_{ij} = 1$ and $a_i \notin B_j$ if $a_{ij} = 0$. Here, (16.4.1) says that the inner product of a row with itself is k and the inner product of two distinct rows of A is λ. In terms of our incidence system S, this says that every element occurs in exactly k blocks and every pair of distinct elements occurs together in λ different blocks. But it does not tell us how many elements are in each block. Let b_i be the number of elements in block B_i, $i = 1, \ldots, v$. Then we have

$$b_1 + b_2 + \cdots + b_v = kv, \tag{16.4.2}$$

$$\frac{b_1(b_1 - 1)}{2} + \cdots + \frac{b_v(b_v - 1)}{2} = \frac{\lambda v(v - 1)}{2}. \tag{16.4.3}$$

The first of these counts the total number of incidences of elements in blocks, which must be kv, since each of v elements is in k blocks. The second counts pairs of elements, since B_i contains $(b_i(b_i - 1)/2)$ unordered pairs and each of the $(v(v - 1)/2)$ unordered pairs of elements occurs together λ times. These equations together yield

$$\begin{aligned}(b_1 - k)^2 + \cdots + (b_v - k)^2 &= \lambda v(v - 1) - (2k - 1)kv + vk^2 \\ &= vk(k - 1) - (2k - 1)kv + vk^2 = 0,\end{aligned} \tag{16.4.4}$$

where we used (16.4.1) to replace $\lambda(v - 1)$ by $k(k - 1)$. But (16.4.4) requires $b_1 = b_2 = \cdots = b_v = k$. Thus, every block contains k objects, and so our incidence system S is a symmetric block design with parameters v, k, λ and our theorem is proved.

The following theorems, due to Ryser [1], are much deeper.

THEOREM 16.4.2. *Suppose $A = (a_{ij})$ is a $v \times v$ matrix of integers and $k(k - 1) = \lambda(v - 1)$ and $AA^T = A^TA = (k - \lambda)I + \lambda J$. Then A or $-A$ is composed entirely of zeros and 1's and is the incidence matrix of a block design.*

Proof: ■ Let s_j be the sum of the jth column of A. Thus,

$$s_j = \sum_{t=1}^{v} a_{tj}. \tag{16.4.5}$$

We have from the value of AA^T,

$$\sum_{j=1}^{v} a_{tj}a_{ij} = \begin{cases} k, & \text{if } i = t, \\ \lambda, & \text{if } i \neq t. \end{cases} \tag{16.4.6}$$

Hence,

$$\sum_{t,j=1}^{v} a_{tj}a_{ij} = k + (v - 1)\lambda = k^2, \tag{16.4.7}$$

whence from (16.4.5),

$$\sum_{j=1}^{v} a_{ij}s_j = k^2, \qquad i = 1, \ldots, v. \tag{16.4.8}$$

Adding (16.4.8) for $i = 1, \ldots, v$, we have

$$\sum_{j=1}^{v} s_j^2 = vk^2. \tag{16.4.9}$$

From A^TA we have for the diagonal values,

$$\sum_{i=1}^{v} a_{ij}^2 = k. \tag{16.4.10}$$

But now, since for any integer $|x| \leq x^2$,

$$|s_j| = \left|\sum_{i=1}^{v} a_{ij}\right| \leq \sum_{i=1}^{v} |a_{ij}| \leq \sum_{i=1}^{v} a_{ij}^2 = k. \tag{16.4.11}$$

This inequality, together with (16.4.9) requires

$$s_j^2 = k^2, \qquad j = 1, \ldots, v. \tag{16.4.12}$$

Thus, the relations in (16.4.11) are all equalities, and in the jth column all nonzero entries have the same sign, and as $|a_{ij}| = a_{ij}^2$, we have $a_{ij} = 0, +1$, or -1. Hence, a column contains either k +1's and $(v - k)$ zeros or k −1's and $(v - k)$ zeros. Since the inner product of distinct columns is the positive integer λ, all columns have the same sign for their nonzero entries, and so either A or $-A$ consists entirely of zeros and 1's and, by our previous theorem, is the incidence matrix of a symmetric block design.

Appealing to Theorem 10.2.3, we see the following.

COROLLARY. *If A is a matrix of integers satisfying* (16.4.1) *and either $AJ = kJ$ or $JA = kJ$, then A is the incidence matrix of a symmetric block design.*

The next theorem is more arithmetical.

THEOREM 16.4.3. *Let A be a matrix of integers satisfying* (16.4.1). *Suppose that $k - \lambda$ is odd and that k and $k - \lambda$ are relatively prime. Then, multiplying the columns of A by $+1$ or -1, appropriately, A becomes the incidence matrix of a symmetric block design.*

Proof: ■ We write $|A|$ for the determinant of A and A_{rs} for the cofactor of the element a_{rs} of A. Then, it is well known that the matrix A^{-1} is given by

$$A^{-1} = [a_{rs}^*], \qquad r, s = 1, \ldots, v, \qquad a_{rs}^* = \frac{A_{sr}}{|A|}. \tag{16.4.13}$$

Here, from (10.2.2),

$$|B| = (k - \lambda)^{v-1}(k + (v - 1)\lambda) = k^2(k - \lambda)^{v-1},$$

and so

$$|A| = \pm k(k - \lambda)^{(v-1)/2}. \tag{16.4.14}$$

From the relation $A^T = B$ and the fact that A is nonsingular, we have

$$A^TA = A^{-1}BA = (k - \lambda)I = \lambda A^{-1}JA\,. \tag{16.4.15}$$

Define, for $i, j = 1, \ldots, v$,

$$s_j = \sum_{w=1}^{v} a_{wj}\,, \qquad t_{ij} = \sum_{w=1}^{v} a_{wi}a_{wj}\,. \tag{16.4.16}$$

If we use (16.4.13) and equate elements in position (m, n) of the two sides of (16.4.15), we find

$$|A|\, t_{mn} = (k - \lambda)\, |A|\, \delta_{mn} + \lambda \sum_{i,j} A_{im}a_{jn}\,, \tag{16.4.17}$$

where δ_{rs} is the Kronecker δ. Here (16.4.8) and (16.4.9) are valid, since their derivation depended only on the value of AA^T. Multiplying (16.4.8) by A_{im} and summing over i gives

$$\sum_{i,j} A_{im}a_{ij}s_j = k^2 \sum_i A_{im}\,. \tag{16.4.18}$$

But always

$$\sum_i A_{im}a_{ij} = \delta_{mj}\, |A|\,. \tag{16.4.19}$$

Hence, (16.4.18) simplifies to

$$|A| \sum_j \delta_{mj}s_j = k^2 \sum_i A_{im} \tag{16.4.20}$$

or

$$|A|\, s_m = k^2 \sum_i A_{im}\,. \tag{16.4.21}$$

Here, using $\sum_j a_{jn} = s_n$ and (16.4.21), if we multiply (16.4.17) by k^2 and divide by $|A|$, we obtain

$$k^2t_{mn} = k^2(k - \lambda)\delta_{mn} + \lambda s_m s_n\,. \tag{16.4.22}$$

Putting the value of $|A|$ in (16.4.21) gives

$$\pm k(k - \lambda)^{(v-1)/2}s_m = k^2 \sum_i A_{im}\,, \qquad m = 1, \ldots, v. \tag{16.4.23}$$

By hypothesis, k and $k - \lambda$ are relatively prime, and so (16.4.23) gives

$$s_m \equiv 0 (\text{mod } k)\,, \qquad m = 1, \ldots, v\,. \tag{16.4.24}$$

Any column of A can be multiplied by -1 without altering the validity of (16.4.1), and we shall suppose that this has been done, so $s_m \geq 0$, $m = 1, \ldots, v$.

From (16.4.24) we write $s_m = ku_m$, $m = 1, \ldots, v$, where $u_m \geq 1$ is an integer. Substituting in (16.4.22) yields

$$t_{mn} = (k - \lambda)\delta_{mn} + \lambda u_m u_n . \tag{16.4.25}$$

Now suppose some $u_i = 0$. Then

$$t_{ii} = k - \lambda + \lambda u_i^2 = k - \lambda . \tag{16.4.26}$$

In this case, $s_i = 0$ and

$$0 = s_i^2 \equiv a_{1i}^2 + \cdots + a_{vi}^2 \equiv k - \lambda (\mathrm{mod}\ 2) . \tag{16.4.27}$$

This conflicts with the hypothesis that $k - \lambda$ is odd. Hence, no u_i is zero. From (16.4.9) we have

$$s_1^2 + s_2^2 + \cdots + s_v^2 = vk^2, \tag{16.4.28}$$

and putting $x_i = ku_i$ gives

$$u_1^2 + u_2^2 + \cdots + u_v^2 = v . \tag{16.4.29}$$

But if no u_i is zero and $u_i \geq 0$, this implies

$$u_1 = u_2 = \cdots = u_v = 1 , \tag{16.4.30}$$

whence

$$s_i = k, \qquad i = 1, \ldots, v . \tag{16.4.31}$$

Since (16.4.25) also gives $t_{ii} = k$, every a_{ij} is zero or 1, and A is an incidence matrix, proving our theorem.

If $k - \lambda$ is even, the conclusion of this theorem may fail to hold. When $\lambda = 1$, many counterexamples exist.

THEOREM 16.4.4. *Suppose $v = n^2 + n + 1$, $k = n + 1$, $\lambda = 1$, the parameters or a finite projective plane. Let $A = [a_{ij}]$ be a $v \times v$ matrix of integers such that*

$$AA^T = nI + J . \tag{16.4.32}$$

Multiplying appropriate columns of A by -1 so that the column sums are non-negative, then A is of one of two types:

Type I. *A is the incidence matrix of a finite projective plane.*

Type II. *n is even, one of the columns has a single zero and $(n^2 + n)$ of $+1$'s, $(n + 1)$ columns consist of $(n + 1)$ of $+1$'s and n^2 zeros, and the remaining columns have sum zero.*

There exists an A of type II for every n, which is the order of a Hadamard matrix, and also for $n = 10$.

Proof: ■ Most of the proof of Theorem 16.4.3 is applicable, since $k = n + 1$ and $k - \lambda = n$ are relatively prime. If n is odd, all the proof applies and A is the incidence matrix of a finite projective plane. Also, if n is even and no u_i is zero, then A yields a finite plane. Here, (16.4.29) becomes

$$u_1^2 + \cdots + u_v^2 = v = n^2 + n + 1. \tag{16.4.33}$$

Now, since for any integer x, $x \leq x^2$,

$$(n + 1)u_j = s_j = a_{1j} + a_{2j} + \cdots + a_{vj} \leq a_{1j}^2 + \cdots + a_{vj}^2 = t_{jj}. \tag{16.4.34}$$

But, by (16.4.25),

$$t_{jj} = n + u_j^2. \tag{16.4.35}$$

The inequality (16.4.34) becomes

$$0 \leq n - (n + 1)u_j + u_j^2 = (1 - u_j)(n - u_j). \tag{16.4.36}$$

Now we have taken the sign of the jth column so that $u_j \geq 0$. If $u_j \leq 1$ for every j, then from (16.4.33), $u_j = 1$ for every j, and as in Theorem 16.4.3, A is the incidence matrix of a finite projective plane. If $u_j > 1$ for some j, then from (16.4.36), we must have $u_j \geq n$. Clearly from (16.4.33), we cannot have $u_j \geq n + 1$, since $(n + 1)^2 > n^2 + n + 1$. Hence, if some $u_j > 1$, then $u_j = n$. We cannot have more than one u_j equal to n in (16.4.33). Let us permute the columns of A so that the unique column with $u_j = n$ is the first column and so $u_1 = n$. Now, from (16.4.33), we have

$$u_2^2 + \cdots + u_v^2 = n + 1. \tag{16.4.37}$$

Hence, there must be $(n + 1)$ additional u's equal to 1, and the remaining u's must be zero. Now we have

$$s_1 = (n + 1)u_1 = n^2 + n$$

and

$$s_j = n + 1 \qquad \text{for} \qquad n + 1\ j\text{'s}$$

and

$$s_j = 0 \qquad \text{for} \qquad n^2 - 1\ j\text{'s}.$$

For our t's, from (16.2.25),

$$s_1 = t_{11} = n + n^2, \tag{16.4.38a}$$

and if $s_j = M + 1$,

$$t_{ij} = n, \quad t_{jj} = n + 1, \qquad \text{if } s_j = 0, \quad t_{jj} = 0, \quad i \neq j, \quad t_{jj} = n. \tag{16.4.38b}$$

Whenever $s_j = t_{jj}$, the jth column consists of s_j of $+1$'s and $(v - s_j)$ zeros.

This completes all parts of the proof except the production of examples of integral matrices of type II. Since the first column comprises $(n^2 + n)$ $+1$'s and one zero, we may place the zero in position (1, 1) without loss of generality. For the columns with $s_j = n + 1$, we have $t_{1j} = n$, and so one of the 1's in this column is in the first row. This gives us $(n + 1)$ of 1's in the first row, and as the inner product of this row with itself is $n + 1$, the remaining entries are zeros. Let n be even and X an $n \times n$ matrix of integers such that the first column of X comprises n of $+1$'s and such that

$$XX^T = nI_n. \tag{16.4.39}$$

Clearly, a Hadamard matrix with its first column normalized to consist of $+1$'s satisfies these requirements. We construct A as follows:

$$A = \left|\begin{array}{c|c|c|c|c}
0 & 1\ 0\ \cdots\ 0 & 1\ 0\ \cdots\ 0 & 1\ 0\ \cdots & 1\ 0\ \cdots\ 0 \\ \hline
1 & & & & \\
1 & & & & \\
\vdots & X & 0 & 0\ \cdots & 0 \\
1 & & & & \\ \hline
1 & & & & \\
1 & & & & \\
\vdots & 0 & X & 0\ \cdots & 0 \\
1 & & & & \\ \hline
 & & & \ddots & 0 \\ \hline
\vdots & 0\ \cdots\ 0\ \cdots & & & \\
 & \vdots \qquad \vdots & & & X \\
1 & 0\ \cdots\ 0\ \cdots & & & \\
1 & & & &
\end{array}\right. \tag{16.4.40}$$

Here A, with the first row and column removed, is the direct sum of $(n + 1)$ copies of X. The 1's in the first row of A are above the first columns of the X's. We verify directly from the properties of X that

$$AA^T = nI + J. \tag{16.4.41}$$

For $n = 10$ and A of order 111, we may take the following as X:

$$K = \begin{bmatrix} 1 & -1 & -1 & -1 & -1 & -1 & 1 & 1 & 1 & 1 \\ 1 & 1 & -1 & -1 & -1 & -1 & -1 & -1 & -1 & -1 \\ 1 & 1 & -2 & 1 & 1 & 1 & 1 & 0 & 0 & 0 \\ 1 & 1 & 1 & 2 & -1 & -1 & 0 & 1 & 0 & 0 \\ 1 & 1 & 1 & -1 & 2 & -1 & 0 & 0 & 1 & 0 \\ 1 & 1 & 1 & -1 & -1 & 2 & 0 & 0 & 0 & 1 \\ 1 & -1 & 1 & 0 & 0 & 0 & 2 & -1 & -1 & -1 \\ 1 & -1 & 0 & 1 & 0 & 0 & -1 & -2 & 1 & 1 \\ 1 & -1 & 0 & 0 & 1 & 0 & -1 & 1 & -2 & 1 \\ 1 & -1 & 0 & 0 & 0 & 1 & -1 & 1 & 1 & -2 \end{bmatrix} \qquad (16.4.42)$$

Eugene Johnsen has found infinitely many type II solutions for $n \equiv 2 \pmod 4$ as well as for $n \equiv 0 \pmod 4$.

Appendix I

Balanced Incomplete Block Designs with from 3 to 15 Replications

The following table gives one or more solutions for a block design $D(v, b, r, k, \lambda)$ with $3 \le r \le 15$ and takes $k \le v/2$. In every case where a solution is known, at least one solution is given. The only parameters omitted are those such as $D(7, 14, 6, 3, 2)$, which exist trivially as multiples of known designs, in this case taking $D(7, 7, 3, 3, 1)$ twice. Taking $D(v, b, r, k, \lambda)$ for t times given a $D(v, tb, tr, k, t\lambda)$. For the most part, these have been taken from Fisher and Yates "Statistical Tables for Biological, Agricultural, and Medical Research," Oliver and Boyd, Edinburgh (1957), for $3 \le r \le 10$, and from a paper by C. R. Rao [1], for $11 \le r \le 15$. These writers did not assign numbers for the designs given by subspaces of finite Euclidean or projective geometries over finite fields. Thus, EG(2, 3) is the Euclidean geometry of dimension 2 over GF(3) and PG(2, 4) is the projective geometry of dimension 2 over GF(4).

These parameters are given in general in (12.2.2) and (12.2.3). In numbers 13 and 31, the complete design is given. In every other case, base blocks are given with respect to some Abelian group of automorphisms. In some instances Bose's notation for the mixed difference approach is used as given in Section 15.3. Number 88 is a good instance of Rao's notation, in which (x, y)mod $(5, 7)$ means that all residues modulo 5 are to be added to x, and all modulo 7 to y, but (x, y) mod$(-, 7)$ means that x is to be fixed and all residues modulo 7 are to be added to y. In every case, ∞ means an element fixed by the automorphism group.

When there is no design, reference is made to the appropriate theorems. In a number of cases, designs are described as residual designs of others. This is the process described in Section 10.1, by which we delete from a symmetric design one block and its elements from all other blocks.

Number	v	b	r	k	λ	Fisher and Yates [1] Number†	Solution
1	7	7	3	3	1		1, 2, 4 mod 7. PG(2, 2).
2	9	12	4	3	1		Residual of 3. EG(2, 3).
3	13	13	4	4	1		0, 1, 3, 9 mod 13. PG(2, 3).
4	6	10	5	3	2	1	Residual of 5.
5	11	11	5	5	2	2	1, 3, 4, 6, 9 mod 11. Type Q (Sec. 11.6).
6	16	20	5	4	1		Residual of 7. EG(2, 4).
7	21	21	5	5	1		3, 6, 7, 12, 14 mod 21. PG(2, 4).
8	10	15	6	4	2	3	Residual of 10.
9	13	26	6	3	1	4	[1, 3, 9]; [2, 5, 6]mod 13. (Theorem 15.3.4.)
10	16	16	6	6	2	5	[(1, 0, 0, 0); (0, 1, 0, 0); (0, 0, 1, 0); (0, 0, 0, 1); (1, 1, 0, 0); (0, 0, 1, 1)]mod (2, 2, 2, 2).
11	25	30	6	5	1		Residual of 12. EG(2, 5).
12	31	31	6	6	1		1, 5, 11, 24, 25, 27 mod 31.
13	8	14	7	3	3	6	1, 2, 3, 4; 1, 2, 7, 8; 1, 3, 6, 8; 5, 6, 7, 8; 3, 4, 5, 6; 2, 4, 5, 7; 1, 4, 6, 7; 1, 2, 5, 6; 1, 3, 5, 7; 2, 3, 5, 8; 3, 4, 7, 8; 2, 4, 6, 8; 1, 4, 5, 8; 2, 3, 6, 7.
14	15	35	7	3	1	7	$[1_1, 4_1, 0_2]$; $[2_1, 3_1, 0_2]$; $[1_2, 4_2, 0_3]$; $[2_2, 3_2, 0_3]$; $[0_1, 0_2, 0_3]$; $[1_3, 4_3, 0_1]$; $[2_3, 3_3, 0_1]$mod 5.
15	15	21	7	5	2	8	Does not exist. Would be residual of 17. (Theorem 16.1.3.)
16	15	15	7	7	3	9	0, 1, 2, 4, 5, 8, 10 mod 15. Type T (Sec. 11.6).
17	22	22	7	7	2	10	Does not exist. (Theorem 10.3.1.)
18	36	42	7	6	1		Does not exist. Would be residual of 19. (Theorem 12.3.3.)
19	43	43	7	7	1		Does not exist. (Theorem 10.3.1.)
20	9	18	8	4	3	11	[0, 1, 2, 4]; [0, 1, 4, 6]mod 9.
21	21	28	8	6	2	12	Does not exist. Would be residual of 23. (Theorem 16.1.3.)
22	25	50	8	4	1	13	[(0, 0), (1, 0), (0, 1), (4, 4)]mod (5, 5); [(0, 0), (2, 0), (0, 2), (1, 1)]mod (5, 5).

† This part of Table 1 is taken from Fisher and Yates, *Statistical Tables for Biological, Agricultural, and Medical Research*, published by Oliver & Boyd Ltd., Edinburgh. By permission of the authors and publishers.

TABLE 1 (continued)

Number	v	b	r	k	λ	Fisher and Yates [1] Number	Solution
23	29	29	8	8	2	14	Does not exist. (Theorem 10.3.1.)
24	49	56	8	7	1		Residual design of 25. EG(2, 7).
25	57	57	8	8	1		1, 6, 7, 9, 19, 38, 42, 49(mod 57). PG(2, 7).
26	10	30	9	3	2	15	[∞, 0, 5]; [0, 1, 4]; [0, 2, 3]; [0, 2, 7]mod 9. In (15.3.17).
27	10	18	9	5	4	16	Residual of 30.
28	16	24	9	6	3	17	Residual of 31. Nonresidual solution. In (16.1.19).
29	19	57	9	3	1	18	[1, 7, 11]; [2, 14, 3]; [4, 9, 6](mod 19). (Theorem 15.3.4.)
30	19	19	9	9	4	19	1, 4, 5, 6, 7, 9, 11, 16, 17(mod 19). Type Q (Section 11.6).
31	25	25	9	9	3	20	*a b c d e f g h i* *n x e q l k c g w* *b h j e s p n l u* *o u l f k i p r c* *c g o m j e v p y* *p i g o t s w x b* *d m x c h j u w r* *q a b p u c x y t* *e d y u w q s o f* *r c s a o w b v n* *f q t j n m i c s* *s k d h c v y t l* *g r m l d a q s p* *t j n g a d o u k* *h t r k q b m e o* *u n h i v y r q g* *i y p w b n k d m* *v w a t i u l m e* *j l w y g r t f b* *w p k v f h a j q* *k s u x m g f b v* *x f v r p t e n d* *l v q b x o d i j* *y e i s r x j k a* *m o f n y l h a x*
32	28	63	9	4	1	21	Elements ∞ and (x, y, z)mod (3, 3, 3). Base [(0, 2, 1); (0, 0, 1); (1, 1, 2); (1, 1, 0)] [(0, 2, 0); (1, 2, 2); (0, 0, 2); (1, 0, 0)] [∞; (0, 1, 1); (1, 1, 1); (2, 1, 1)] All mod (3, 3, 3). Adding (0, 0, 0), (1, 0, 0), and (2, 0, 0) to the base gives a complete replication, with the last block fixed. (Also Theorem 15.3.6.)

Number	v	b	r	k	λ	Fisher and Yates [1] Number	Solution
33	28	36	9	7	2	22	Residual of 34.
34	37	37	9	9	2	23	1, 7, 9, 10, 12, 16, 26, 33, 34 mod 37. Type B (Theorem 11.6.5).
35	46	69	9	6	1	24	Solution unknown.
36	64	72	9	8	1		Residual of 37. EG(2, 8).
37	73	73	9	9	1		1, 2, 4, 8, 16, 32, 37, 55, 64 mod 73. PG(2, 8).
38	21	70	10	3	1	25	[0, 1, 13]; [0, 4, 10]; [0, 16, 19]mod 21 and [0, 7, 14]mod 21 period 7. Theorem 15.3.3. Also (15.4.16).
39	21	30	10	7	3	26	Residual of 40.
40	31	31	10	10	3	27	$[1_1, 6_1, 2_2, 5_2, 3_3, 4_3, 3_4, 5_4, 6_4, \infty_1]$mod 7. $[2_1, 5_1, 3_2, 4_2, 1_3, 6_3, 3_4, 5_4, 6_4, \infty_2]$mod 7. $[3_1, 4_1, 1_2, 6_2, 2_3, 5_3, 3_4, 5_4, 6_4, \infty_3]$mod 7. $[1_1, 2_1, 4_1, 1_2, 2_2, 4_2, 1_3, 2_3, 4_3, 0_4]$mod 7. $[0_1, 1_1, 2_1, 3_1, 4_1, 5_1, 6_1, \infty_1, \infty_2, \infty_3]$. $[0_2, 1_2, 2_2, 3_2, 4_2, 5_2, 6_2, \infty_1, \infty_2, \infty_3]$. $[0_3, 1_3, 2_3, 3_3, 4_3, 5_3, 6_3, \infty_1, \infty_2, \infty_3]$.
41	36	45	10	8	2	28	Does not exist. Would be residual of 43. (Theorem 16.1.3.)
42	41	82	10	5	1	29	[1, 37, 16, 18, 10]; [8, 9, 5, 21, 39]mod 41. In (15.3.12).
43	46	46	10	10	2	30	Does not exist. (Theorem 10.3.1.)
44	51	85	10	6	1	31	Solution unknown.
45	81	90	10	9	1		Residual of 46. EG(2, 9).
46	91	91	10	10	1		0, 1, 3, 9, 27, 49, 56, 61, 77, 81 mod 91. PG(2, 9).

TABLE 1 (continued)

Number	v	b	r	k	λ	Rao [1] Number†	Solution
47	12	44	11	3	2	32	[0, 1, 3]; [4, 5, 9]; [2, 8, 6]; [∞, 7, 10]mod 11. Second solution: [0, 1, 3]; [0, 1, 4]; [0, 2, 6]; [∞, 0, 5]mod 11. Also (15.3.17).
48	12	33	11	4	3	33	[0, 1, 3, 7]; [2, 4, 9, 10]; [∞, 5, 6, 8]mod 11.
49	12	22	11	6	5	34	[0, 1, 3, 7, 8, 10]; [∞, 0, 5, 6, 8, 10]mod 11.
50	23	23	11	11	5	35	1, 2, 4, 6, 8, 9, 12, 13, 16, 18 mod 23. Type Q (Section 11.6).
51	45	99	11	5	1	36	Elements (x, y, z)mod (3, 3, 5). [(0, 1, 0), (0, 2, 0), (1, 0, 2), (2, 0, 2), (0, 0, 1)]mod (3, 3, 5), [(2, 1, 0), (1, 2, 0), (2, 2, 2), (1, 1, 2), (0, 0, 1)]mod (3, 3, 5), [(0, 0, 0), (0, 0, 1), (0, 0, 2), (0, 0, 3), (0, 0, 4)]mod (3, 3, –).
52	45	55	11	9	2	37	Solution unknown. Residual of 53.
53	56	56	11	11	2	38	Solution unknown.
54	100	110	11	10	1	39	Solution unknown. Residual of 55. EG(2, 10).
55	111	111	11	11	1	40	Solution unknown. PG(2, 10).
56	13	26	12	6	5	41	[0, 1, 3, 6, 7, 11]; [0, 1, 2, 3, 7, 11]mod 13.
57	19	57	12	4	2	42	[0, 1, 3, 12]; [0, 1, 5, 13]; [0, 4, 6, 9]mod 19.
58	21	42	12	6	3	43	[0, 2, 10, 15, 19, 20]; [0, 3, 7, 9, 10, 16]mod 21.
59	22	33	12	8	4	44	Solution unknown.
60	25	100	12	3	1	45	[0, 1, 3]; [0, 4, 13]; [0, 5, 11]; [0, 7, 17]mod 25. A second solution: [(0, 1), (4, 1), (1, 3)]; [(1, 0), (3, 3), (1, 2)]; [(3, 2), (2, 1), (0, 2)]; [(1, 1), (2, 4), (2, 0)]mod (5, 5).
61	33	44	12	9	3	46	Solution unknown.
62	34	34	12	12	4	47	Does not exist. (Theorem 10.3.1.)
63	37	111	12	4	1	48	Elements $(x, y) x \equiv 0, 1, 2 \bmod 3$, $y \equiv 0, \ldots, 10 \bmod 11$. Also $y = \infty$ and $(x, y) = (\infty, \infty)$. [(0, 0), (0, 1), (1, 2), (1, 5)]; [(0, 1), (0, 3), (0, 8), (1, 0)], [(0, ∞), (0, 7), (1, 5), (2, 1)]mod (3, 11), [(∞, ∞), (0, 0), (1, 0), (2, 0)]mod (–, 11), [(0, ∞), (1, ∞), (2, ∞), (∞, ∞)].

Number	v	b	r	k	λ	Rao [1] Number	Solution
64	45	45	12	12	3	49	Solution unknown.
65	55	66	12	10	2	50	Does not exist. Would be residual of 67. (Theorem 16.1.3.)
66	61	122	12	6	1	51	Solution unknown.
67	67	67	12	12	2	52	Does not exist. (Theorem 10.3.1.)
68	121	132	12	11	1		Residual of 69. EG(2, 11).
69	133	133	12	12	1		1, 8, 9, 11, 25, 37, 69, 88, 94, 99, 103, 121 mod 133. PG(2, 11).
70	27	117	13	3	1		[0, 1, 22]; [0, 2, 8]; [0, 3, 14]; [0, 7, 17]mod 26. [∞, 0, 13]mod 26 period 13. Lines in EG(3, 3).
71	27	39	13	9	4		Residual of 75. Planes in EG(3, 3).
72	27	27	13	13	6	53	[(0, 0, 1), (1, 0, 0), (1, 2, 0), (1, 1, 1), (2, 0, 2), (1, 1, 0), (1, 0, 2), (0, 2, 0), (0, 2, 1), (1, 2, 1), (2, 1, 1), (0, 2, 2), (2, 2, 1)]mod (3, 3, 3). Type *Q* (Section 11.6).
73	40	130	13	4	1		[0, 1, 26, 32]; [0, 7, 19, 36]; [0, 3, 16, 38]mod 40. [0, 10, 20, 30]mod 40 period 10. Lines in PG(3, 3).
74	40	52	13	10	3	54	Solution unknown.
75	40	40	13	13	4		1, 2, 3, 5, 6, 9, 14, 15, 18, 20, 25, 27, 35 mod 40. Planes in PG(3, 3).
76	53	53	13	13	3	55	Does not exist. (Theorem 10.3.1.)
77	66	143	13	6	1	56	Solution unknown.
78	66	78	13	11	2	57	Solution unknown. Residual of 79.
79	79	79	13	13	2	58	Solution unknown.
80	144	156	13	12	1	59	Solution unknown. Residual of 81. EG(2, 12).
81	157	157	13	13	1	60	Solution unknown. PG(2, 12).
82	15	42	14	5	4	61	[0, 1, 4, 9, 11]; [0, 1, 4, 10, 12]; [∞, 0, 1, 2, 7]mod 14. Double the unsolvable case 15.
83	15	35	14	6	5	62	[$\infty, 0_0, 0_1, 1_1, 2_1, 4_1$]; [$\infty, 0_1, 0_0, 6_0, 5_0, 3_0$]. [$1_0, 2_0, 4_0, 0_1, 1_1, 3_1$]; [$2_0, 3_0, 5_0, 0_1, 1_1, 3_1$] [$0_0, 4_0, 5_0, 0_1, 1_1, 3_1$]mod 7.

† This part of Table 1 is taken from C. R. Rao: "A Study of BIB Designs with Replications 11 to 15," *Sankhyā*, **23** (1961), 117–127. By permission of the author and the editors of *Sankhyā*.

TABLE 1 (continued)

Number	v	b	r	k	λ	Rao [1] Number	Solution
84	27	77	14	4	2	63	$[0_0, 3_0, 9_0, 10_0]$; $[0_0, 0_1, 2_1, 7_1]$; $[0_0, 0_1, 9_1, 10_1]$ $[0_0, 2_0, 5_1, 8_1]$; $[0_0, 3_0, 4_1, 7_1]$; $[0_0, 4_0, 3_1, 9_1]$ $[0_0, 5_0, 2_1, 6_1]$mod 11.
85	22	44	14	7	4	64	Solution unknown. Double the unsolvable case 17.
86	29	58	14	7	3	65	[1, 7, 16, 20, 23, 24, 25]; [2, 3, 11, 14, 17, 19, 21]mod 29.
87	36	84	14	6	2	66	[0, 1, 3, 5, 11, 23]; [0, 5, 8, 9, 18, 24]mod 35. and [∞, 0, 7, 14, 21, 28]mod 35 period 7 taken twice. Double the unsolvable case 18.
88	43	86	14	7	2	67	Elements (x, y) $x = 0, \ldots, 4$ mod 5 and $x = \infty$, $y = 0, \ldots, 6$ mod 7 and $y = \infty$. [(∞, 0), (0, 1), (0, 6), (1, 5), (1, 2), (2, 3), (2, 4)]mod (5, 7). [(∞, 0), (0, 1), (0, 6), (3, 5), (3, 2), (1, 3), (1, 4)]mod (5, 7). [(∞, 0), (∞, ∞), (0, 0), (1, 0), (2, 0), (3, 0), (4, 0)]mod (–, 7), taken twice. [(∞, 0), (∞, 1), (∞, 2), (∞, 3), (∞, 4), (∞, 5), (∞, 6)], taken twice. Double the unsolvable case 19.
89	78	91	14	12	2	68	Does not exist. Would be residual of 91. (Theorem 16.1.3.)
90	85	170	14	7	1	69	Solution unknown.
91	92	92	14	14	2	70	Does not exist. (Theorem 10.3.1.)
92	11	55	15	3	3	71	[0, 1, 3]; [0, 1, 5]; [0, 2, 7]; [0, 1, 8]; [0, 3, 5]mod 11.
93	13	39	15	5	5	72	[0, 1, 2, 4, 8]; [0, 1, 3, 6, 12]; [0, 2, 5, 6, 10]mod 13.
94	16	80	15	3	2	73	[0, 1, 3]; [0, 3, 8]; [0, 2, 12]; [0, 1, 7]; [0, 4, 9]mod 16. Second solution: $[0_1, 1_1, 2_1]$; $[0_1, 2_1, 5_1]$; $[0_0, 7_0, 0_1]$; $[1_0, 6_0, 0_1]$; $[2_0, 5_0, 0_1]$; $[3_0, 4_0, 0_1]$; $[1_0, 7_0, 0_1]$; $[2_0, 6_0, 0_1]$; $[3_0, 5_0, 0_1]$; $[0_0, 0_1, 4_1]$;mod 8.

Number	v	b	r	k	λ	Rao [1] Number	Solution
95	16	48	15	5	4	74	[0, 1, 2, 4, 7]; [0, 1, 8, 5, 10]; [0, 1, 3, 7, 11]mod 16.
96	16	40	16	6	5	75	[0, 1, 3, 5, 9, 12]; [0, 1, 2, 3, 6, 12]mod 16. [0, 8, 1, 9, 2, 10]mod 16 period 8.
97	16	30	15	8	7		[∞, 0, 1, 2, 7, 9, 12, 13]; [3, 4, 5, 6, 8, 10, 11, 14]mod 15.
98	21	35	15	9	6	76	[$0_0, 1_0, 2_0, 4_0, 0_1, 1_1, 2_1, 4_1, 2_2$]; [$0_0, 6_0, 5_0, 3_0, 6_2, 4_2, 3_2, 2_2, 0_1$]; [$1_0, 6_1, 5_1, 3_1, 6_2, 4_2, 3_2, 2_2, 0_0$]; [$4_0, 1_0, 3_0, 0_1, 2_1, 6_1, 4_2, 1_2, 2_2$]; [$0_0, 2_0, 6_0, 4_1, 1_1, 3_1, 4_2, 1_2, 2_2$]mod 7.
99	26	65	16	6	3	77	[∞, (0, 0), (1, 3), (2, 1), (3, 4), (4, 2)]mod (−, 5), [∞, (0, 0), (1, 2), (2, 4), (3, 1), (4, 3)]mod (−, 5), [∞, (0, 0), (1, 0), (2, 0), (3, 0), (4, 0)]mod (−, 5), [(0, 1), (0, 4), (1, 2), (1, 3), (2, 1), (2, 4)]mod (5, 5), [(0, 1), (0, 4), (2, 2), (2, 3), (3, 2), (3, 3)]mod (5, 5).
100	28	42	15	10	5	78	Solution unknown.
101	31	155	15	3	1		[0, 1, 18]; [0, 2, 5]; [0, 4, 10]; [0, 8, 20]; [0, 9, 16]mod 31. Lines in PG(4, 2). Steiner triple system.
102	31	93	15	5	2	79	[1, 2, 4, 8, 16]; [3, 6, 12, 24, 17]; [9, 18, 5, 10, 20]mod 31.
103	31	31	15	15	7		1, 2, 4, 5, 7, 8, 9, 10, 14, 16, 18, 19, 20, 25, 28 mod 31. Type Q (Section 11.6). 1, 2, 3, 4, 6, 8, 12, 15, 16, 17, 23, 24, 27, 29, 30 mod 31. Type H_6 (Section 11.6).
104	36	36	15	15	6	80	Solution unknown.
105	43	43	15	15	5	81	Solution unknown.
106	46	69	15	10	3	82	Solution unknown.
107	56	70	15	12	3	83	Solution unknown.
108	61	183	15	5	1	84	[1, 9, 20, 58, 34]; [4, 36, 19, 49, 14]; [16, 22, 15, 13, 56]mod 61.
109	71	71	15	15	3	85	Solution unknown.

TABLE 1 (concluded)

Number	v	b	r	k	λ	Rao [1] Number	Solution
110	76	190	15	6	1	86	Solution unknown.
111	91	195	15	7	1	87	[0, 10, 27, 28, 31, 43, 50]; [0, 11, 20, 25, 49, 55, 57]mod 91. [0, 13, 26, 39, 52, 65, 78]mod 91 period 13.
112	91	105	15	13	2	88	Does not exist. Would be residual of 113. (Theorem 16.1.3.)
113	106	106	15	15	2	89	Does not exist. (Theorem 10.3.1.)
114	136	204	15	10	1	90	Solution unknown.
115	196	210	15	14	1	91	Does not exist. Would be residual of 116. (Theorem 12.3.3.)
116	211	211	15	15	1	92	Does not exist. (Theorem 10.3.1.)

Appendix II

Hadamard Matrices of the Williamson Type

t	n	$W_1^2 + W_2^2 + W_3^2 + W_4^2$	W_1	W_2	W_3	W_4
3	12	$1^2 + 1^2 + 1^2 + 3^2$	1	1	1	$1 - 2\omega_1$
5	20	$1^2 + 1^2 + 3^2 + 3^2$	1	1	$1 - 2\omega_1$	$1 - 2\omega_2$
7	28	$1^2 + 3^2 + 3^2 + 3^2$	1	$1 - 2\omega_1$	$1 - 2\omega_2$	$1 - 2\omega_3$
7	28	$1^2 + 1^2 + 1^2 + 5^2$	1	1	$1 + 2\omega_1 - 2\omega_2$	$1 + 2\omega_3$
9	36	$3^2 + 3^2 + 3^2 + 3^2$	$1 - 2\omega_1$	$1 - 2\omega_2$	$1 - 2\omega_3$	$1 - 2\omega_4$
9	36	$1^2 + 1^2 + 3^2 + 5^2$	1	$1 - 2\omega_2 + 2\omega_1$	$1 - 2\omega_4$	$1 + 2\omega_3$
			1	1	$1 - 2\omega_2$	$1 + 2\omega_1 + 2\omega_3 + 2\omega_4$
11	44	$1^2 + 3^2 + 3^2 + 5^2$	$1 + 2\omega_1 - 2\omega_2$	$1 - 2\omega_4$	$1 - 2\omega_5$	$1 + 2\omega_3$
13	52	$1^2 + 1^2 + 1^2 + 7^2$	1	1	$1 + 2\omega_1 - 2\omega_4 + 2\omega_5 - 2\omega_6$	$1 - 2\omega_2 - 2\omega_3$
			1	$1 + 2\omega_4 - 2\omega_5$	$1 + 2\omega_6 - 2\omega_1$	$1 - 2\omega_2 - 2\omega_3$
13	52	$3^2 + 3^2 + 3^2 + 5^2$	$1 - 2\omega_2$	$1 - 2\omega_4$	$1 - 2\omega_1 - 2\omega_3 + 2\omega_5$	$1 + 2\omega_6$
13	52	$1^2 + 1^2 + 5^2 + 5^2$	$1 + 2\omega_4 - 2\omega_3$	$1 + 2\omega_6 - 2\omega_2$	$1 + 2\omega_1$	$1 + 2\omega_5$
15	60	$1^2 + 3^2 + 5^2 + 5^2$	1	$1 - 2\omega_5$	$1 + 2\omega_6$	$1 + 2\omega_1 - 2\omega_2 + 2\omega_3 + 2\omega_4 - 2\omega_7$
			$1 - 2\omega_1 + 2\omega_7$	$1 - 2\omega_3$	$1 + 2\omega_2$	$1 + 2\omega_4 + 2\omega_5 - 2\omega_6$
			$1 + 2\omega_6 - 2\omega_4$	$1 + 2\omega_5 - 2\omega_1 - 2\omega_3$	$1 + 2\omega_7$	$1 + 2\omega_2$
15	60	$1^2 + 1^2 + 3^2 + 7^2$	1	1	$1 - 2\omega_1 - 2\omega_5 + 2\omega_7$	$1 - 2\omega_3 - 2\omega_6 + 2\omega_2 - 2\omega_4$
17	68	$3^2 + 3^2 + 5^2 + 5^2$	$1 - 2\omega_2$	$1 - 2\omega_8$	$1 - 2\omega_1 + 2\omega_5 + 2\omega_6$	$1 - 2\omega_4 + 2\omega_3 + 2\omega_7$
17	68	$1^2 + 3^2 + 3^2 + 7^2$	$1 + 2\omega_6 + 2\omega_7 - 2\omega_5 - 2\omega_3$	$1 - 2\omega_2$	$1 - 2\omega_8$	$1 - 2\omega_1 - 2\omega_4$
			1	$1 + 2\omega_6 - 2\omega_5 - 2\omega_4$	$1 + 2\omega_7 - 2\omega_1 - 2\omega_3$	$1 - 2\omega_8 - 2\omega_2$
17	68	$1^2 + 3^2 + 3^2 + 7^2$	1	$1 + 2\omega_5 - 2\omega_4 - 2\omega_2$	$1 + 2\omega_3 - 2\omega_1 - 2\omega_8$	$1 - 2\omega_7 - 2\omega_6$
19	76	$1^2 + 5^2 + 5^2 + 5^2$	1	$1 + 2\omega_1 + 2\omega_4 - 2\omega_2$	$1 + 2\omega_8 + 2\omega_6 - 2\omega_3$	$1 + 2\omega_7 + 2\omega_9 - 2\omega_5$
			$1 + 2\omega_9 + 2\omega_5 - 2\omega_4 - 2\omega_3$	$1 + 2\omega_8 + 2\omega_2 - 2\omega_7$	$1 + 2\omega_6$	$1 + 2\omega_1$
			1	$1 + 2\omega_9 + 2\omega_8 - 2\omega_3$	$1 + 2\omega_7 + 2\omega_4 - 2\omega_5$	$1 + 2\omega_1 + 2\omega_6 - 2\omega_2$
19	76	$3^2 + 3^2 + 3^2 + 7^2$	None			

Appendix II (continued)

t	n	$W_1^2 + W_2^2 + W_3^2 + W_4^2$	W_1	W_2	W_3	W_4
19	76	$1^2 + 1^2 + 5^2 + 7^2$	1	1	$1 + 2\omega_1 + 2\omega_8 - 2\omega_3$	$1 + 2\omega_2 + 2\omega_6 - 2\omega_9 - 2\omega_7$ $- 2\omega_5 - 2\omega_4$
			$1 + 2\omega_8 - 2\omega_2$	$1 + 2\omega_7 - 2\omega_4$	$1 + 2\omega_6 + 2\omega_3 - 2\omega_9$	$1 - 2\omega_1 - 2\omega_5$
			$1 + 2\omega_4 - 2\omega_8$	$1 + 2\omega_2 - 2\omega_5$	$1 + 2\omega_1$	$1 + 2\omega_7 - 2\omega_9 - 2\omega_6 - 2\omega_3$
21	84	$3^2 + 5^2 + 5^2 + 5^2$	$1 - 2\omega_7$	$1 + 2\omega_3 + 2\omega_5 - 2\omega_8$	$1 + 2\omega_6 + 2\omega_4 - 2\omega_2$	$1 + 2\omega_9 + 2\omega_1 - 2\omega_{10}$
21	84	$1^2 + 1^2 + 1^2 + 9^2$	$1 - 2\omega_3 + 2\omega_2$	$1 - 2\omega_6 + 2\omega_{10}$	$1 - 2\omega_9 + 2\omega_8$	$1 + 2\omega_1 + 2\omega_5 + 2\omega_4 - 2\omega_7$
			1	1	$1 + 2\omega_9 + 2\omega_7 - 2\omega_6$ $- 2\omega_5$	$1 + 2\omega_1 + 2\omega_8 + 2\omega_4 + 2\omega_2$ $- 2\omega_{10} - 2\omega_3$
			$1 + 2\omega_9 - 2\omega_3$	$1 + 2\omega_8 - 2\omega_{10}$	$1 + 2\omega_4 - 2\omega_5$	$1 + 2\omega_7 + 2\omega_2 + 2\omega_1 - 2\omega_6$
21	84	$1^2 + 3^2 + 5^2 + 7^2$	$1 + 2\omega_5 - 2\omega_4$	$1 + 2\omega_{10} + 2\omega_2 - 2\omega_9$ $- 2\omega_8 - 2\omega_6$	$1 + 2\omega_1$	$1 - 2\omega_7 - 2\omega_3$
			$1 + 2\omega_9 - 2\omega_5$	$1 + 2\omega_2 - 2\omega_{10} - 2\omega_4$	$1 + 2\omega_7 + 2\omega_6 - 2\omega_8$	$1 - 2\omega_1 - 2\omega_3$
			$1 + 2\omega_8 - 2\omega_6$	$1 + 2\omega_2 - 2\omega_{10} - 2\omega_4$	$1 + 2\omega_7 + 2\omega_5 - 2\omega_9$	$1 - 2\omega_1 - 2\omega_3$
23	92	$1^2 + 1^2 + 3^2 + 9^2$	$1 + 2\omega_{11} + 2\omega_9 - 2\omega_8$ $- 2\omega_4$	$1 + 2\omega_5 - 2\omega_7$	$1 - 2\omega_3 + 2\omega_1 - 2\omega_{10}$	$1 + 2\omega_2 + 2\omega_6$
23	92	$3^2 + 3^2 + 5^2 + 7^2$	None			
25	100	$1^2 + 3^2 + 3^2 + 9^2$	$1 + 2\omega_6 - 2\omega_{11}$	$1 - 2\omega_1 + 2\omega_3 - 2\omega_{12}$	$1 - 2\omega_7 + 2\omega_4 - 2\omega_9$	$1 + 2\omega_5 + 2\omega_{10} + 2\omega_2 - 2\omega_8$
25	100	$5^2 + 5^2 + 5^2 + 5^2$	$1 + 2\omega_1 + 2\omega_9 - 2\omega_6$	$1 + 2\omega_7 + 2\omega_{12} - 2\omega_8$	$1 + 2\omega_5 + 2\omega_2 - 2\omega_4$	$1 + 2\omega_{10} + 2\omega_{11} - 2\omega_3$
25	100	$1^2 + 1^2 + 7^2 + 7^2$	1	1	$1 + 2\omega_{12} + 2\omega_6 - 2\omega_7$ $- 2\omega_5 - 2\omega_3 - 2\omega_2$	$1 + 2\omega_9 + 2\omega_8 - 2\omega_1 - 2\omega_{11}$ $- 2\omega_{10} - 2\omega_4$
			$1 + 2\omega_3 - 2\omega_7$	$1 + 2\omega_4 - 2\omega_1$	$1 + 2\omega_8 - 2\omega_{11} - 2\omega_{10}$ $- 2\omega_9$	$1 + 2\omega_6 - 2\omega_{12} - 2\omega_5 - 2\omega_2$
27	108	$1^2 + 1^2 + 9^2 + 5^2$	1	1	$1 + 2\omega_{12} + 2\omega_7 + 2\omega_5$ $- 2\omega_9 - 2\omega_3 + 2\omega_4$	$1 + 2\omega_{13} + 2\omega_{10} + 2\omega_8 + 2\omega_6$ $- 2\omega_1 - 2\omega_{11} - 2\omega_2$
29	116	$1^2 + 3^2 + 5^2 + 9^2$	$1 + 2\omega_{12} + 2\omega_6 + 2\omega_2$ $- 2\omega_{11} - 2\omega_9 - 2\omega_4$	$1 + 2\omega_{10} + 2\omega_7$ $- 2\omega_8 - 2\omega_3 - 2\omega_5$	$1 + 2\omega_1$	$1 + 2\omega_{13} + 2\omega_{14}$
37†	148	$1^2 + 7^2 + 7^2 + 7^2$	1	$1 + 2\alpha_5 - 2\alpha_0 - 2\alpha_1$	$1 + 2\alpha_8 - 2\alpha_3 - 2\alpha_4$	$1 + 2\alpha_2 - 2\alpha_6 - 2\alpha_7$
43‡	172	$1^2 + 1^2 + 1^2 + 13^2$	$1 + 2\alpha_0 - 2\alpha_2$	$1 + 2\alpha_3 - 2\alpha_1$	$1 + 2\alpha_4 - 2\alpha_6$	$1 + 2\alpha_5$

† $\alpha_j = \omega_{2^j} + \omega_{2^{9+j}}$.

‡ $\alpha_j = \omega_{3^j} + \omega_{3^{7+j}} + \omega_{3^{14+j}}$.

Bibliography

Albert, A. A.
[1] "On non-associative division algebras," *Trans. Amer. Math. Soc.*, **72** (1952), 296–309.
[2] *Fundamental Concepts of Modern Algebra.* Chicago: University of Chicago Press, 1956.

Baumert, L. D.
[1] "Hadamard matrices of Williamson type," *Math. of Comp.*, **19** (1965), 442–447.

Baumert, L. D., S. W. Golomb, and M. Hall, Jr.
[1] "Discovery of an Hadamard matrix of order 92," *Bull. Amer. Math. Soc.*, **68** (1962), 237–238.

Baumert, L. D., and Marshall Hall, Jr.
[1] "A new construction for Hadamard matrices," *Bull. Amer. Math. Soc.*, **71** (1965), 169–170.

Bhattacharya, K. N.
[1] "A note on two-fold triple systems," *Sankhyā*, **6** (1943), 313–314.
[2] "A new balanced incomplete block design," *Sci. Cult.*, **9** (1944), 508.

Blumenthal, L. M.
[1] *A Modern View of Geometry.* San Francisco: W. H. Freeman & Co., 1961.

Bose, R. C.
[1] "On the construction of balanced incomplete block designs," *Ann. Eugenics*, **9** (1939), 353–399.

Bose, R. C., E. T. Parker, and S. Shrikhande
[1] "Further results on the construction of mutually orthogonal Latin squares and the falsity of Euler's conjecture," *Can. J. Math.*, **12** (1960), 189–203.

Bose, R. C., and S. Shrikhande
[1] "On the construction of sets of mutually orthogonal Latin squares and the falsity of a conjecture of Euler." *Trans. Amer. Math. Soc.*, **95** (1960), 191–209.

BRUCK, R. H.
[1] "Difference sets in a finite group," *Trans. Amer. Math. Soc.*, **78** (1955), 464–481.

BRUCK, R. H., and H. J. RYSER
[1] "The nonexistence of certain finite projective planes," *Can. J. Math.*, **1** (1949), 88–93.

CHOWLA, S., P. ERDÖS, and E. G. STRAUS
[1] "On the maximal number of pairwise orthogonal Latin squares of a given order," *Can. J. Math.*, **12** (1960), 204–208.

CHOWLA, S., and H. J. RYSER
[1] "Combinatorial problems," *Can. J. Math.*, **2** (1950), 93–99.

COLE, F. N., A. S. WHITE, and L. D. CUMMINGS, JR.
[1] "Complete classification of triad systems on fifteen elements," *Mem. Nat. Acad. Sci.*, **14** (1925), Second Memoir, 89.

CONNOR, W. S.
[1] "On the structure of balanced incomplete block designs," *Ann. Math. Stat.*, **23** (1952), 57–71.

DANTZIG, G.
[1] "Maximization of a linear function of variables subject to linear inequalities," Chap. XXI, *Activity Analysis of Production and Allocation* (T. C. Koopmans, ed.). New York: John Wiley & Sons, Inc., 1951.

DEAN, R.
[1] *Elements of Abstract Algebra.* New York: John Wiley & Sons, Inc., 1966.

DEBRUIJN, N. G.
[1] "A combinatorial problem," *Nederl. Akad. Wetensch., Proc.* **49,** 758–764; Indagationes Math., **8** (1946), 461–467.

DIANANDA, P. H.
[1] "On non-negative forms in real variables some or all of which are non-negative," *Proc. Cambridge Philos. Soc.*, **58** (1962), 17–25.

DICKSON, L. E.
[1] "Cyclotomy, higher congruences, and Waring's problem," *Amer. J. Math.*, **57** (1935), 391–424.

DILWORTH, R. P.
[1] "A decomposition theorem for partially ordered sets," *Ann. Math.*, (2) **51** (1950), 161–166.

DULMAGE, A. L., D. M. JOHNSON, and N. S. MENDELSOHN
[1] "Orthomorphisms of groups and orthogonal Latin squares, *I*," *Can. J. Math.*, **13** (1961), 356–372.

EHLICH, H.
[1] "Neue Hadamard-Matrizen," *Arch. Math.*, **16** (1965), 34–36.

ERDÖS, P., and I. KAPLANSKY
[1] "The asymptotic number of Latin rectangles," *Amer. J. Math.*, **68** (1946), 230–236.

ERDÖS, P., and G. SZEKERES
[1] "A combinatorial problem in geometry," *Compositio Math.*, **2** (1935), 463–470.

FARKAS, J.
[1] "Uber die Theorie der einfachen Ungleichungen," *J. Reine Angew. Math.*, **124** (1902), 1–24.

FISHER, R. A.
[1] "An examination of the different possible solutions of a problem in incomplete blocks," *Ann. Eugenics*, **10** (1940), 52–75.

FISHER, R. A., and F. YATES
[1] *Statistical Tables for Biological, Agricultural, and Medical Research*, 2nd ed. London: Oliver and Boyd Ltd., 1943.

FLOOD, M.
[1] "On the Hitchcock distribution problem," *Pacific J. Math.*, **3** (1953), 369–386.

FRANKLIN, J.
[1] "Sur le développement du produit infini $(1 - x)(1 - x^2) \cdots$." *C. R. Acad. Fr.*, **92** (1881), 448–450.

GALE, D., H. W. KUHN, and A. W. TUCKER, JR.
[1] "Linear programming and the theory of games," *Activity Analysis of Production and Allocation* (T. C. Koopmans, ed.). New York: John Wiley & Sons, Inc., 1951, 317–329.

GOLDBERG, K.
[1] Personal communication.

GOOD, I. J.
[1] "Normal recurring decimals," *J. London Math. Soc.*, **21** (1946), 167–169.

HALL, MARSHALL, JR.
[1] "A survey of difference sets," *Proc. Amer. Math. Soc.*, **7** (1956), 975–986.
[2] "Characters and cyclotomy," *Proc. Symposia in Pure Math., Amer. Math. Soc.*, **8** (1965), 31–43.

HALL, MARSHALL, JR., and H. J. RYSER

[1] "Cyclic incidence matrices," *Can. J. Math.*, **3** (1951), 495–502.

[2] "Normal completions of incidence matrices," *Amer. J. Math.*, **76** (1954), 581–589.

HALL, M. JR., and J. D. SWIFT

[1] "Determination of Steiner triple systems of order 15," *Math. Tables Aids Comput.*, **9** (1955), 146–156.

HALL, PHILIP

[1] "On representatives of subsets," *J. London Math. Soc.*, **10** (1935), 26–30.

[2] "The Eulerian functions of a group," *Quart. J. Math., Ox. Series*, **7** (1936), 134–151.

HANANI, H.

[1] "The existence and construction of balanced incomplete block designs," *Ann. Math. Stat.*, **32** (1961), 361–386.

HARDY, G. H., and S. RAMANUJAN

[1] "Asymptotic formulae in combinatorial analysis," *Proc. London Math. Soc.*, (2) **17** (1918), 75–115.

HARDY, G. H., and E. M. WRIGHT

[1] *An Introduction to the Theory of Numbers.* Oxford: Clarendon Press, 1938.

HUGHES, D. R.

[1] "A class of non-Desarguesian projective planes," *Can. J. Math.*, **9** (1957), 378–388.

JOHNSEN, EUGENE

[1] "Matrix rational completions satisfying generalized incidence equations," *Can. J. Math.*, **17** (1965), 1–12.

JONES, B. W.

[1] *The Arithmetic Theory of Quadratic Forms.* Carus Mathematical Monograph No. 10, Mathematical Association of America (1950).

KIRKMAN, REV. THOMAS

[1] "On a problem in combinations," *Camb. and Dublin Math. J.*, **2** (1847), 191–204.

KÖNIG, D.

[1] *Theorie der endlichen und unendlichen Graphen.* New York: Chelsea Publishing Co., 1950.

LEHMER, D. H.

[1] "On the Hardy-Ramanujan Series for the partition function," *J. London Math. Soc.*, **12** (1937), 171–176.

MacNeish, H. F.
[1] "Euler squares," *Ann. Math.*, **23** (1922), 221–227.

Mann, H. B.
[1] "The construction of orthogonal Latin squares," *Ann. Math. Stat.*, **13** (1942), 418–423.

Mitchell, H. H.
[1] "On the generalized Jacobi-Kummer cyclotomic function," *Trans. Amer. Math. Soc.*, **17** (1916), 165–177.

Moore, E. H.
[1] "Concerning triple systems," *Math. Ann.*, **43** (1893), 271–285.

Ore, Oystein
[1] "On coset representatives in groups," *Proc. Amer. Math. Soc.*, **9** (1958), 665–670.

Paley, R. E. A. C.
[1] "On orthogonal matrices," *J. Math. Phys.*, **12** (1933), 311–320.

Parker, E.
[1] "On collineations of symmetric designs," *Proc. Amer. Math. Soc.*, **8** (1957), 350–351.
[2] "Construction of some sets of mutually orthogonal Latin squares," *Proc. Amer. Math. Soc.*, **10** (1959), 946–949.

Rao, C. Radhaskrishna
[1] "A study of BIB designs with replications 11 to 15," *Sankhyā*, **23** (1961), 117–127.

Rademacher, H.
[1] "A convergent series for the partition function $p(n)$," *Proc. Nat. Acad. Sci.*, **23** (1937), 78–84.

Ramsey, F. P.
[1] "On a problem of formal logic," *Proc. London Math. Soc.*, 2nd series, **30** (1930), 264–286.

Reiss, M.
[1] "Über eine Steinersche combinatorische Aufgabe welche in 45sten Bande dieses Journals, Seite 181, gestellt worden ist," *J. reine u. angew. Math.*, **56** (1859), 326–344.

Rota, G. C.
[1] "On the foundations of combinatorial theory, I. Theory of Möbius functions," *Z. Wahrscheinlichkeitstheorie und Verw. Gebiete*, **2** (1964), 340–368.

Ryser, H. T.
[1] "Matrices with integer elements in combinatorial investigations," *Amer. J. Math.*, **74** (1952), 769–773.

SCARPIS, V.

[1] "Sui determinanti di valore massimo," *Rend. R. Ist. Lombardo Sci. e Lett.*, (2) **31** (1898), 1441–1446.

SHRIKHANDE, S. S.

[1] "The impossibility of certain symmetrical balanced incomplete block designs," *Ann. Math. Stat.*, **21** (1950), 106–111.

SINGER, JAMES

[1] "A theorem in finite projective geometry and some applications to number theory," *Trans. Amer. Math. Soc.*, **43** (1938), 377–385.

STEINER, J.

[1] "Combinatorische Aufgabe," *J. reine angew. Math.*, **45** (1853), 181–182.

TARRY, G.

[1] "Le problème des 36 officiers," *C. R. Assoc. Fr. Av. Sci.*, **1** (1900), 122–123; **2** (1901), 170–203.

TUCKER, A.

[1] "Combinatorial theory underlying linear programs," *Recent Advances in Mathematical Programming* (Graves and Wolfe, eds.). New York: McGraw-Hill Book Company, 1963, 1–16.

TUKEY, JOHN

[1] "Convergence and uniformity in topology," *Annals of Math. Studies, No. 2*, Princeton University Press (1940).

VEBLEN, O., and J. H. M. WEDDERBURN

[1] "Non-Desarguesian and non-Pascalian Geometries," *Trans. Amer. Math. Soc.*, **8** (1907), 379–388.

VEBLEN, O., and J. W. YOUNG

[1] *Projective Geometry*, Vol. 1. Boston: Ginn and Company, 1910.

WEISNER, L.

[1] "Abstract theory of inversion of finite series," *Trans. Amer. Math. Soc.*, **38** (1935), 474–484.

WHITEMAN, A. L.

[1] "A family of difference sets," *Illinois J. Math.*, **6** (1962), 107–121.

WILLIAMSON, J.

[1] "Hadamard's determinant theorem and the sum of four squares," *Duke Math. J.*, **11** (1944), 65–81.

[2] "Note on Hadamard's determinant theorem," *Bull. Amer. Math. Soc.*, **53** (1947), 608–613.

Witt, E.
[1] "Theorie der quadratischen Formen in beliebigen Körpern," *J. für die r. u. ang. Math.*, **176** (1937), 31–44.

Zassenhaus, H.
[1] "Uber endliche Fastkörper," *Abh. Math. Sem. Hamburg*, **11** (1936), 187–220.

Index

A B C D E F G H I J 5 4 3 2 1 7 0 6 9 8 7